W9-CRB-662

Springer Series in Optical Sciences   Volume 56

Edited by Theodor Tamir

Springer Series in Optical Sciences

Volumes 1–41 are listed on the back inside cover

D. Sayre   M. Howells   J. Kirz
H. Rarback (Eds.)

X-Ray Microscopy II

Proceedings of the International Symposium,
Brookhaven, NY, August 31 – September 4, 1987

With 306 Figures

Springer-Verlag Berlin Heidelberg New York
London Paris Tokyo

DAVID SAYRE, Ph. D.

IBM T. J. Watson Research Center,
Yorktown Heights, NY 10598, USA

MALCOLM HOWELLS, Ph. D.

Center for X-Ray Optics,
Lawrence Berkeley Laboratory,
University of California at Berkeley,
1 Cyclotron Rd., Berkeley, CA 94720, USA

Professor JANOS KIRZ, Ph. D.

Department of Physics,
State University of New York,
Stony Brook, NY 11794, USA

HARVEY RARBACK, Ph. D.

National Synchrotron Light Source,
Brookhaven National Laboratory,
Upton, NY 11973, USA

ISBN 3-540-19392-8 Springer-Verlag Berlin Heidelberg New York
ISBN 0-387-19392-8 Springer-Verlag New York Berlin Heidelberg

Library of Congress Cataloging-in-Publication Data. X-ray microscopy II : proceedings of the International Symposium. Brookhaven, NY, August 31–September 4, 1987 / D. Sayre, M. Howells, J. Kirz (eds.). p. cm.–(Springer series in optical sciences ; v. 56) Based on papers presented at the International Symposium on X-ray Microscopy held at Brookhaven National Laboratory. 1. X-ray microscope–Congresses. I. Sayre, David. II. Howells, Malcolm R. III. Kirz, J. (Janos), 1937-. IV. International Symposium on X-ray Microscopy (2nd : 1987 : Brookhaven, N.Y.) V. Series. QH212.X2X23 1988 535'.332–dc 19 88-20043

Printing: Druckhaus Beltz, 6944 Hemsbach/Bergstr.
Binding: J. Schäffer GmbH & Co. KG, 6718 Grünstadt.
2154/3150-543210 – Printed on acid-free paper

Preface

This volume is based on papers presented at the International Symposium on X-Ray Microscopy held at Brookhaven National Laboratory, Upton NY, August 31–September 4, 1987. Previous recent symposia on the subject were held in New York in 1979, Göttingen in 1983 and Taipei in 1986.

Developments in x-ray microscopy continue at a rapid pace, with important advances in all major areas: x-ray sources, optics and components, and microscopes and imaging systems. Taken as a whole, the work presented here emphasizes three major directions: (a) improvements in the capability and image-quality of x-ray microscopy, expressed principally in systems attached to large, high-brightness x-ray sources; (b) greater access to x-ray microscopy, expressed chiefly in systems employing small, often pulsed, x-ray sources; and (c) increased rate of exploration of applications of x-ray microscopy.

The number of papers presented at the symposium has roughly doubled compared with that of its predecessors. While we are delighted at this growth as a manifestation of vitality and rapid growth of the field, we did have to ask the authors to limit the length of their papers and to submit them in camera-ready form. We thank the authors for their contributions and for their efforts in adhering to the guidelines on manuscript preparation.

In our capacity as organizers of the Symposium we acknowledge with thanks the financial support received from the U.S. National Science Foundation; the Center for X-ray Optics, Lawrence Berkeley Laboratory; the National Synchrotron Light Source, Brookhaven National Laboratory; and the Physics Department, State University of New York at Stony Brook. We are grateful to Ms. Judy Thompson who served as Conference Secretary with skill and dedication.

D. Sayre   M. Howells
J. Kirz   H. Rarback

Contents

Part II   **X-Ray Optics and Components**

Part III   X-Ray Microscopes and Imaging Systems

Part V	Summary of Session on Future X-Ray Microscopy Facilities

Introduction

D. Sayre[1], *M. Howells*[2], *J. Kirz*[3], *and H. Rarback*[4]

[1]IBM Research Center, Yorktown Heights, NY 10598, USA
[2]Center for X-ray Optics, Lawrence Berkeley Laboratory,
  Berkeley, CA 94720, USA
[3]Department of Physics, State University of New York,
  Stony Brook, NY 11794, USA
[4]National Synchrotron Light Source,
  Brookhaven National Laboratory, Upton, NY 11973, USA

X-ray microscopy, as the name implies, deals with high resolution image formation using x-rays. The papers in this volume highlight various facets of the field, emphasizing more the latest developments than their historical origins [1]. (The proceedings of previous symposia on this subject can be found in [2-4].) Although individual contributions tend to emphasize only one area of activity, we feel that the mosaic that emerges from these tiles gives nevertheless a rather complete picture. To provide a framework, the volume is organized into five parts.

The papers in Part I describe developments in x-ray sources. High brightness sources, such as undulators operating on electron storage rings, are beginning to have a major impact. X-ray lasers are approaching the wavelength and power level where they will take on an important role in flash imaging. While these sources appear to be intrinsically large and costly, smaller laboratory sources are being developed, which may eventually point the way to off-the-shelf systems that could become more widely available.

Developments in optical elements and other components form the subject of the papers in Part II. The largest group of contributions deal with the fabrication and properties of zone plates. New efforts are reported to improve their resolution and efficiency, and to extend their useful wavelength range. Reflective optics has been used successfully for creating a hard x-ray mircoprobe, and prospects for further applications involving multilayers seem especially bright. This section also includes several papers that present new developments and new understanding in the area of x-ray detection and on the nature of x-ray interactions.

A great variety of instruments and imaging systems are described in Part III. There are new and improved scanning, imaging, projection and contact microscopes, image conversion, microtomography, fluorescence, diffraction, and holographic instruments. Many of these operate with synchrotron radiation, but a significant number use smaller installations. Plans for future imaging systems are also discussed, as well as some of the ultimate limitations.

Although the development of the instruments must precede their use, it is particularly encouraging that a growing number of scientists are already using x-ray microscopy, or are seriously thinking about potential applications. The papers in Part IV deal with applications to biology, radiobiology, geology, and materials science. Some of the work makes use of the elemental sensitivity, other papers illustrate that the ability to view specimens in their natural environment at improved resolution leads to new insights.

Significant investments are being planned in synchrotron radiation sources worldwide (e.g. Advanced Light Source in Berkeley, other labs at Argonne, Grenoble, Trieste, Beijing, Hefei, Taiwan). Part V is the summary of an evening session that took place during the Symposium to formulate the special needs of the microscopist in relation to these future facilities.

Literature

1. The origins of the field are traced in V.E. Cosslett, W.C. Nixon: X-Ray Microscopy (Cambridge, 1960). More recent and less detailed historical accounts may be found in M. Howells, J. Kirz, D. Sayre, G. Schmahl: Phys. Today 38(8), 22 (1985), and in J. Kirz, H. Rarback: Rev. Sci. Instrum. 56, 1 (1985).
2. Ultrasoft X-ray Microscopy: Its Application to Biological and Physical Sciences, ed. by D.F. Parsons, Ann. N.Y. Acad. Sci., Vol. 342 (New York Academy of Sciences, New York 1980)
3. X-Ray Microscopy, ed. by G. Schmahl, D. Rudolph, Springer Ser. Opt. Sci., Vol. 43 (Springer, Berlin, Heidelberg 1984)
4. X-Ray Microscopy: Instrumentation and Biological Applications, ed. by P.-C. Cheng, G.-J. Jan (Springer, Berlin, Heidelberg 1987)

Part I

X-Ray Sources

Experience with Synchrotron Radiation Sources *

S. Krinsky

National Synchrotron Light Source, Brookhaven National Laboratory,
Upton, NY 11973, USA

The development of synchrotron radiation sources is discussed, emphasizing
characteristics important for x-ray microscopy. Bending magnets, wigglers
and undulators are considered as sources of radiation. Operating experience at the National Synchrotron Light Source on the VUV and XRAY storage
rings is reviewed, with particular consideration given to achieved current
and lifetime, transverse bunch dimensions, and orbit stability.

1. DEVELOPMENT OF SYNCHROTRON RADIATION FACILITIES

Throughout the world synchrotron radiation facilities are being built and
operated as sources of ultraviolet and x-radiation for research in the
basic and applied sciences [1]. The capacities of existing facilities are
being strained by the ever increasing number of scientists using synchrotron radiation, and large new facilities are being planned in the United
States, Europe, Japan, and elsewhere. Early work with synchrotron radiation was performed parasitically on synchrotrons and storage rings designed
and operated for high energy physics. Over the last decade, new facilities
designed specifically for synchrotron radiation research have been successfully constructed and operated. Examples are the NSLS and ALADIN in the
United States, the SRS at Daresbury in England, BESSY in Germany, SUPERACO
in France, and the PHOTON FACTORY in Japan. At the new dedicated facilities, advances in storage ring design together with improvements in beamline instrumentation have resulted in a significant increase in the quality
of the radiation sources available to experimenters.

Of particular importance has been the increase of spectral brightness of
the new sources. Brightness [2] is the proper figure of merit for experiments requiring photon beams with small angular divergence incident upon
small samples, and in experiments utilizing spatially coherent radiation
[3]. Spectral brightness is defined as the number of photons per unit
source area, per unit solid angle, per unit bandwidth. In order to increase the brightness of the radiation sources in a storage ring, one must
reduce the volume of phase space occupied by the electron beam. Accelerator physicists use the term "emittance" to refer to the area in horizontal
or vertical phase space occupied by the electron beam. The condition for
optimizing the luminosity for colliding beam experiments in high energy
physics requires large horizontal emittance, whereas high brightness radiation sources result from small emittance. Using existing technology, it
was possible to design low emittance storage rings dedicated to synchrotron

---

*This work was performed under the auspices of the U.S. Department of
Energy.

4

radiation research, with significantly higher brightness than the sources available on high energy physics machines. The benefits expected from providing higher brightness sources have been realized, e.g. reduction of deleterious effects due to aberrations in beamline optics, higher resolution, improved spatial coherence, and this has motivated the development of the next generation of synchrotron radiation sources, which are being designed to have even smaller electron beam emittances and, consequently, even higher brightness than the existing facilities.

Another important development leading to enhanced radiation sources has been the use of insertion devices [4], see Fig. 1. Early work with synchrotron radiation was done exclusively using the radiation emitted from electrons bent in circular arcs in the dipole magnets of a storage ring. More recently, improved sources have been obtained by placing special magnets called "wigglers" in the straight sections (insertions) of the storage ring. These magnets produce along the electron trajectory a magnetic field alternating in polarity, causing the electron to wiggle transversely. If we assume the wiggler magnetic field to be approximately sinusoidal,

$$B = B_w \sin (2\pi z/\lambda_w) \; ,$$

then the resulting angular deviation of the electron trajectory can be written

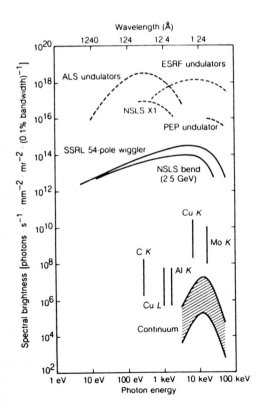

Fig. 1. Comparison of spectral brightness of some conventional X-ray sources, and synchrotron sources from bending magnets and insertion devices

5

$$x' = \frac{K}{\gamma} \sin (2\pi z/\lambda_w) ,$$

where λ_w is the period length of the static wiggler magnetic field, γ is the electron energy measured in units of its rest mass, and

$$K = 0.93 \ B_w(T) \ \lambda_w(cm)$$

is a dimensionless parameter determining the maximum angular deviation.

Electrons passing through a wiggler magnet radiate due to the transverse acceleration they experience. When $K \ll 1$, the radiated spectrum is peaked about the fundamental wavelength

$$\lambda_1 = \frac{\lambda_w}{2\gamma^2} \left(1 + \frac{K^2}{2}\right) ,$$

and there is little intensity at the higher harmonic wavelengths $\lambda_k = \lambda_1/k$ ($k = 2, 3, 4, ...$). As K is increased toward unity, the third harmonic becomes important, and for large K many harmonics have significant intensity.

The radiated spectral intensity from a bending magnet is characterized by a critical wavelength $\lambda_c(\text{Å}) = 18.6/B(T)E^2(\text{GeV})$, where B is the magnetic field strength and E the electron energy. The spectral intensity is roughly constant for $\lambda > \lambda_c$, and falls off rapidly for $\lambda < \lambda_c/3$, see Fig. 2. When an insertion device is operated with K large, the radiated spectrum is similar to that of synchrotron radiation from an arc source, and we speak of "wiggler" radiation. Wigglers provide an enhancement in the radiated flux per unit solid angle, relative to an arc source. Also, by designing the wiggler to operate at higher magnetic field than the arc sources, one can obtain a source of harder photons.

When the insertion device is operated with K small, there is a high intensity at only a few harmonics, Fig. 2, and one speaks of "undulator" radiation. Although there is no clear dividing line between the undulator and wiggler regimes, there is a difference between undulator radiation and that from an arc source. Consider turning down the magnetic field strength on the undulator. The number of radiated photons decreases as K^2, but the first harmonic wavelength approaches a limit, $\lambda_1 = \lambda_w/2\gamma^2$, independent of K.

BENDING MAGNET

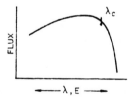

UNDULATOR

$$\lambda_p^m = \frac{\lambda_u}{2m\gamma^2}\left(1 + \frac{K^2}{2}\right)$$

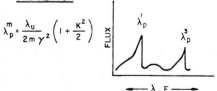

Fig. 2. Qualitative features of radiated spectrum from bending magnet and undulator sources

6

Hence, in the limit of K very small, there is the emission of only a few photons, but they are still hard. In contrast, when the magnetic field in an arc source is reduced, the radiation becomes softer. Storage rings operating with electron energy between 0.5-1GeV are appropriate for the production of ultraviolet and soft x-rays from bending magnets and wigglers, and ultraviolet radiation from undulators. Facilities in this class are the NSLS VUV-RING, ALADIN, BESSY and SUPERACO. Storage rings operating with electron energies between 1.5-3GeV can deliver soft x-rays from undulators, and hard x-rays from bending magnets and wigglers. Examples are the NSLS XRAY RING, DARESBURY, THE PHOTON FACTORY, and the planned rings at Berkeley, Trieste, and BESSY II in Germany. Hard x-rays from undulators can be produced from electrons with energy 6-8 GeV. At present, CESR at Cornell, and PEP at SLAC can run in this energy range. New facilities dedicated to synchrotron radiation research are planned in the 6-8 GeV range at Argonne and Grenoble, and also in Japan.

The reduction of the emittance of a storage ring necessary to achieve high brightness requires an increase in the size and cost of the machine. The emittance is roughly proportional to $\gamma^2\theta^3$, where θ is the bend angle per dipole magnet. The larger the electron energy γ, the smaller θ must be to achieve low emittance. Small bend angle per dipole is accompanied by increased numbers of quadrupole magnets to focus the electron beam, leading to increased ring circumference.

In addition to the development of high brightness sources there has been recent interest in producing compact storage rings for industrial applications of x-ray lithography [5]. These machines are being designed to have a critical wavelength near 10Å, and although the emittance should not be too large, it need not be exceptionally low. Therefore, these rings can be smaller and less expensive than the machines being built for synchrotron radiation facilities. In Japan, Germany and the United States, there are two approaches to the design of compact sources. One approach uses superconducting magnets with fields of about 4T, with the ring circumference ≈12m and the electron energy ≈600 MeV. The second approach uses conventional magnets operating at about 1.6T, with ring circumference ≈30m and electron energy ≈1GeV. Compact storage rings may one day be useful for x-ray microscopy techniques not requiring high brightness.

2. PERFORMANCE OF THE NSLS STORAGE RINGS

The storage rings at the NSLS were specifically designed for use as dedicated synchrotron radiation sources. Their basic parameters as achieved in normal operations, are summarized in Table 1.

The lifetime of the XRAY ring corresponds to an exponential decay of the beam current due to scattering off of the residual gas in the vacuum chamber. At 200 ma, the pressure is about 2nT, and the observed lifetime is 10-20 hrs. The XRAY ring is normally run with 25 consecutive electron bunches and a gap of 5 empty buckets. Leaving a gap in the bunch distribution is found to alleviate ion trapping.

In contrast, the lifetime in the VUV ring is determined not by the vacuum, but by scattering between electrons in the same bunch (Touschek scattering). The Touchek lifetime is inversely proportional to the bunch density and, therefore, is reduced as the transverse dimensions are decreased. In the VUV ring, when the vertical emittance of the electron beam is reduced using skew quadrupoles, one observes that the lifetime

Table 1. NSLS Storage Rings

	X-RAY RING	VUV RING
Energy	2.5 GeV	0.75 GeV
Circumference	170 m	51 m
Critical Wavelength	2.5 Å	25 Å
Current	200 ma	700 ma
Lifetime (1/e)	10-20 hrs	2-4 hrs
Emittance ϵ_x	10^{-7} m-rad	1.5×10^{-7} m-rad
ϵ_y	10^{-9} m-rad	1.5×10^{-9} m-rad
Typical Source Size σ_x	0.35 mm	0.40 mm
σ_y	0.12 mm	0.25 mm

decreases. Also, the lifetime depends only on the individual bunch currents, and not on the average current. The VUV ring is normally run with five consecutive bunches and a gap of four empty buckets. Again the gap helps eliminate ion trapping. A 4th harmonic cavity is planned to increase the bunch length and thus increase the beam lifetime.

Of key importance to the optimum utilization of the high-brightness sources is the achievement of a stable orbit with movements small compared to the dimensions of the electron beam. At present orbit motions of 50-100μ are observed, and much effort is aimed at identifying and eliminating the source of these variations. There is an active R&D program to develop improved detectors of orbit position using synchrotron radiation. Key issues are intensity dependence of the position measurements and long term stability. New high resolution RF detectors for the pick-up electrodes in the storage rings are also being designed. A vigorous program is underway to develop and implement orbit feedback systems, and a prototype system was successfully operated on X17T for the mini-undulator discussed at this conference. With the feedback 5μ stability was achieved, [6] as illustrated in Fig. 3.

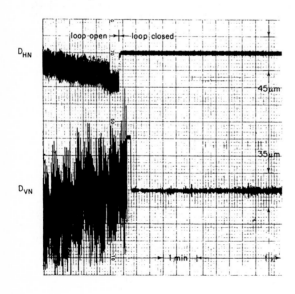

Fig. 3. Improvement of orbit stability resulting from closing the feedback loop on X17T

In conclusion, the operation of the VUV and XRAY rings has been in accordance with the expectations based on the original design. The experimental program has grown rapidly, and in the near future there will be ≈30 beamlines on the VUV ring and ≈50 on the XRAY ring. Work on x-ray microscopy will continue to be one of the most exciting programs at the NSLS, and the high quality of the source required for this work will stimulate further advances in the design of synchrotron radiation facilities.

ACKNOWLEDGEMENT

I wish to thank Harvey Rarbach for several useful discussions.

REFERENCES

1. The development of synchrotron radiation sources is reviewed by H. Winick: In Scientific American (Nov. 1987), p 88.

2. The importance of optimizing the brightness of synchrotron radiation sources was emphasized by G.K. Green: Spectra and Optics of Synchrotron Radiation, Brookhaven National Lab Report, BNL 50522 (1976).

3. Coherence of synchrotron radiation is discussed by D. Attwood, K. Halbach, and K.J. Kim: Science 288, 1265 (1985).

4. See e.g. S. Krinsky, M.L. Perlman and R.E. Watson: In Handbook on Synchrotron Radiation, Vol. 1, edited by E.E. Koch (North Holland, 1983); and S. Krinsky: IEEE Trans. Nucl. Sci. NS-30, 3078 (1983).

5. See e.g. R.P. Haelbich, J.P. Silverman, and J.M. Warlaumont: Nucl. Inst. Methods 222, 291 (1984); and A.D. Wilson: In SPIE Vol. 537 Electron-Beam, X-ray, and Ion-Beam Techniques for Submicron Lithographies IV (1985), p 85.

6. R.J. Nawrocky, L. Ma, H.M. Rarbach, O.V. Singh and L.H. Yu: In Proc. of the 1987 Particle Accelerator Conference, Washington, D.C., March 15-19, 1987.

The NSLS Mini-Undulator

H. Rarback[1], *C. Jacobson*[2], *J. Kirz*[2], *and I. McNulty*

[1]National Synchrotron Light Source, Brookhaven National Laboratory,
Upton, NY 11973, USA
[2]Physics Department, State University of New York,
Stony Brook, NY 11794, USA

1. INTRODUCTION

The soft x-ray microscopy program at the NSLS depends critically on the coherence properties of the source. The coherent flux is directly proportional to the brightness [1]; therefore our programs benefit directly from improvements in that area. Work on scanning microscopy, holography and diffraction all started on the bending magnet beamline U15, which was the brightest source of soft x-rays available until the first soft x-ray undulator started operation at the Photon Factory [2].

We first became aware of the dramatically increased source brightness that undulators offer [3] when, at Krinsky's suggestion, Howells started to think seriously about building an undulator based beamline [4]. The Soft X-ray Undulator project at the NSLS is based on this initiative, and is to be completed in 1988. The long gestation period made us rather impatient, but fortunately we had a chance to get a taste of what lies ahead: Hastings realized that if a prototype undulator were built with only 10 periods rather than 37, the existing hardware on the NSLS X-ray ring would be able to handle the output power, and such a device could be made operational in a short time. A ten period "mini-undulator" was then rapidly constructed and installed, along with a temporary beamline and experimental station. This endeavor required the dedicated effort of most of the NSLS staff, and of our collaborators at the Lawrence Berkeley Laboratory. It is a tribute to their talents and hard work that within nine months of the decision to build the "mini", we observed the first monochromatic light from it.

2. CHARACTERISTICS of the MINI-UNDULATOR

The mini-undulator is a hybrid device with samarium cobalt permanent magnets and steel poles, and a period length of 8 cm. The structure is designed to fit outside the vacuum chamber; therefore, the minimum poleface gap is 33 mm, which corresponds to a magnetic deflection parameter $K = 2.3$. The gap can be changed by the experimenter, providing a practical tuning range for the first harmonic of the undulator between 1.7 and 6.0 nm, with the NSLS X-ray ring operating at 2.5 GeV. The ends of the undulator have correction coils to compensate for a nonzero net field integral $\int B_y \, dz$; in practice, no significant beam steering errors were observed with the current in the coils set to a nominal fixed value. After a substantial period of machine studies it was determined that other beamlines around the ring saw no significant changes when the gap of the undulator was changed during regular operations.

The beam that undulators produce is highly collimated in space and peaked in wavelength. The total power output is modest (80 watts in our worst case scenario, with 250 mA circulating current and $K = 2.3$), but the peak power density on any element that intercepts the beam at normal incidence is 350 W/cm^2 at a distance of 10 m from the source under similar conditions. This strong collimation is of course what is so attractive about the source, but it also requires careful planning of the beamline. The x-ray beam points along the direction of the electron beam inside the undulator. In fact, the beam divergence is dominated by the electron beam divergence, the intrinsic emission angle of the radiation being much smaller than the angular spread of the beam in the horizontal direction.

3. COMPONENTS of the BEAMLINE

One particular problem associated with the sharply collimated nature of undulator radiation is the difficulty of initial alignment. It is not easy to find the beam, and it is accompanied by the tails of the radiation from the bending magnets at the two ends of the straight section. Bending magnet radiation has a strong component in the visible, that fills the aperture with blue light. To render the sharply peaked soft x-ray component visible, we clamped an alumina-coated plate to a water cooled paddle. The alumina fluoresces reddish purple when illuminated by soft x-rays. The paddle was oriented at 45° to the beam, and observed through a viewport. It was mounted on a manipulator that was stepping motor driven so the paddle could be inserted into or removed from the beam.

The paddle made it possible to steer the beam in approximately the right direction. For precise alignment we used a system of position monitors [5]. The vertical position sensors consisted of a pair of vertical molybdenum blades that were water cooled via heatpipes, yet electrically isolated. The blades were connected to electrometers to measure the photoemission current due to the top and bottom edges of the beam. The blades were held fixed relative to each other, but were mounted on a motorized manipulator. Hence, one could determine the height of the beam by moving the pair vertically until the signals were equal, and horizontally until the sum of the signals was a maximum. The system for determining the horizontal position was identical, but for a 90° rotation around the beam direction. The noise level and sensitivity of the system was sufficient to detect variations in beam position at the monitor of the order of 10 μm.

The position monitors were also used to keep the beam position fixed after the initial alignment. Signals derived from the monitors were used in a feedback system that steered the electron beam through the undulator with the aid of two pairs of steering magnets in each plane [6]. This system was essential for the experimental program. It operated with a bandwidth of about 10 Hz, and stabilized the beam on the specimen against short term variations as well as long term drift. The remaining instability that will require further attention is the higher frequency oscillation of the beam that affects time-sequential experiments such as scanning microscopy [7].

Another area that requires particular attention is the safety of personnel and equipment. The surfaces that may be exposed to the primary beam must have adequate cooling, for they may not only distort, but could also fail mechanically due to the power incident upon them. In addition, as the source is located in a long straight section of the storage ring rather than in a short arc of a bending magnet, undulator radiation is accompanied

11

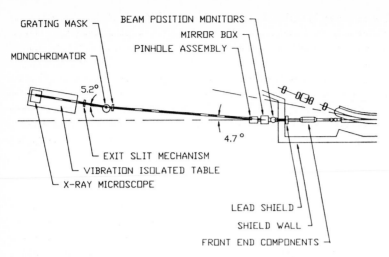

Fig. 1. X17B mini-undulator beamline at the NSLS

by gamma rays from bremsstrahlung created as the electron beam interacts with residual gas. We used a water cooled mirror and a heatpipe cooled pinhole to handle these problems. The mirror was designed and fabricated at the Lawrence Berkeley Laboratory [8]. It is a metal mirror with carefully designed near-surface cooling, mounted on a manipulator with stepping motor driven adjustments in three directions. Its optical performance under load was examined by interferometric methods, and no measurable distortions were found [9]. The mirror deflected the beam by 80 mrad in the horizontal plane, with an energy cutoff at about 1.0 keV due to its nickel coating. The gamma rays were not deflected, but instead were stopped in a stack of lead bricks placed in the straight-through direction.

In normal operation the beam was time shared-with another group [10] switching every 12 hours. The switch involved removing the mirror of one group and inserting that of the other. In practice this operation turned out to be relatively simple, and could be accomplished in less than 15 minutes. The switching also required a special interlock system to assure that the valves associated with the beamline using the beam be open, as the poppets of some of the valves were not appropriately cooled.

The heatpipe-cooled pinhole in our beamline absorbed the bulk of the power, yet transmitted most of the coherent flux [11]. The 300 μm pinhole was formed in a massive copper block by pressing a tungsten wire into the soft copper. It was mounted from a stepping motor driven manipulator to facilitate the positioning of the pinhole. The heatpipe transferred the power from the UHV chamber to cooling fins outside the vacuum.

Past the position monitor, the mirror and the pinhole, the beam intensity could be monitored using an aluminum photodiode mounted on a rotary feedthrough. The next optical element was the toroidal grating monochromator. A set of four masks in front of the grating defined the exposed area. The grating itself was made of fused silica coated with gold. It had a groove spacing of 600 lines/mm, and was mounted on a sine bar driven by a stepping motor. The grating was set to deflect horizontally by 85 mrad. The effective entrance slit of the monochromator was the cooled pinhole, 6.8 m

12

upstream, while an adjustable exit slit was located 0.84 m downstream of the grating. The measured resolving power was in excess of 500 at 400 eV. A second aluminum photodiode downstream of the exit slit made it possible to optimize the performance of the beamline and to monitor the alignment.

From the storage ring to the exit slit of the monochromator, the beam was in an ultrahigh vacuum environment. To gain flexibility in the experimental area, we separated it from the rest of the beamline by using a 150 nm thick aluminum foil. As the region downstream of the foil was still at good vacuum, the foil did not have to withstand significant forces, yet it made it possible to make numerous changes without baking this section of the system.

4. PERFORMANCE

Because of the remarkable collimation properties of undulators, their performance is extremely sensitive to the properties of the electron beam. Consequently, the mini-undulator was four orders of magnitude brighter than a similar undulator at Daresbury, and two orders of magnitude brighter than a 60 period undulator at the Photon Factory before both storage rings were shut down for upgrades to lower emittance lattices. The standard approaches for incorporating emittance effects into the well-understood theory of zero-emittance undulator radiation either fail to explain undulator output at wavelengths other than those of odd harmonics [12], or are rather costly in terms of the computer time used [13]. Over the past two years JACOBSEN [14] has developed a suite of computer programs (*Sensible Modeling of Undulator Throughput, or SMUT*) which make use of the fact that the electron beam spatial and angular distributions are both Gaussian, so their combined footprint at a specified distance from the undulator is also described by a Gaussian distribution. This allows one to integrate the zero-emittance undulator radiation pattern over the electron beam distribution in a reasonably efficient manner.

Measurements of the undulator output were made both with the photodiode located behind the exit slit, and with a gas filled ionization chamber further downstream. To make the comparison between the calculation and the measurement quantitative, the beamline efficiency had to be taken into account. This involved knowing the reflectivity of the mirror, taken from the compilation of HENKE [15], and the efficiency of the grating, calculated by PADMORE [16]. Overall, the efficiency varies between 2% and 11% in the wavelength range 1.4–5.0 nm. Details of the theory and philosopy of the calculations, as well as the comparison with measurements in the case of the mini-undulator, have been presented elsewhere [17].

The angular size of the electron beam, small as it is at the NSLS X-ray ring, is still considerably larger than the intrinsic divergence of the undulator beam. As a result, the on-axis spectrum begins to resemble the angle-integrated undulator output: the radiation is not as sharply peaked in wavelength or in spatial extent as the naive theory would predict. The spectral peaks are shifted to longer wavelengths, and the angular distribution depends little on whether one looks at the fundamental or one of the higher harmonic peaks. The measurements are in very good agreement with the calculations, except that the observed intensity of the third harmonic radiation is somewhat below the predictions. An example of the on-axis undulator spectrum is shown in fig. 2. This value of K maximizes the output at the long wavelength side of the nitrogen K-edge, ideal for scanning microscopy in an air environment.

13

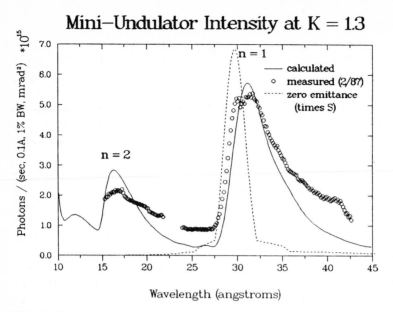

Fig. 2. Comparison between calculated and measured on-axis spectra, at $K = 1.3$; the scale factor S is a function of the actual beam emittance [17]

The spatially coherent flux detected at the experimental area was in the range between 10^7 and 10^8 photons / (sec · 100 mA · 0.1% BW), a value sufficiently large to enable viewing it directly on a surface coated with P-31 phosphor. This flux is two orders of magnitude higher than what we had at the bending magnet beamline U15, although the beamline was not optimized for this undulator. We look forward to the use of the full undulator on beamline X1, which we anticipate will produce another order of magnitude improvement in coherent flux.

5. ACKNOWLEDGEMENTS

We would like to thank the many people who helped make the mini-undulator possible. In particular, we acknowledge Jerry Hastings, Malcolm Howells and Sam Krinsky for essential contributions to the conception of the undulator; John Galayda and Ben Craft for the expeditious commissioning of the device; Hank Hsieh, Tom Oversluizen, Payman Mortazavi, and Martin Woodle for the engineering they provided; Nasif Iskander and Harald Ade for taking some of the intensity data; Howard Padmore for making the grating efficiency calculations; and all of the NSLS staff who made the "mini" a reality.

The Stony Brook effort is supported by the NSF grant **BBS8618066**; the NSLS is supported by the DOE under contract **DE-AC02-76CH00016**.

REFERENCES

1. A.M. Kondratenko, A.N. Skrinsky: *Opt. Spectrosc.* **42**, 189 (1975)

2. H. Maezawa, Y. Suzuki, H. Kitamura, T. Sasaki: *Applied Optics* **25**, 3260 (1986)

3. D.T. Attwood, K.-J. Kim: "Partial Coherence and Spectral Brightness at X-ray Wavelengths", this volume

4. M.R. Howells, J. Kirz, S. Krinsky: "A Beam Line for Experiments with Coherent Soft X-rays", *Brookhaven National Laboratory Informal Report* **BNL32519** (1982)

5. P. Mortazavi, M. Woodle, H. Rarback, D. Shu, M.R. Howells: *Nucl. Instr. Meth.* **A246**, 389 (1986)

6. R.J. Nawrocky, J. Bittner, H. Rarback, L. Ma, P. Siddons, L.-H. Yu: "Automatic Beam Steering in the NSLS Storage Rings Using Closed Orbit Feedback", *Proc. Conference on Synchrotron Radiation Instrumentation (Madison, 1987)*, to be published in Nucl. Instr. Meth.

7. H. Rarback, H. Ade, C. Jacobsen, J. Kirz, I. McNulty: "The Stony Brook / NSLS Scanning Microscope", this volume

8. R. DiGennero, B. Gee, J. Guigli. H. Hogrefe, M.R. Howells, H. Rarback: "A Water-Cooled Mirror System for Synchrotron Radiation", *Proc. Conference on Synchrotron Radiation Instrumentation (Madison, 1987)*, to be published in Nucl. Instr. Meth.

9. S.-N. Qian, H. Rarback, D. Shu, P. Takacs: "In-Situ Shearing Interferometry of National Synchrotron Light Source Mirrors", *Proc. Conference on Metrology: Figure and Finish, SPIE* **749**, 30 (1987)

10. A collaboration consisting of the University of Pennsylvania, Exxon Research and Engineering Co., and BNL staff operated the X17A line. For related work, see for example, I.-W. Lyo, R. Murphy, E.W. Plummer, D. Sondericker, W. Eberhardt: *NSLS Annual Report 1986,* p. 73

11. D. Shu, P. Mortazavi, H. Rarback, M.R. Howells: *Nucl. Instr. Meth.* **A246**, 417 (1986)

12. S. Krinsky: *IEEE Trans. Nucl. Sci.* **NS-30**, 3078 (1983)

13. R. Tatchyn, A.D. Cox, S. Qadri: *Proc. International Conference on Insertion devices for Synchrotron Sources, SPIE* **582**, 47 (1985)

14. C. Jacobsen, H. Rarback: *Proc. International Conference on Insertion devices for Synchrotron Sources, SPIE* **582**, 201 (1985)

15. B.L. Henke, P. Lee, T.J. Tanaka, R.L. Shimabukuro, B.K. Fujikawa: *Atom. Data and Nucl. Data Tables* **27**, 1 (1982)

16. Computer program kindly provided by H. Padmore, Daresbury SRS

17. H. Rarback, C. Jacobsen, J. Kirz, I. McNulty: "The Performance of the NSLS Mini-Undulator", *Proc. Conference on Synchrotron Radiation Instrumentation (Madison, 1987)*, to be published in Nucl. Instr. Meth.

Partial Coherence and Spectral Brightness at X-Ray Wavelengths

D. Attwood and Kwang-Je Kim

Center for X-Ray Optics, Lawrence Berkeley Laboratory,
University of California, Berkeley, CA 94720, USA

Microscopic examination of physical and biological systems requires short wavelength radiation which can be focused to very small dimensions. Indeed it is possible in the limit to both "see" and "write" features approaching the radiation wavelength λ, and furthermore to identify particular chemical elements through their characteristic electronic transitions. Sources of radiation which can deliver a substantial flux of photons to very small dimensions are said to be of high brightness. Sources of high brightness radiate a large photon flux (photons/second), from a small area · solid angle product $(\Delta A \cdot \Delta \Omega)$. If, in addition, the photons are emitted within a spectrally narrow region, $\Delta \lambda$, the radiation is said to be of high spectral brightness. The focusability of such radiation is, however, set by additional attributes of spatial coherence and, of course, by available optics [1]. The emphasis in this paper is on radiation which is not only of high spectral brightness, but which has additional attributes which permit it to be focused to dimensions approaching that set by its finite wavelength, λ. With a near perfect lens [1] such radiation could be focused to dimensions approaching the Rayleigh limit, 1.2 Fλ, where F is the lens F number. Spatially coherent radiation, capable of being focused to wavelength limited dimensions, is often referred to as "diffraction limited". Figure 1 illustrates the concept of source brightness and focusability.

If one considers the propagation of such radiation from its source, to and through requisite optics, it is convenient to describe the radiation in terms of field properties along and transverse to the local propagation direction. It is the transverse properties that determine focusability. When the fields are perfectly correlated everywhere along transverse contours, the radiation is said to be of full transverse coherence, equivalent to being "diffraction limited". Phase correlations in the propagation direction are related to longitudinal or "temporal" coherence. Assuming sufficient transverse coherence properties, the ability to form interference patterns ("fringes") in interferometers, or in holographic recordings, is often described in terms of the longitudinal coherence length [2],

$$\ell = \lambda^2/\Delta\lambda, \tag{1}$$

where $\Delta\lambda$ is the spectral width of the radiation.

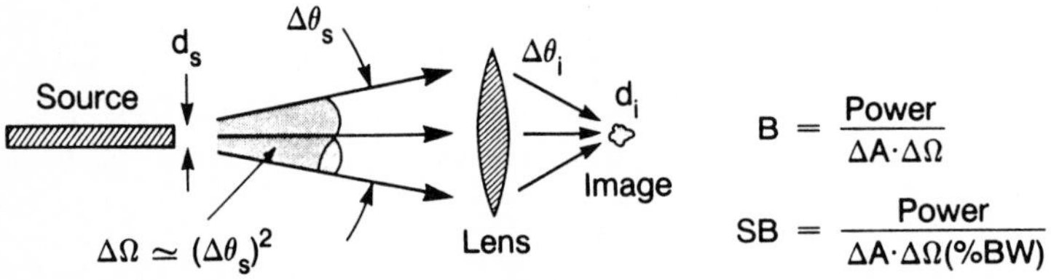

Fig. 1. The left side of the illustration shows a source of aperture d_s radiating into a cone of full angle $\Delta\theta_s$. The brightness of a radiation source measures the power, or photon flux, per unit area and per unit solid angle. For small angles $\Delta\Omega \simeq (\Delta\theta_s)^2$. Brightness is an important measure because it is a conserved quantity in a perfect optical system. For a perfect lens the notion that "brightness is conserved" leads us to $d_s \cdot \theta_s = d_i \cdot \Delta\theta_i$. In a real optical system lens and mirror efficiencies, imperfections, and aberrations modify this statement.

16

As described by WOLF [2], the only true source of coherent radiation is a point source which oscillates with a purely sinusoidal motion for all time. Thus all physical sources, at best, generate partially coherent radiation, having finite transverse and longitudinal coherence properties. This paper deals with the emergence of partially coherent radiation at x-ray wavelengths--radiation suitable for early experimentation with x-ray microscopy, x-ray microholography, and interferometry. When considering phase uniformity, it is of value to clearly understand the wavelength imposed limitations to transverse coherence. For a Gaussian intensity distribution one finds the far field spatial and angular widths, measured at $1/e^2$ intensity points, to be given by [3]

$$\frac{d_{1/e^2}}{2} \cdot \theta_{HW1/e^2} = \frac{\lambda}{\pi}, \tag{2}$$

which represents a situation of full transverse coherence, where the phase at any point on an appropriate surface transverse to the propagation direction is unambiguously related to the phase at any other point on that transverse surface. With a proper lens, radiation obeying eq. (2) could be focussed to "diffraction limited" dimensions, set only by the finite wavelength λ, and the lens $F^{\#}$, e.g. to a diameter of 1.2 $F\lambda$. When the radiation is not diffraction limited, that is eq. (2) is not met, $d \cdot \theta \gg \lambda$, the focussing properties are not as good. As illustrated in Fig. 2, the wavefront is no longer spherical, no longer characterized by a single propogating mode, but rather is multi-mode in nature.

Two particularly interesting sources of partially coherent radiation at short wavelengths, each at the forefront of research on very high brightness radiation, are magnetic undulators on modern electron storage rings, and plasma driven x-ray lasers. Undulators, as illustrated in Fig. 3, are periodic magnet structures traversed by relativistic electrons, typically of energy 1-10 GeV. The radiation from the resultant oscillations is Lorentz and Doppler shifted to very short wavelengths in the ultraviolet and x-ray spectral regions. The radiation cone is greatly narrowed, to units measured in microradians, by these same relativistic effects. If the electrons are maintained within a bunch of small lateral extent and limited random angular motions--that is, within a very small phase spare volume--the resultant radiation can have very good coherence properties. The interesting feature of new storage rings is that the electron beam is contained in an extremely small phase space--or "emittance" as the accelerator community refers to it--allowing the generation of diffraction limited radiation down to soft x-ray wavelengths. For a ring of electron energy $E = \gamma m_0 c^2$, with a periodic magnet structure (undulator) of N periods and periodicity λ_u, the radiation wavelength is given by

$$\lambda_x = \frac{\lambda_u}{2\gamma^2} \left(1 + \frac{K^2}{2}\right), \tag{3}$$

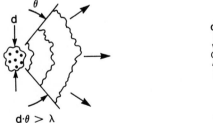

$$d \cdot \theta > \lambda$$

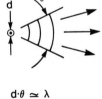

$$d \cdot \theta \simeq \lambda$$

Fig. 2. Diffraction limited radiation is characterized by a size angle product $d \cdot \theta$ just equal to the wavelength. According to the uncertainty principle radiation characterized by that wavelength and radiation cone (θ) could not be used to infer spatial detail of the source any smaller than d. The resultant phase contours are "wrinkle-free" and excellent for phase-sensitive interference experiments like holography, interferometry, and wavelength limited microscopy. A source characterized by $d \cdot \theta > \lambda$ generates a radiation field characterized by irregular phase contours, is not near the uncertainty or wavelength limit, and in general is not as useful for phase sensitive applications.

17

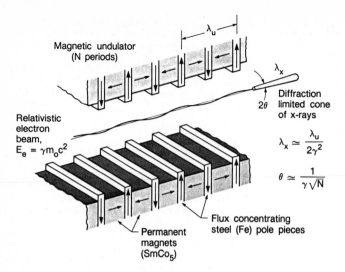

Fig. 3. Undulator radiation is emitted when relativistic electrons traverse a periodic magnet structure. Near diffraction limited radiation results when the electron bunch is contained within a very small phase space, e.g., within a small transverse dimension and within a very narrow angular (trajectory) cone. Modern relativistic storage rings strive to contain the electrons to a phase space volume comparable to that set by diffraction limits at the radiation wavelength, e.g., $(d \cdot \theta)_{e'} \simeq (d \cdot \theta)_{x\text{-ray}} \simeq \lambda$.

where K is the magnetic deflection parameter, and where the radiation cone half angle is

$$\theta_{1/2} \simeq \frac{1}{\gamma\sqrt{N}} \cdot \qquad (4)$$

For a 1.5 GeV machine $\gamma \simeq 3,000$, N is of order 100, $K \simeq 0(1)$, and λ_u is typically measured in centimeters. For best performance the storage ring must have an emittance less than or equal to the phase space of diffraction limited radiation at the wavelength generated. Comparing to eq. (2) this requires that the electron beam be contained in a phase-spare such that [4,5,6]

$$\pi \sigma \sigma' < \frac{\lambda}{4}, \qquad (5)$$

where σ and σ' are the rms measure of electron beam spatial and angular extent, e.g. $\pi \sigma \sigma'$ is the elliptical phase space area. For radiation generated at wavelengths shorter than the phase-space matching condition, eq. (5), the fraction of photon flux or power which is spatially coherent decreases by λ^2, corresponding to the two transverse directions. Another factor λ accrues from decreased coherence length (see eq. 1), so that at decreasingly short wavelengths, coherent power generally declines in proportion to λ^3.

The second major source of very high brightness radiation at short wavelength is that of x-ray lasers. Short wavelength lasers took a giant step in the last three years with the successful plasma driven laser experiments at Lawrence Livermore National Laboratory, [7,8,9] Princeton University, [10,11,12] and more recently, at the Naval Research Laboratory [13]. In these experiments a high power visible (2ω Nd, LLNL) or infrared (CO_2, Princeton) laser is used to form a high temperature, high density plasma of highly ionized atoms. In the Livermore experiments, initially in selenium, the heated atoms collide frequently, stripping off electrons, until a balance is reached between the energy of the colliding ions, and the energy to remove one more electron. Generally an ionization bottleneck can be arranged at some closed shell, as in neon-like selenium, Se^{+24}, where a large fraction (10-30%) of a single ionic configuration is achieved. This is set by choice of the atom (z-dependent ionization energies for the closed shell) and choice of laser intensity, which controls temperature. Collisional pumping of these closed shell ions can then result in lasing as the ion is excited to energy levels for which allowed transitions and decay times result in a population inversion of the proper time scale (matched to the laser-plasma heating and disassembly time) and density to result in observable gain. Multimegawatt outputs in sub-nanosecond pulses have been observed for $3p{\rightarrow}3s$ transitions at 206 and 209 Å in Se plasmas. Lesser outputs have been achieved in yttrium at 154 Å, and molybdenum at 131 Å, using the same scheme of collisionally pumping on neon-like ions. More recent results at Livermore [8] have involved nickel-like atoms, such as europium, where lasing has been observed at 71 Å.

The Princeton results [10,11,12] involve CO_2 laser generation of a magnetically confined carbon plasma. In this case the carbon atoms are fully stripped, and then allowed to recombine with electrons as the plasma expands and cools. As the electron cascades down towards the ground state, it too has achieved lasing on a $3{\rightarrow}2$ transition, at 182 Å. Very high peak powers have also been achieved with these recombination lasers. In both recombination and collisionally pumped lasers, plasma heating and expansion lead to relatively large phase-space volumes of the lasing medium, so that the radiation field is far from diffraction limited, e.g. multimode in nature. Very interesting schemes have been proposed to provide mode control, albeit at a concomitant price in output power. One such idea, an offshoot of earlier ideas on "dot spectroscopy" of laser-plasmas, involves imbedding a very thin lasant plasma column within a larger plasma medium [14]. Figure 4 illustrates the concept. Efforts to achieve shorter wavelengths, moving towards the biologically interesting water window (23 to 44 Å), continue to make progress, but are limited by problems of energetics (ion stripping and excitation), and the avoidance of non-thermal heating processes at higher laser intensities.

Comparison of the various radiation sources is problematical as each has its own advantages and thus is suitable to differing applications. For instance some applications are critically dependent on exposure time or peak intensity, while others are not. Most applications require a certain minimum number of photons within a specified wavelength range or coherence length. Some are critically dependent on reaching a specified atomic or molecular transition energy–and perhaps tuning through that resonance–while others are not. For phase sensitive applications, which may demand a diffraction limited phase front, coherence is thus an issue; for other applications simple brightness, spectral brightness (within a specified relative bandwidth), or photon flux may suffice. Thus in Figs. 5, 6, 7 and 8 we present graphs of both peak (single pulse) and average values of spectral brightness and coherent power. Achieved values are shown as solid lines or circular data points. Anticipated performance is shown as dashed lines. Typical repetition rates for storage rings are in the 100 MHz range.

Acknowledgment

The authors are pleased to acknowledge contributions from their student colleagues N. Iskander, N. Wang, E. Insko and E. Bollt. This work was sponsored by the U.S. Department of Energy, Office of Basic Energy Sciences under contract DE-AC03-76SF00098, and the U.S. Department of Defense, Air Force Office of Scientific Research, under contract T-49620-87-K-0001.

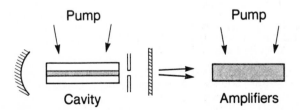

Fig. 4. A scheme for including transverse mode control in a laser-plasma pumped x-ray laser (following reference [14]). The lasant material is contained in a thin filament surrounded by hydrodynamically similar plasma. The pinhole and mirror would limit gain to the lowest order mode. Mirror reflectivities and multi-pass gain duration are areas for future development.

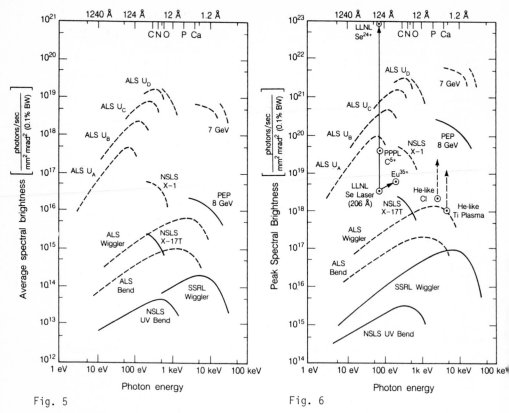

Fig. 5   Fig. 6

Figs. 5-8. Spectral brightness as a function of photon energy is shown for several sources of short wavelength radiation. Average and peak (single pulse) quantities are shown for both. Shown are various sources of synchrotron radiation lasers and laser heated plasmas. Graphs of average and peak coherent power are shown separately. Coherent power is taken to be diffraction limited and of the longitudinal coherence specified e.g. one micron or 100 microns.

References

1. E. Hecht, A. Zajac: Optics (Addison-Wesley, Reading, Mass. 1979).
2. M. Born, E. Wolf: Principles of Optics, ed. 6, ch. 10 (Pergamon, New York, 1983).
3. A. Siegman: Lasers (Univ. Science Books, Mill Valley, CA 1986), ch. 16.
4. D. Attwood, K. Halbach, K-J. Kim: Science 228, p. 1265, 14 June 1985.
5. D. Attwood, K-J. Kim, N. Wang, N. Iskander: Jour. de Phys. (Paris), Coll. C6, supp. 10, 47, October 1986; X-ray Lasers, P. Jaegle and A. Sureau, Ed. (Editions de Physique, 91944 Les Ulis Cedex, France 1986).
6. K-J. Kim: Nucl. Instr. Met., A246, 71, 1986.
7. D.L. Mathews et al.: Phys. Rev. Lett. 54, 110, 1985; M.D. Rosen et al.: Phys. Rev. Lett. 54, 106, 1985.
8. B.J. McGowen et al.: J. Appl. Phys. 61, 5243, 15 June 1987.
9. B.J. McGowen et al.: Phys. Rev. Lett. (to be published).

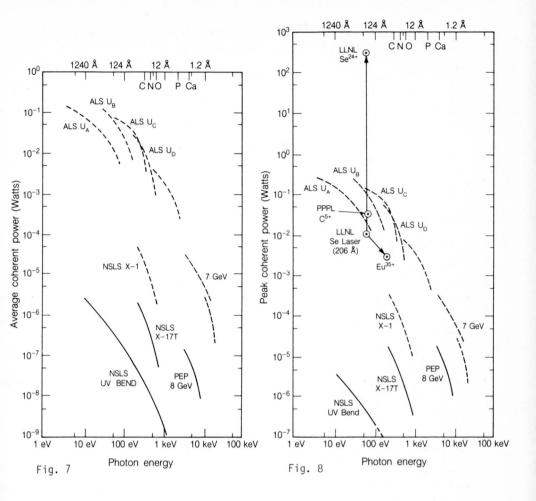

Fig. 7

Fig. 8

Photon energy

10. S. Suckewer et al.: Phys. Rev. Lett. 55, 1753, 21 October 1985.
11. S. Suckewer et al.: Phys. Rev. Lett. 57, 1004, 25 August 1986.
12. C.H. Skinner et al: Proc. SPIE Symp. (San Diego, 1987).
13. T.N. Lee, E.A. McLean, R.C. Elton: Phys. Rev. Lett. 59, 1185, 14 September 1987.
14. M.D. Rosen, J.E. Trebes, D.L. Matthews: Comments Plasma Phys. Controlled Fusion, 10, 245 (1987).

A Plasma Focus as Radiation Source for a Laboratory X-Ray Microscope

W. Neff[1], J. Eberle[1], R. Holz[1], F. Richter[1], and R. Lebert[2]

[1]Fraunhofer Institute for Laser Engineering and Technology,
Steinbachstraße, D-5100 Aachen, Fed. Rep. of Germany
[2]Rheinisch-Westfälische Technische Hochschule Aachen,
Lehrstuhl für Lasertechnik, Steinbachstraße, D-5100 Aachen,
Fed. Rep. of Germany

1. Introduction

To open the X-ray microscopy to a wider range of applications there is need for a user oriented tool for routine biological investigations, which is located in the laboratories of biologists. The requirements are defined by the specimen to be examined:

- They are entire cells in culture, nuclei or other organells, wet and thick (0.1 - 10 μm).
- There should be only short delay between preparation and experiment.
- The specimen should stay under atmospheric conditions and room temperature during examination.
- In order to eliminate blurring due to the motion of the cell and due to radiation damage the exposure time should be as short as possible.

Many of the points mentioned above are met by future microscopes at dedicated beam lines at storage rings. Because of some problems of the more practical rather than principle kind there is need for a compact laboratory type microscope.

The question is, what kind of X-ray microscope can be realized for use in a biological laboratory. The answer depends on the availability of sources bright enough and optics with improved efficiency especially developed to match the emission characteristic of a compact plasma source. All the optical schemes not needing especially coherent radiation can be used for laboratory microscopes as are used for microscopes at storage rings /1,2,3/.

Contact microscopy images have been obtained for thin specimens in many laboratories not using synchrotron radiation. The diffraction-limited resolution is Δx < 50 nm. As an X-ray sensitive resist has to be processed and to be examined with a scanning electron microscope (SEM) artefacts might be introduced during processing /4,5,6/.

With spatially incoherent radiation with a small spectral bandwidth imaging microscopy can be realized /7/. Using a CCD-array as X-ray detector and a pulsed plasma source the image could be obtained directly with nanosecond exposure time.

Ideally, soft X-ray sources for microscopy are intense and tunable. It is important to get the highest spectral brilliance possible. Sources discussed for a laboratory-type microscope are plasma sources generated by gas discharges and laser-generated plasmas. They produce line radiation hopefully with the advantage that no or only a coarse monochromator is necessary and are bright enough to achieve one pulse nanosecond exposure. That sources should be compact, perhaps less expensive and small enough to be used in the biological laboratory /8,9/.

2. The Laboratory-Type Microscope

The size of a laboratory-type X-ray microscope should be comparable with the size of an electron microscope. Figure 1 shows a scheme of the instrument with a plasma focus as a pulsed plasma source. In the first version it is planned to do the experiments in the "water window" region. Intended lines are the resonance lines of hydrogen- and heliumlike ionized nitrogen located around 2.48 nm. Plasma sources for shorter wavelengths around 1 nm especially developed for X-ray lithography can be used in future instruments with phase contrast microscopy /10/.

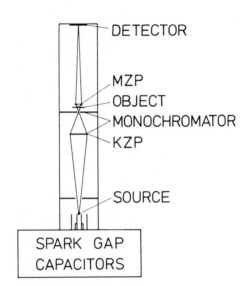

DETECTOR

MZP
OBJECT
MONOCHROMATOR
KZP

SOURCE

SPARK GAP
CAPACITORS

Fig. 1: Schematic set-up of a laboratory-type microscope. The size is planned to be comparable with an electron microscope, about 3 meters high. The source is located about 0.6 meter above floor level

The optical arrangement has been designed and is under construction by the Göttingen Group. The source is demagnified onto a pinhole with a condenser zone plate thus representing a linear monochromator with a spectral resolution of about $\lambda/\Delta\lambda > 75$. Calculations of the modulation transfer function based on the emission characteristics of our source prove that bandwidth sufficient for the desired resolution /11/. The sample is situated in a chamber at atmospheric pressure. The micro zone plate is placed some 100 µm above the sample and is adjusted in an optical transmission microscope arrangement. The image is recorded by film or by real time imaging with CCD /12/. Working with a magnification of about 100 a resolution of 50 nm should be possible. With improved efficiencies of the zone plates and X-ray detectors the expected plasma emission of about 1 J into the full solid angle from a source with a diameter of 200 µm should be sufficient to achieve an exposure with one pulse. With the short exposure time radiation damage does not disturb the image and there is no blurring due to molecular movements.

3. A Plasma Focus as Radiation Source for Soft X-Rays

Plasma focus devices /13,14/ consist of a low inductance capacitive energy storage switched to a gas discharge between two coaxial electrodes. The electrodes are electrically separated by a cylindrical insulator (Fig. 2). A triggered spark gap is used as switching device. The space between the electrodes is filled with several millibars of the working gas (i.e. hydrogen, nitrogen, neon). After applying high voltage across the electrodes a surface discharge along the insulator occurs. A plasma sheath is formed which lifts off from the insulator surface (Fig. 2/I). Due to the fast rising current the energy density of the driving magnetic field is sufficient to provide a plasma expansion as a shock wave (II). Its velocity at the end of the electrode system is about 200 times the speed of sound in the undisturbed gas. After leaving the inter electrode space the plasma is focused towards the axis and a modified Z-pinch is formed (III). The final velocity just before formation of the cylindrical focus plasma reaches 10^7 cm/s.

For several nanoseconds the power density reaches values up to 10^{14} W/cm². Using hydrogen or deuterium as working gas particles are accelerated to energies up to several MeV due to plasma wave interactions /15,16/. In convential plasma focus machines these phenomena have only little reproducibility from discharge to discharge. Improvement of the reproducibility of the plasma focus operating with hydrogen has been achieved by an external control of the ignition phase. The modifications resulted in a higher degree of symmetry and homogeneity of the plasma which leads to smaller pinch radii and higher power densities (Fig. 3a).

The main work with plasma focus machines was done with deuterium as working gas to examine fusion reactions occurringin the plasma column. Operating with light elements the only radiation mechanism in the soft X-ray region is bremsstrahlung of electrons in the electric field of the ions and the recombination radiation continuum which leads to a poor spectral brightness. Higher spectral brightness of the pinched plasma column can only be achieved by using the emission lines of highly ionized elements. Early work with plasma foci to study highly ionized states of high Z elements were done by PEACOCK et al. /17/. They observed line radiation of hydrogen- and heliumlike neon and argon. The device has not been optimized for the use of heavy gases.

The application of the source determines the filling gas that has to be used. For X-ray lithography some work has been done with neon and krypton which results in a plasma emission around 1 nm wavelength. For X-ray microscopy one can choose nitrogen or carbondioxide to produce an emission in the "water-window" range. The lines that can be used are:

2.8787 nm	NITROGEN VI	$1s^2 \longrightarrow 1s2p$	4.0268 nm	CARBON V
2.4898 nm	NITROGEN VI	$1s^2 \longrightarrow 1s3p$	3.4973 nm	CARBON V
2.4781 nm	NITROGEN VII	$1s \longrightarrow 2p$	3.3736 nm	CARBON VI

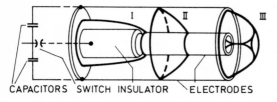

CAPACITORS SWITCH INSULATOR   ELECTRODES

Fig. 2: Schematic drawing of the plasma focus device. The moving plasma sheath is shown in three stages of the discharge (Ignition:I; Rundown: II; Pinch:III)

Experiments with pulsed plasma sources with carbon dioxide and nitrogen have been done /18/. No work has been reported on plasma focus devices operating with these gases.

4. Plasmafocus in Nitrogen

Gas discharges operating in the pressure and electric field regime characteristic for plasma focus devices tend to be inhomogeneous and filamentary. Investigations have shown that inhomogeneities produced during the first few nanoseconds are preserved during the whole discharge and lead to poor plasma focusing. The problem was solved for hydrogen (Fig. 3 a) :
- It is necessary to produce a sufficient number of initial electrons in the ignition phase. This can be done by field emission from a cylindrical knife edge around the insulator.
- The material of the insulator and the roughness of its surface are selected in order to the electron emission coefficient. A high emission coefficient results in homogeneous plasma sheath along the insulator surface by avalanche processes.
- These mechanisms need a high power input which is provided by a low inductive, triggered spark gap arrangement designed especially for plasma focus devices.

The dynamics of the plasma is controlled in order to sweep the whole gas out of the electrode system by complete ionization of the sheath and to obtain minimum sheath thickness to get maximum compression and power density in the pinch.

Plasma focus operation in nitrogen is different from the operation in hydrogen or noble gases:

- The Townsend coefficient for nitrogen is larger. Hence parasitic discharges in the volume between the electrodes have a higher growth rate.
- The other kind of gas adsorbed on the insulator may influence the breakdown mechanism.

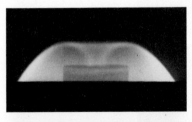

a

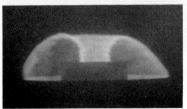

b

Fig. 3: Image converter images of the plasma sheath during collapse. Exposure time is 10 ns.
a) Controlled operation in hydrogen
b) Operation in nitrogen

- The rundown phase is modified due to a high number of internal degrees of freedom such as different stages of ionization and excitation.
- It is expected that radiation processes play a role for the shock wave structure. UV - radiation may preionize the neutral gas in front of the shock wave and influence the shock thickness.

The problems resulting from these differences represent a major challenge and are not yet completely solved. Up to now it is not possible to achieve the same degree of homogeneity and symmetry of the plasma sheath as in discharges with hydrogen (Fig. 3b).

5. Experimental Results

Using a pinhole camera with a 5 µm beryllium foil and typical pinhole diameter of 15 µm the X-ray emitting region was recorded. The side on view shows a cylindrical X-ray source with a diameter of d = 400 µm and a length of about 1 = 5 mm (Fig. 4). For microscopy the source is used end on. The whole length of the source is used by the optics if the spectral resolution of the monochromator is below $\lambda/\Delta\lambda = g/l = 150$; where g is the distance from the source to the condenser zone plate.

The emission time of the radiation was measured with a microchannel plate detector having a temporal resolution of about 500 ps. The emission lasts about 8 nanoseconds with a rise time below 1 nanosecond (Fig. 5). With a streak camera no magnetic confinement of the pinch was observed. So it is possible to estimate the duration of the emission to 4 nanoseconds in good agreement with the measured size of the emitting region (d=400 µm) and the measured sheath velocity of 10^7 cm/s. Reduction of the sheath thickness leads to a smaller source and a shorter emission time.

Using a 2 m grazing incidence spectrograph the emission spectrum from one single discharge was recorded. In order to observe the spectrum of one pulse a resolution of $\lambda/\Delta\lambda = 130$ was selected to get sufficient illumi-

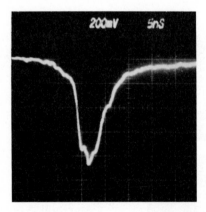

Fig. 4 (left): Side on pinhole image of the source behind 5 µm beryllium

Fig. 5 (right): Time function of the plasma emission measured with a micro-channel plate detector

nation. The grating spectrograph was not calibrated, so that the emitted energy could only be determined from a spectrum taken with a crystal spectrograph. Fitting the grazing incidence spectrum to the calibrated crystal spectrum in the wavelength region around 1.9 nm the emitted energy of the spectral line at 2.4791 nm is 0.5 J into 4π solid angle.

Figure 6 shows the measured spectra. The photometric density of the photoplate is corrected taking into account the spectral efficiencies and transmissions. The spectrum consists of the Lyman series and recombination continuum of the most abundant helium- and hydrogenlike ions. For X-Ray microscopy we want to use the strongest line at 2.49 nm that consists of two lines not resolved here: The Lyman-α line of hydrogenlike nitrogen and the $1s2p-1s^2$ transition of heliumlike nitrogen mentioned above. The separation of these lines $\Delta\lambda=0.0117$ nm corresponds to a bandwidth of $\lambda/\Delta\lambda > 200$. This is large enough to use them as one line with the optics planned /11/.

The line widths in Figure 6 are determined by the experimental setup. The real line widths are influenced by the following broadening mechanisms occurringin plasma sources:

Atomic	$\lambda/\Delta\lambda$	$=$	$4.8 \cdot 10^8 \cdot (Z^4 \ \lambda/nm)^{-1}$	$= 8.0 \cdot 10^4$
Stark	$\lambda/\Delta\lambda$	$=$	$4.1 \cdot 10^{16} \cdot (n_i/cm^{-3})^{-2/3}$	$= 3.5 \cdot 10^3$
Doppler	$\lambda/\Delta\lambda$	$=$	$1.3 \cdot 10^4 \cdot M/(kT_i/eV)$	$= 1.5 \cdot 10^3$

Z : Nuclear charge $= 7$ λ : Wavelength $= 2.4$ nm
n_i: Ion density $< 10^{19}$ cm^{-3} M : Ion mass $= 14$

Using one line with the resulting bandwidth of $\lambda/\Delta\lambda > 1000$ high resolution can be achieved without additional narrowband filtering.

The spectral characteristics are strongly influenced by the pressure of the filling gas and the discharge voltage. Taking the spectra from Fig. 6 we changed the pressure from 3 mbar to 1.5 mbar. Thereby the continuous background was decreased by a factor of two. The pressure and voltage

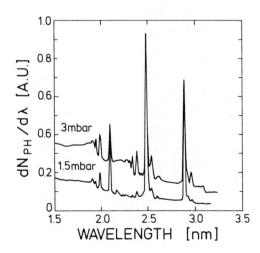

Fig. 6: Emission spectra of the nitrogen X-ray source at different filling gas pressures taken with a 2 m grazing incidence spectrograph.

regime has to be chosen in order to optimize the emission of the Lyman-α line of hydrogenlike nitrogen and to suppress the continuum radiation.

6. Summary and Conclusions

It has been demonstrated that a source for a laboratory-type microscope can be realized on the basis of a plasma focus device.
- The size of the planned laboratory X-ray microscope with the plasma focus source will be similar to the size of an electron microscope.
- The source emits about 10^{16} photons per discharge into $4\,\pi$ solid angle in the Lyman-α line of hydrogenlike nitrogen. That is just about a factor of two below the energy necessary for single pulse exposure.
- The emission time is below 10 nanoseconds. That might be short enough to avoid undesired blurring.
- A coarse monochromator with a bandwith of $\lambda/\Delta\lambda$ = 75 is sufficient to obtain a resolution of 50 nm. To achieve even higher resolution with future zone plates one doppler broadened line with $\lambda/\Delta\lambda$ = 1000 might be used.

7. Acknowledgements

This work has been supported by the German Federal Minister for Research and Technology (BMFT) under the contract number 13N5329A/0.

8. References

1. M. Howells, J. Kirz, D. Sayre, G. Schmahl: Physics Today, August 1985
2. J. Kriz, D. Sayre: "Prospects and Problems in X-Ray Microscopy", in X-Ray Microscopy, ed. G.Schmahl und D. Rudolph, (Springer, Berlin 1984)
3. J. Kirz, H. Rarback: Rev. Sci. Instrum. 56 (1), 1 (1985)
4. R. Feder, V. Banton, D. Sayre, J.L. Costa, M.G. Baldini, N.K. Kim, Science 227, 63 (1985)
5. R.W. Eason, P.C. Cheng, R. Feder, A.G. Michette, R.J. Rossner, F.O. Neil, Y. Owando, P.T. Rumsby, M.J. Shaw, I.C.E. Turuc: Opt. Acta 33 (4), 501 (1986)
6. R.J. Rossner, R. Feder, A. Ng, F. Adams, M. Caldarolo, P. Celliers, P.C. Cheng, L. Da Silva, D. Parfeniuk, R.J. Speer: J. Microsc. 144 (2), 5 (1986)
7. D. Rudolph, B. Niemann, G. Schmahl, O. Christ: "The Göttingen X-Ray Microscope and X-Ray Microscopy Experiments at the BESSY Storage Ring", in X-Ray Microscopy, eds. G. Schmahl und D. Rudolph, (Springer, 1984)
8. D.T. Attwood, K-J. Kim:"Overview of Soft X-Ray Sources",this volume
9. F.O. Neill, Y. Owandano, I.C.E. Turuc, A.G. Michette, C.H. Hills, A.M. Rogoyski:"Low Power Pulsed Lasers as Plasma Sources for Soft X-Rays" in AIP Conf. Proc. No. 147, 354 (1986)
10. G. Schmahl, D. Rudolph and P. Guttmann,"Phase Contrast X-Ray Microscopy - Experiments at the BESSY Storage Ring", this volume
11 D. Rudolph, B.Niemann, G.Schmahl and J. Thieme: "Status of a Laboratory Microscope", this volume
12 W. Mayer-Ilse: "Applications of Charge Coupled Detectors", this volume
13. J.W. Mather: "Dense Plasma Focus", in Methods of Experimental Physics , ed. by R.H. Loveberg, H.R. Griem, Vol.9B (Academic Press, New York, London 1971) p.187

14. N.V. Filippov, T.I. Filippova, V.P. Vinogradov, Nucl. Fusion suppl. $\underline{2}$, 577 (1962)
15. G. Herziger: "X-Ray Emission from a 1 kJ Plasma Focus" , in X-Ray Microscopy, ed. G.Schmahl und D. Rudolph, (Springer, Berlin 1984)
16. H. Krompholz, G. Herziger, "Phenomena of Selforganization in Dense Plasma", in Chaos and Order in Nature, ed. H. Haken, Springer Ser. Syn., Vol.11 (Springer, Berlin, Heidelberg 1983)
17. N.J. Peacock, R.J. Speer, M.G. Hobby: J. Phys B $\underline{2}$, 798 (1969)
18. P. Choi, A.E. Dangor and C. Deeney:"A Small Gas Puff Z-Pinch Source", in Soft X-Ray Optics and Technology ed. E.E. Koch and G. Schmahl, SPIE Vol. $\underline{733}$, 52 1986

Lawrence Livermore National Laboratory
Soft X-Ray Laser Program*

J. Trebes, S. Brown, E.M. Campbell, N.M. Ceglio, D. Eder, D. Gaines, A. Hawryluk, C. Keane, R. London, B. MacGowan, D. Matthews, S. Maxon, D. Nilson, M. Rosen, D. Stearns, G. Stone, and D. Whelan

Lawrence Livermore National Laboratory, University of California, P.O. Box 5508, L-473, Livermore, CA 94550, USA

1.0 Abstract

The Laboratory soft x-ray laser program at the Lawrence Livermore National Laboratory's Nova laser facility has been pursuing soft x-ray laser research with the goal of developing x-ray laser physics and the necessary technology for producing usable x-ray lasers. For the microscopy community, this means developing short wavelength (λ < 44 Å), high power (10^8 W), highly coherent lasers to be used in microscopy and holography applications. Significant progress toward this goal has been achieved. A variety of x-ray lasers based on the collisional pumped Ne-like laser scheme have been developed and characterized in great detail. A new collisionally pumped laser utilizing Ni-like ions has been demonstrated with laser wavelengths as short as 50 Å. This new scheme is scalable to sub 44 Å wavelengths. Preliminary ionization balance experiments have also begun for recombination pumped lasers which should also readily scale to sub 44 Å wavelengths. A laser cavity has been developed and demonstrated using multi-layer x-ray mirrors and beam-splitters. This offers the possibility of improved laser coherence and efficiency. An applications beamline was constructed and used in a photo-ionization physics experiment in conjunction with Bell Laboratory. This was an important demonstration of the ability to make existing x-ray lasers usable and available for the general science community. Finally, x-ray holography using an x-ray laser has been demonstrated.

2.0 Neon-Like Lasers

The neon-like lasers developed at LLNL are produced when a powerful optical laser pulse ionizes an exploding foil geometry target.[1] The population inversion is produced by both collisional excitation and di-electronic recombination during the optical laser drive pulse.[1,2] The typical irradiance on the foil is 5 x 10^{13} W/cm^2 of 0.5 μm light in a 500 ps gaussian pulse. With a selenium coated plastic foil, this produces laser emission at five wavelengths ranging from 182 and 263 Å. The dominant emission occurs at 206 and 209 Å. Output powers of several megawatts have been measured on these lines.[3] The output intensity at 206 Å as a

---

*Work performed under the auspices of the U.S. Department of Energy by the Lawrence Livermore National Laboratory under contract number W-7405-ENG-48.

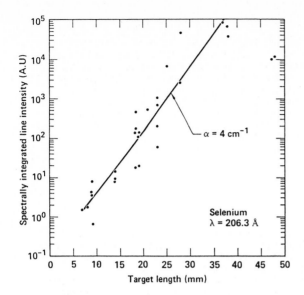

Figure 1. Measured intensity as a function of length for the Se laser.

function of length is shown in Fig. 1. The laser emission increases more than five orders of magnitude above the spontaneous emission level as the laser length is increased. The divergence of the Se laser has been measured to be 10 mrad x 20 mrad with some time-dependent structure due to refraction of the x-rays in the time evolving density profile of the laser amplifier.[3] The laser source size has also been measured as a function of length.[4] Initially, the source size has a diameter of 250 μm. As the laser length is increased, the size decreases to a dimeter of 50 μm. With this source size, the output energy of 150 μJ from a 4 cm long laser is greater than the predicted saturation output energy of 60 μJ. Wavelength scaling using the Ne-like scheme has been demonstrated down to wavelengths as short as 106.4 Å in Ne-like molybdenum.[5] Unfortunately, the power and energy required to pump this system for producing wavelengths below 44 Å are beyond the capabilities of existing visible light lasers.

3.0 Nickel-Like Lasers

In an effort to achieve sub 44 Å lasers, Ni-like lasers are being developed. Nickel-like lasers offer scalability to sub 44 Å wavelengths with power requirements which can be met by existing visible light lasers such as the Nova Laser at LLNL. The Ni-like laser is an analog to the Ne-like laser except 4-4 transitions are used as laser transitions instead of the 3-3 transitions in the Ne-like lasers. The 4-4 transitions in Ni-like ions have a shorter wavelength for a given ionization potential than due to 3-3 transitions in Ne-like systems. This results in less power being required for lasing to be achieved. Lasing at 66 and 71 Å has been demonstrated in Ni-like Eu.[6] The measure gain coefficients are about 1 cm^{-1} with gain length products of up to 3.8 being obtained. Evidence for gain at 50.5 Å has been obtained in Ni-like Yb.[6] Experiments using Ni-like tungsten with laser emission at 43 Å are planned.

4.0 Laser Cavities

Laser cavity development is important for the improved efficiency, divergence, and coherence of soft x-ray lasers. Multi-pass amplification of a soft x-ray laser has been demonstrated with a cavity consisting of a spherical multi-layer mirror and a flat multi-layer beamsplitter placed at either end of the Se laser (also see N. Ceglio, these proceedings).[7] Both double and triple-pass amplification experiments were conducted with time resolved measurements of single, double, and triple-pass laser emission being obtained. In the double-pass experiments, the second pass was about seven times more intense than the initial amplified spontaneous emission. In the triple-pass experiments, the second pass was about 20 times more intense than the first pass, while the third pass is comparable to the first pass.

5.0 Recombination Lasers

Recombination lasers, such as those demonstrated by Suckewer, et al.,[7] and Key et al.,[8] have simple atomic physics and a rapid scaling to short wavelengths. Hydrogen-like Al has predicted gains of 5 cm on the 3-2 transition at 39 Å using a drive laser intensity of 3×10^{15} W/cm^2. The output pulselength is predicted to be 10 ps in duration.[9] Lithium-like Cr is predicted to also have laser emission at 39 Å with a drive laser power of 3×10^{14} W/cm^2. Ionization balance and spectral identification experiments are underway as a preliminary step to producing sub 44 Å lasing using recombination laser systems.[11]

6.0 Applications Beamline

The Se x-ray laser wavelengths (~200 Å), pulse length (~200 ps), and output energy (0.5 mJ) are well matched to test the population inversion physics for a 372 Å photo-ionization laser in Na^{+1} proposed by W. Silfvast and O. Wood.[12] In order to utilize the Se laser, an applications beamline was constructed to collimate, relay, and focus the x-ray laser beam into a Na cell.[13] A schematic drawing of the system is shown in Fig. 2. The beamline consisted of two spherical multi-layers and a thin Al filter. The spherical mirrors had a peak reflectivity of ~26% at 208 Å and a radius curvature of 4 m. The narrow bandpass (~10%) of the mirrors was used to eliminate the background emission due to the non-lasing processes. The 1000 Å thick Al filter was used to eliminate stray 0.5 μm visible light from the drive laser scattered by the Se exploding foil target. A measured spot size of 300 μm in diameter was obtained using this optics system. This measurement was done with the optics slightly defocused. This spot size is consistent with a spot size of 100 μm in diameter at best focus. The Se laser beamline has been used in a preliminary photo-ionization experiment with argon.[14] Experiments with Na are planned.

7.0 Holography

X-ray holography offers the possibility of obtaining high resolution (<500 Å) three-dimensional images of live biological specimens on timescales short compared to biological processes. This requires high power, highly coherent, sub 44 Å lasers. With the demonstration of Ni-like lasers, the demonstration of laser cavities, and the possibility of oscillator-amplifier x-ray laser chains,[15] such high quality lasers

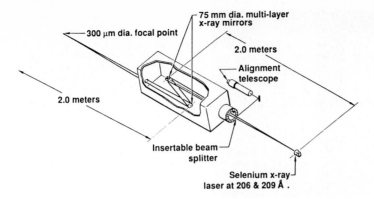

● Two multi-layer mirrors focus the x-ray laser
  to a 300 μm dia. point.

● The laser can be pointed with a 75 μrad accuracy
  with this mirror configuration.

Figure 2. Schematic of the x-ray laser application beamline.

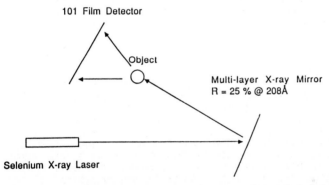

Figure 3. Schematic arrangement of the modified Gabor holography geometry showing the narrow bandpass x-ray mirror.

may soon be available for holography of biological specimens. As a first step toward the goal of making holograms of biological specimens, a proof of principle x-ray laser holography experiment has been demonstrated.[16]

The Se laser at 206 and 209 Å was used in a modified Gabor geometry. A schematic of the system is shown in Fig. 3. An ultra-smooth (< 1 Å rms roughness), ultra-flat (λ/100) narrow bandpass multi-layer mirror is used to relay the x-ray laser beam to the holography object. The narrow bandpass (10%) eliminates high energy background. X-ray film was used as a detector. An Al filter was used to protect the film from stray light. The primary object was an 8 μm diameter carbon fiber.

The results are shown in Fig. 4. The top portion shows a false gray scale image of the hologram fringe pattern with the film response decon-

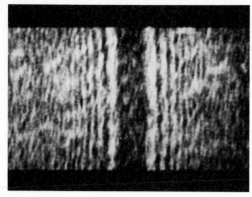

a

Figure 4. a) Fringe pattern from hologram of 8 μm wire; b) average intensity as a function of position in the fringe pattern; c) calculated intensity as a function of position for a hologram of an 8 μm wire.

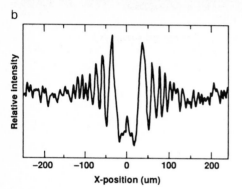

b

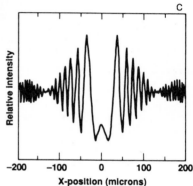

c

volved numerically. The middle portion shows the average intensity across the data as a function of position, the bottom portion shows a calculated fringe pattern for the wire. Agreement is quite good except at large distances from the center of the hologram where the fringe spatial frequency is too high compared to the spatial resolution of the film. Holograms of three-dimensional gold bar figures were also obtained and reconstructions using a He-Ne laser were successfully produced of both the wire and the three-dimensional structure. Alternative holography systems which offer potentially higher resolution and the requirements for making holograms of biological specimens with sub 44 Å lasers are being developed.

8.0 Conclusion

The LLNL soft x-ray laser program is now focussing on the development of sub 44 Å laser schemes which can be scaled to high power and high coherence. The technologies for this such as cavities, oscillator amplifier designs and x-ray optics are being developed. There is also an initial effort in developing x-ray laser holography as a useful tool in high resolution microscopy.

9.0 References

1. D. Matthews, et al.: Phys. Rev. Lett. $\underline{54}$, 110 (1985)
2. M. Rosen, et al.: "On the Dynamics of Collisional Excitation X-ray Lasers," to be published
3. D. Matthews, et al.: J. Optical Soc., to be published
4. D. Whelan, et al.: SPIE
5. B. J. MacGowan, et al.: "Observation of X-Ray Amplification in Neon Like Molybdenum," Lawrence Livermore National Laboratory Report No. UCRL-95154 (1986)
6. B. J. MacGowan, et al.: "Demonstration of Soft X-ray Amplification in Nickel Like Ions," submitted to Phys. Rev. Lett.
7. N. Ceglio, et al.: "Multi-Pass Amplification of Soft X-rays in a Laser Cavity," submitted to Optics Letters
8. S. Suckewer, et al.: Phys. Rev. Lett. $\underline{55}$, 1753 (1985)
9. M. Key, et al.: J. de Physique $\underline{C6}$, 71 (1986)
10. D. Eder, et al.: "Recombination X-ray Lasers Using H-Like Magnesium and Aluminum," submitted to JOSA B
11. C. Keane, et al.: to be published
12. W. Silfvast, et al.: Soft Wavelength Coherent Radiation: Generation and Applications, D. Attwood and J. Bokor, eds. (AIP, NY 1986)
13. D. Nilson, et al.: to be published
14. W. Silfvast, O. Wood, et al.: to be published
15. M. Rosen, et al.: Comm. Plasma Phys. "Controlled Fusion," $\underline{10}$ 245 (1987)
16. J. Trebes, et al,: Science Magazine, October 23, 1987

X-Ray Laser Sources for Microscopy

C.H. Skinner, D.E. Kim, A. Wouters, D. Voorhees, and S. Suckewer

Princeton University, Plasma Physics Laboratory Princeton, NJ 08544, USA

Progress and prospects in soft X-ray laser development at Princeton are presented. A comparison to plasma and synchrotron sources is made with a view to applications in microscopy.

1. Introduction

A significant development since the publication of the proceedings on the previous International Symposium on X-ray Microscopy[1] has been the development of soft X-ray lasers. Following the first joint announcement by groups at Princeton University and Lawrence Livermore National Laboratory at the American Physical Society Meeting in Boston in November 1984, progress in increasing the output power and range of soft X-ray lasing wavelengths has been rapid. These sources are beginning to be used in several applications. In this article we will review progress and prospects in soft X-ray laser development at Princeton and compare the performance to synchrotron and plasma sources with a view to applications in microscopy. First results from a contact microscope based on the soft X-ray laser are presented in a separate article in this volume[2].

2. The Princeton Soft X-ray Laser

In the Princeton Soft X-Ray Laser population inversion and gain is generated in a rapidly recombining plasma confined in a magnetic field. A commercial CO_2 laser (maximum energy 1 kJ, duration 70 nsec) is focussed onto a carbon disc located in a strong (up to 90 kG) magnetic field, creating a carbon plasma of sufficient temperature that the electrons are stripped off most of the carbon ions. After the laser pulse the plasma cools rapidly by radiation losses and fast three-body recombination preferentially populates the upper excited levels in hydrogen-like carbon, CVI (see Fig. 1). The $m = 2$ level is rapidly depopulated by the strong $2 \to 1$ radiative transition and in this way a population inversion is built up between level $n = 3$ and $m = 2$ leading to amplified spontaneous emission at 182 Å. Gain at shorter wavelengths has been measured using lithium like ions i.e. AlXI(154 Å) and SiXII (129 Å).

The strong confining magnetic field provides several advantages. It enables the electron density to be controlled to the optimum value and forms the plasma to a long thin geometry suitable for a laser. The geometry is also suited to fast radiation cooling, and the radiation cooling can be enhanced by the addition of high Z materials.

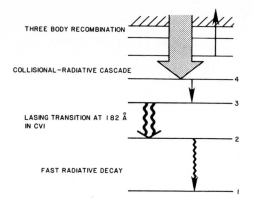

THREE BODY RECOMBINATION

COLLISIONAL—RADIATIVE CASCADE

LASING TRANSITION AT 182 Å
IN CVI

FAST RADIATIVE DECAY

Fig. 1 Principle of generation
of population inversion and
lasing action in hydrogen-like
ions in a rapidly recombining
plasma.

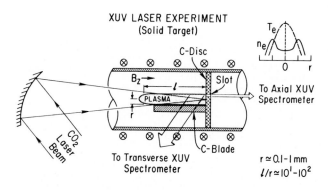

XUV LASER EXPERIMENT
(Solid Target)

C-Disc

B_z

Slot

PLASMA

To Axial XUV
Spectrometer

CO_2 Laser Beam

C-Blade

To Transverse XUV
Spectrometer

$r \simeq 0.1-1$ mm
$l/r \simeq 10^1-10^2$

Fig. 2 Cross section
of target chamber.
Plot of radial densi-
ty and temperature
profile in plasma.

The CO_2 laser is focussed into a target chamber inside a solenoidal
magnet that is surrounded by an array of spectrometers. The spectrometers
are absolutely calibrated and measure the soft X-ray gain by measuring
the directionality of the emitted radiation. In a long thin plasma
most of the stimulated emission is along the plasma axis and the gain
is measured by comparing the axial to transverse intensity.

Figure 2 shows the target in detail. A carbon blade attached to
the carbon disc helps to generate an elongated plasma. It also helps
to cool the plasma by thermal conduction and additional radiation losses.
An important feature illustrated here is the radial profile of the plasma.
The plasma pressure is balanced by the magnetic field pressure so that
on axis there is a high temperature due to heating by the laser and
corresponding low density. In the outer cooler regions the density
is higher and it is in these outer regions that the conditions are most
favorable for fast recombination. The gain is generated in an annular
region around the center of the cylindrical plasma. An off-axis slot
in the target transmits the stimulated emission to the axial spectrometer.
The system has a very accessible output beam, can be fired every few
minutes and operated by only two people.

Gain is generated on the 3-2 transition in CVI at 182 Å. The 4-2 transition at 135 Å emits mostly spontaneous emission so one way to measure

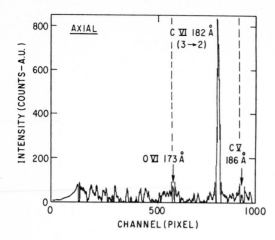

Fig. 3 Axial spectra in the region near 182 Å from a carbon disk target with four carbon blades. The laser energy was 500 J.

gain is by the axial enhancement of the 182 Å emission compared to the spontaneous, isotropic 135 Å emission. A second, independent way is to use the spectrometer calibrations to measure the relative intensity in the two directions. Reference 3 describes gain measurements of gL = 6.5 corresponding to an axial enhancement of the 182 Å emission of a factor of 100. This is illustrated in the spectra in Fig. 3, showing the very high intensity of the 182 Å emission in the axial direction. Measurements of higher values of gain, gL $\approx$ 8, with enhancement of up to 500 of the 182 Å line are presented in Ref. 4.

Usually the axial emissions are collected by a grazing incidence mirror and focussed onto the entrance slit of the axial spectrometer. Since the mirror is made by bending a piece of float glass the optical quality is not ideal and the spatial distribution of the light at the slit contains a mix of angular and spatial information of the emission from the target. By scanning the axial instrument back and forth in the transverse direction we can measure the relative divergence of lasing compared to non-lasing lines. Figure 4 shows the lasing line at 182 Å highly peaked in comparison to the relatively constant spontaneous emission lines. To get an absolute measure of the divergence the grazing incidence mirror was removed and the axial spectrometer scanned across the soft X-ray beam. A beam divergence of 5 mrad was measured at magnetic fields of 35 and 50 kG [4].

More recently we have been generating gain at shorter wavelengths in lithium-like ions by using aluminum and silicon targets. The approach is analagous to that for carbon and Fig. 4 shows the relative divergence of the AlXI 154 Å line with a gain-length of gL $\approx$ 3-4, and the relative divergence of the SiXII line at 129 Å with an estimated gain-length of gL $\approx$ 1-2.

3. Two Laser Approach to Wavelengths Significantly Below 100 Å

The advantage of illumination in the 24 Å-44 Å wavelength region is well known in microscopy. However we wish to point out it is not the only region in which contrast may be obtained. At 182 Å the ratio between the absorption cross sections for oxygen and carbon is a factor of three

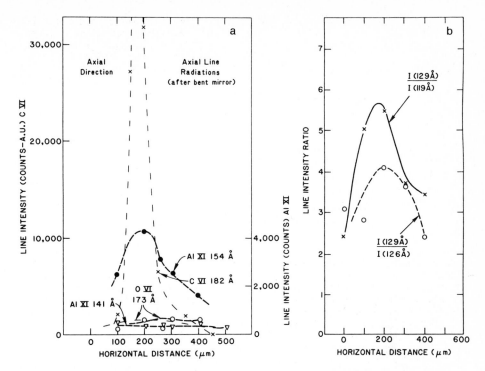

Fig. 4(a) Relative divergence of the AlXI 154 Å stimulated emission. The spectral line intensities of the AlXI 154 Å, and OVI 173 Å lines are shown as a function of the transverse position of the axial spectrometer. For comparison, relative divergence measurements of the CVI 182 Å stimulated emission compared to the OVI 173 Å emission are shown (light dashes).

(b) Relative divergence of the SiXII 129 Å stimulated emission. The line intensity ratios SiXII 129 Å/119 Å and SiXII 129 Å/126 Å are shown as a function of the transverse position of the axial spectrometer.

which should provide sufficient contrast for biological imaging. Extension of the operating range of current soft X-ray lasers to the 24 Å-44 Å range is a significant technical challenge due to the severe demands placed on pump laser power at shorter wavelengths (see e.g. [5]). The equipment cost may be reduced however by using a very short pulse laser[6]. A picosecond laser pulse length is also well matched to the radiative time scales in candidate ions.

An experiment to investigate a two-laser approach to laser action at wavelengths significantly below 100 Å has been constructed at Princeton with the first experiments beginning in Fall 1987. The basic approach is to split the task of creating a soft X-ray gain between two pump lasers (Fig. 5). First a CO_2 or Nd laser creates a highly ionized plasma containing ions of the appropriate ionization stage confined in a magnetic field. Then a powerful picosecond laser ($I \sim 10^{18}$ W/cm^2) populates a

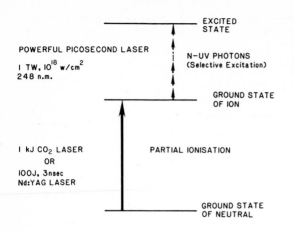

Fig. 5 One scheme considered for generating a population inversion in the "Two
Laser Approach".

selected excited state in the ion by multiphoton excitation. At these
laser powers very high order multiphoton transitions are efficiently
excited and are expected to selectively populate a specific state,
generating a population inversion. Candidates include inner shell and
doubly excited transitions in argon-like and krypton-like ions. This
approach was stimulated by the observation of stimulated emission at
931 Å by multiphoton excitation of an inner shell transition in neutral
krypton [7].

4. Cavity Development

We are also working on cavity development with the aim of increasing
the brightness of our soft X-ray laser. Without mirrors the laser beam
divergence is governed solely by the plasma geometry - any ray that
can pass through the whole gain region can be amplified and the divergence
at present, is 5-10 mrad. With a properly designed laser cavity, the
divergence could be near diffraction limited with a potential increase
in brightness of up to 10^6. To properly establish the cavity modes,
however, many passes through the gain medium are needed and a long
duration gain is necessary. The Princeton 182 Å laser with gain duration
10-30 nsec is ideally suited to cavity development.

Early work at Princeton using newly developed multilayer mirrors
in a double pass arrangement resulted in a factor of two intensity
increase due to the amplification of the stimulated emission in the
second pass[8]. Mirror alignment posed severe difficulties, however
a new experimental set designed to overcome these problems has been
constructed and cavity experiments are planned for the near future.

5. Comparison to Undulator and Plasma Sources

A comparison of laser brightness and coherent power to that of synchrotron
bending magnets and undulators is given elsewhere in this volume[9].
Increases in brightness of several orders of magnitude are anticipated
in future soft X-ray lasers with the implementation of soft X-ray
cavities. For applications in which peak powers or nanosecond exposures
times are required (e.g. flash imaging of live cells) laser sources

are ideal. Applications requiring a high average power are more suited to undulators so the two types of devices are complementary in terms of applications.

Table 1 compares the parameters of the Princeton X-ray Laser to the plasma focus source at Aachen[10]. While the total energy emitted in the plasma focus is higher than the soft X-ray laser, the light is emitted over 4 π solid angle so that the energy per steradian for the laser is more than two orders of magnitude higher. In addition the plasma focus emission extends over a wide wavelength range and a monochromator is necessary with a reduction of useable intensity of one to two orders of magnitude. However the plasma focus does have the advantage (not yet matched by present X-ray lasers) of operating in the water window (24 Å-44 Å) wavelength region.

Table 1

	Aachen Plasma Focus (spontaneous emission)	Princeton Soft X-Ray Laser (stimulated emission)
Energy	$\approx$ 500 mJ	1-3 mJ
Solid Angle	4 π	5 mrad
Energy/Steradian	40 mJ.sr^{-1}	4,000-12,000 mJ.sr^{-1}
Duration	5 nsec	10-30 nsec
Bandwidth	continuum & lines (monochromator necessary)	$\frac{\lambda}{\Delta\lambda} \approx 10^5$
Wavelength	25-29 Å	182 Å (154 Å, 129 Å)

Acknowledgements

We would like to thank H. Furth, J. Hirschberg, E. Valeo, J.L. Schwob, for helpful discussions, L. Meixler and D. Dicicco for technical advice; and Andrew Schuessler, J. Schwarzmann, and Steve Cranmer for target preparation and technical support. This work was made possible by financial support from the U.S. Department of Energy, Basic Energy Sciences, the U.S. Air Force Office of Scientific Research and NRL/SDIO.

References

1. G. Schmahl and D. Rudolph: X-Ray Microscopy, Proceedings of the International Symposium, Goettingen, Federal Republic of Germany, September 14-16, 1983, Springer Series in Optical Sciences 43, Springer Verlag 1984.

2. D. Dicicco, L. Meixler, C.H. Skinner, S. Suckewer, J. Hirschberg, and E. Kohen: Soft X-Ray Laser Microscopy, this volume

3. S. Suckewer, C.H. Skinner, H. Milchberg, C. Keane, and D. Voorhees: Phys. Rev. Lett. 55, 1753 (1985).

4. S. Suckewer, C.H. Skinner, D. Kim, E. Valeo, D. Voorhees, and A. Wouters: Phys. Rev. Lett. <u>57</u>, 1004 (1986).

5. M.D. Rosen, R.A. London, and P.L. Hagelstein: <u>On the Scaling of Ne-like X-ray Laser Schemes to Short Wavelength</u>, submitted to Physics of Fluids 1987.

6. C.W. Clark, M.G. Littman, R. Miles, T.J. McIlrath, C.H. Skinner, S. Suckewer, and E. Valeo: J. Opt. Soc. Am. B <u>3</u>, 371 (1986).

7. K. Boyer, H. Egger, T.S. Luk, H. Pummer, and C.K. Rhodes: J. Opt. Soc. Am. B <u>1</u>, 3, (1984).

8. C. Keane, C.H. Nam, L. Meixler, C.H. Skinner, S. Suckewer, and D. Voorhees: Rev. Sci. Instrum. <u>57</u>, 1296 (1986).

9. D. Attwood: <u>Overview of Soft X-ray Sources</u>, this volume.

10. W.J. Neff: <u>Plasma Focus as Radiation Scource for Soft X-ray</u>, this volume.

Laser Plasma Soft X-Ray Source: A Dedicated Small Source at the Central Laser Facility

A. Damerell, E. Madraszek, F. O'Neill, N. Rizvi, R. Rosser, and P. Rumsby

SERC Central Laser Facility, Rutherford Appleton Laboratory, Chilton, Didcot, Oxon OX11 0QX, UK

1. ABSTRACT

A chamber dedicated for use as a laser-produced plasma soft X-ray source is being commissioned at the Central Laser Facility, RAL. The laser used to drive the source is the Nd-glass Vulcan laser [1]. By splitting off the beam at an early stage of amplification, enough power is obtained for a useful soft X-ray source and a repetition rate of a shot every 2 minutes is possible, without interfering with the other users of the laser. By operating in this way, it should be possible to use the chamber in the low duty cycle mode favoured by biologists. The target chamber has provision for X-rays to be relayed vertically down using a grazing incidence mirror, so that exposures will be possible outside of the vacuum with horizontally placed recording medium.

2. INTRODUCTION

Successful experiments in single shot soft X-ray contact microscopy and lithography have performed using laser-produced plasma soft X-ray sources at the Central Laser Facility of the Rutherford Appleton Laboratory by ROSSER et al. [2] and EASON et al. [3]. Encouraged by these, a number of scientists have tried to use the source to do serious biological work, as reported by STEAD et al. [4]. The long delay time (20 minutes) between shots, necessitated by the cooling requirements of the main disc amplifiers of the Vulcan laser have made these attempts irksome. However, if the beam is diverted before these amplifiers, considerable energy (20J IR) is still available. This is sufficient to create soft X-ray emitting plasmas capable of single shot exposure of high resolution soft X-ray resists. These shots are available at 2 minute intervals and do not affect the other users of the laser. By making the laser available in this mode, it will allow microscopists and lithographers to make many more exposures, and to make them when it suits them rather than trying to fit into other users' schedules. The freedom to take exposures when the object to be viewed is ready is particularly important for wet specimen work. To realize this potential a new target area has been set up and a dedicated microscopy/lithography chamber designed. The chamber will have both airlocks for rapid direct exposures and a permanent relay toroid with vacuum window for in-atmosphere exposures.

3. PRESENT STATUS

The target area has been constructed and is fully interlocked. The optics that switch the beam out of main laser chain are all in place and operational, as are the frequency doubling crystal and the optics that focus onto target. Over a hundred shots have already been fired into the area and 2J of frequency doubled green radiation in a 1 nsec pulse has been obtained on target.

The vacuum chamber is a temporary one, with the final one due for delivery in November. The target mounting mechanism, and focussing lens assembly are all in the process of being manufactured and are also due for delivery in November. By January 1988 it is anticipated that the new system will be operational.

When the new chamber is in use an extensive program of source and filter characterization is planned as a prelude to serious biological studies.

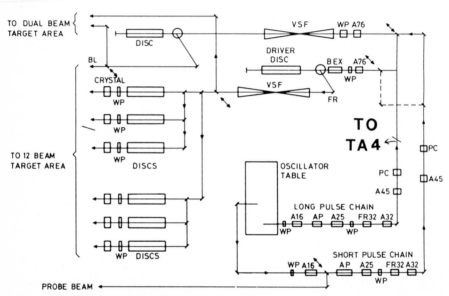

THE VULCAN LASER

Fig. 1

One of the successes of the program to date has been the development of inexpensive toroidal mirrors made by means of bending sections of precision glass tubing as reported by FALDON [5]. By reducing the cost of this optical component by a factor of between 5 and 10, it may become economic to have more than one relay station operating at a time.

4. Conclusion

In 1988, a source that satisfies most of the biological users most urgent requirements, such as exposures available to suit the user, low duty cycle, access at short notice and an exposure chamber allowing horizontal specimen stages, will be available at the Central Laser Facility. This should enable a thorough evaluation of the technique of contact X-ray microscopy to be made and so identify the areas of useful application.

5. References

1. I N Ross et al.: IEEE J. Quantum Electron. $\underline{17}$, 1653 (1981)
2. R J Rosser, K G Baldwin: In Annual Report to the Laser Facility Committee, Rutherford Appleton Laboratory, RAL-84-049, A4.3, (1984)
3. R W Eason, P C Cheng, R Feder, A Michette, R J Rosser, F O'Neill, Y Owadano, P Rumsby, M Shaw: Optica Acta, $\underline{33}$, 501 (1986)
4. T Stead et al.: in this volume.
5. M Faldon: MSc Thesis, Imperial College, University of London (1987)

Soft X-Ray Laser Microscopy

D. DiCicco[a], *L. Meixler*, *C.H. Skinner*, *S. Suckewer*, *J. Hirschberg*[b],
and *E. Kohen*[b]

Princeton University, Plasma Physics Laboratory Princeton,
NJ 08544, USA

1. Introduction

Microscopes based on soft X-ray lasers possess unique advantages in bridging the gap between high resolution electron microscopy of dehydrated, stained cells and light microscopy at comparatively low resolution of unaltered live cells. The high brightness and short pulse duration of soft X-ray lasers make them ideal for flash imaging of live specimens.

The Princeton soft X-ray laser is based on a magnetically confined laser produced carbon plasma. Radiation cooling after the laser pulse produces rapid recombination which produces a population inversion and high gain. A full account is given in a companion paper in this volume [1]. The important characteristics of the laser beam produced by this device are 1 to 3 mJ of 18.2 nm radiation in a 10 to 30 nsec pulse with a divergence of 5 mrad. The 18.2 nm wavelength, while outside the water window, does provide a factor of 3 difference in absorption coefficients between oxygen and carbon.

2. Status of Microscopy at Princeton

Figure 1 shows a schematic of the Princeton soft X-ray laser experiment. Multichannel XUV spectrometers are normally used to monitor X-ray emission from the plasma. During microscopy experiments the soft X-ray beam is diverted 20 deg. via an astigmatic spectacle lens which serves as a rudimentary toroidal grazing incidence mirror. Figures 2a and 2b give views of the mirror, the positioning system, and the rear portion of the environmental cell. The translators are remotely controlled and allow us to steer the X-ray beam to the environmental cell. The alignment is optimized by temporarily placing a PIN diode detector at the environmental cell position.

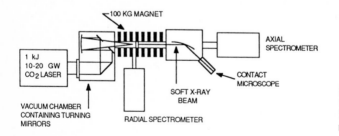

Fig. 1 Schematic of the Princeton Soft X-Ray Laser Experiment

Fig. 2. a) View of the grazing incidence mirror, b) View of the microscope

47

Our environmental cell design follows the arrangement used by Feder et al. [2]. A silicon nitride window serves as the vacuum interface. The window is 200 μm x 200 μm x 120 nm thick and is coated with 100 nm of aluminum. The Al acts as a UV rejection filter and also lends some mechanical support to the membrane. Initial experiments have been performed in order to evaluate the performance of the system without the complications involved in handling live specimens. The first of this series used a piece of # 100 wire mesh in place of a living cell. Images of this mesh were recorded on Kodak 101 film and on P(MMA co MAA) resist and may be seen in figs. 3a and 3b.

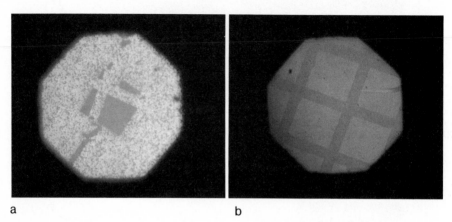

a   b

Fig. 3. a) Image of mesh on Kodak 101 with Si_3N_4 window and Al filter (one laser shot); b) image of mesh on P(MMA co MAA) (one laser shot)

Both of the above images were generated with one laser shot. They differ only in fact that the resist image was obtained without the use of the aluminum coated silicon nitride window so a contribution from UV light from the plasma cannot be ruled out at this time. The P(MMA co MAA) resist was developed in equal parts of methyl iso-butyl ketone and isopropanol. Images obtained on resist with the window in place were too faint to be clearly identifiable using a metallurgical microscope and an effort is being made to observe these images in an SEM. In the near future the rudimentary grazing incidence mirror will be replaced by a diamond turned ellipsoidal mirror of much superior optical quality. We anticipate a two orders of magnitude increase in soft X-ray intensity at the environmental chamber which should enable the resist to be well exposed.

3. Microscope Development

In future work we plan to use the contact microscope to examine live specimens. The cells will be placed or grown on a suitable resist-coated substrate. This would be brought into contact with the window and exposed with the laser beam. Subsequently, the resist would be ultrasonically cleaned, developed, and examined either by phase or electron microscopy.

In addition, a new type of soft X-ray laser microscope, which has already been constructed, will be installed on the soft X-ray laser in the near future. Called COXRALM (Composite Optical X-Ray Laser Microscope), this device is an inverted phase contrast microscope with the capability of observing UV induced fluorescence combined with the option of contact micrograph generation via flash soft X-ray exposure. COXRALM, which is a collaborative effort by Biologists and Physicists, will provide the advantage of being able to observe the specimen up until the time of X-ray exposure. This will directly address the question of specimen condition at exposure and aid in the interpretation of contact micrographs.

Acknowledgements

We would like to thank P.C. Cheng for helpful advice regarding technique and sample preparation and T. Bennett for help with photography. We would also like to thank IBM Watson Research Center for providing the silicon nitride windows and the copolymer resist. This work was made possible by financial support from the U.S. Department of Energy, Basic Energy Sciences, the U.S. Air Force Office of Scientific Research and NRL/SDIO.

(a) PXL Inc., 1-H Princeton Corporate Plaza, Deerpark Drive, Monmouth Jct., New Jersey 08852

(b) Physics Department, University of Miami, Coral Gables, Florida 33124

References

1. C.H. Skinner, D.E. Kim, A. Wouters, D. Voorhees, and S. Suckewer: X-Ray Laser Sources for Microscopy, this volume.

2. R. Feder, J.W. McGowan, and D. Shinozaki: Examining the Submicron World, Plenum Publishing Company, (1986).

3. S. Suckewer, C.H. Skinner, H. Milchberg, C. Keane, and D. Voorhees: Phys. Rev. Lett. 55, 1753 (1985).

4. G. Schmahl and D. Rudolph: X-Ray Microscopy, Proceedings of the International Symposium, Gottingen, Federal Republic of Germany, Sept. 14-16, 1983, Springer Series in Optical Science 43, Springer Verlag 1984.

5. E. Spiller: High Resolution Soft X-Ray Optics, Proceedings of the SPIE, Brookhaven, New York, Nov. 18-20, 1981, SPIE 316, 1981.

Laser Plasma Soft X-Ray Sources: Overcoming the Debris Problem by Means of Relay Optics

C. Hills[1], *R. Feder*[2], *A. Ng*[3], *R. Rosser*[4], *and R. Speer*[5]

[1]King's College London, The Strand, London WC2R 2LS, UK
[2]River Road, Hyde Park, NY 12538, USA
[3]University of British Columbia, Physics Department,
   Vancouver V6T 2A6, Canada
[4]Rutherford Appleton Laboratory, Laser Division,
   Chilton, Didcot, Oxon OX11 0QX, UK
[5]Imperial College, Blackett Laboratory,
   London SW7 2BZ, UK

1. ABSTRACT

Experiments at both the Physics Department, University of British Columbia and the Laser Division, Rutherford Appleton Laboratory have demonstrated that grazing incidence toroidal mirrors can be used to relay the soft X-rays from laser produced plasmas to regions away from the debris associated with these sources. This arrangement allows experiments in lithography and microscopy to be performed without damage to the masks or images. At UBC it was possible to relay the soft X-rays through a vacuum retaining silicon nitride window and perform exposures of PMMA copolymer photoresist in helium at atmospheric pressure.

2. INTRODUCTION

Laser-produced plasmas are very intense sources of soft X-rays and are potentially useful for lithography and microscopy as demonstrated by YAAKOBI et al. [1] and O'NEILL et al. [2]. In particular, as shown in MICHETTE et al. [3], single shot exposures can be made with nanosecond long exposures which allow images of live biological specimens to be made. A major disadvantage has been the debris associated with the expanding plasma. Experiments at both the University of British Columbia's SMIRF laser, reported in ROSSER et al [4], and the Rutherford Appleton Laboratory's Vulcan laser, reported in HILLS et al [5], have shown that it is possible to do contact microscopy/lithography on high resolution soft x-ray resists using toroidal relay mirrors. These mirrors eliminate the debris problem, allowing specimens to be recovered. It also means that permanent soft X-ray transmitting windows can be used, greatly simplifying specimen handling.

3. EXPERIMENT

Figure 1 shows the experimental arrangement used at UBC. The soft X-ray source was a gold plasma produced by focussing 2J of a frequency doubled, 2nsec long pulse from a Nd-glass laser. The laser used was a

50

commercially available Quantel NG34 model which emits 1.06μ radiation.
The energy density at focus was approximately 1×10^{13}W/cm². The relay
optics was a grazing incidence toroidal grating used at zero order as
produced and tested by SPEER et al.[6]. The toroid had a minor radius of
5.65mm and a major radius of 2m. The conjugates were 210mm apart with a
3 degree grazing incidence angle. The set up at the Rutherford Appleton
Laboratory was similar, using 30J, 1 nsec frequency doubled pulses from
the Vulcan Nd:Yag laser and a larger toroid, giving a relay distance of
1m. The focal spot at RAL was larger, giving the same power density on
the gold target of 1×10^{13}W/cm².

Figure 2 shows a single shot exposure of a cell-covered electron
microscope grid, supplied by P C Cheng of IBM, Yorktown Heights Research
Centre, NY, taken at UBC using the relay optic. The exposure was
sufficiently intense to allow development of the PMMA recording material
to be done in one minute with a 1:1 solution of MIBK:IPA. The developed
photoresist is shown viewed through a Nomarski optical microscope.
Similar results were achieved at RAL.

Figure 3 shows how a permanent soft X-ray transparent, vacuum
retaining window was set up at UBC. The window was a 250 by 250 μm
silicon nitride foil of 0.12 μm thickness, supplied by IBM, Yorktown
Heights Research Centre, NY. This is sufficient to withstand 1

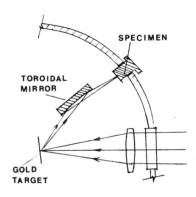

SPECIMEN

TOROIDAL MIRROR

GOLD TARGET

Fig 1. A plan view of a quadrant of the
experimental chamber used at the SMIRF
laser facility, Physics Department,
University of British Columbia

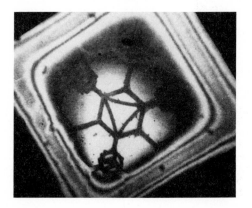

Fig 2. Optical micrograph of soft
X-ray exposed photoresist. The
resist is on a 1mm by 1mm silicon nitride window, for TEM
viewing. The imaged hexagonal
grid measures 200 μm across the
flats of the hexagon.

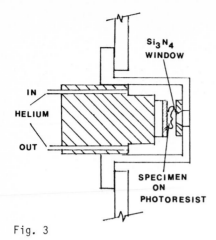

Fig. 3

Fig. 4

Fig 3. A close up view of the specimen holder, showing the silicon nitride vacuum retaining window.

Fig 4. Optical micrograph of a PMMA copolymer resist, exposed in helium at atmospheric pressure. The imaged grid measures 200 μm across the flats of the hexagon.

atmosphere pressure difference, yet transmit 80% of the flux at 3.5 nm. Careful alignment allowed the image of the laser-produced plasma to be relayed through this aperture onto the photoresist which was in a Helium atmosphere.

Figure 4 shows a single shot image of a cell-covered electron microscope grid taken in a one atmosphere of helium environment, using the relay toroid at UBC. As before the exposure was sufficient to allow an image after only one minute of development of the PMMA recording medium in the standard 1:1 MIBK:IPA solution. However, it can be seen that the image has very little contrast, showing that it is on the limits of acceptable exposure. A slightly higher power laser or a better relay optic would be of benefit. Both at UBC and at RAL, higher laser powers are available.

4. Conclusion

Toroidal grating relay optics have been shown to be a way of overcoming the debris problem normally associated with laser-produced plasmas, enabling them to be used as sources for non-destructive soft X-ray microscopy and lithography. Although the images shown here contain biological specimens, they are merely used as test objects to demonstrate the feasibility of the technique and are not intended to show anything of intrinsic biological interest. A dedicated microscopy/lithography source is currently being commissioned at RAL, and it is hoped that this can be used to do work of real biological significance.

5. References

1. B Yaakobi, H Kim, J M Soures, H W Deckman, J Dunsmuir: Appl Phys Lett 43, (1983)
2. F O'Neill, G M Davis, M C Gower, I C E Turcu, M Lawless, M Williams: In X-Rays from Laser Plamas, SPIE conference, San Diego (1987)
3. A G Michette, P C Cheng, R W Eason, R Feder, F O'Neill, Y Owadano, R J Rosser, P Rumsby, M J Shaw: J Physics D, 19, 363 (1986)
4. R J Rosser, R Feder, A Ng, F Adams, P Celliers, R Speer, to be published in Applied Optics, (1987)
5. C P B Hills, R J Rosser, A Ridgeley, T W Ford, A D Stead: In Annual Report of the Central Laser Facility, RAL-87-041, (1987)
6. R J Speer, D Turner, R L Johnson, D Rudolph, G Schmahl: Appl Opt 13, 1258 (1974)

Status of the Taiwan Light Source

Gwo-Jen Jan

Instrumentation and Control Group Synchrotron
Radiation Research Center,
Department of Electrical Engineering,
National Taiwan University, Taipei 10764, Taiwan, Rep. of China

1. Introduction

The establishment of the Synchrotron Radiation Research Center under the Executive Yuan was approved by the government of the Republic of China on July 9, 1983. The major parts of Synchrotron Radiation Facility have been reviewed and advised by the Technical Review Committee (TRC) and decided by the Board of Directors (BOD) during this period. The detailed information from the TRC and BOD was presented at the '86 X-ray microscopy conference [1]. The technical aspects of the Taiwan Synchrotron Radiation facility (Taiwan Light Source) and its current progress as well as the organization of the technical and user divisions of SRRC are described.

2. Organization of Technical and User Division

The overall organization of SRRC has been described at the conference [1]. The technical and user's training divisions are described here. The division and associate division heads as well as the group leader of each technical group is shown on the organization chart in Fig.1.

It is the responsibility of each technical group to develop and implement the facilities for its components of the TLS facility. The User's Training division is responsible for recruiting and training new people as in-house users. This division is also planning and setting up the experimental facility of TLS as well as organizing the user meetings and workshops for Synchrotron Radiation-related research. In this paper we will concentrate on the technical construction and progress within the last two years.

3. Status and progress of SRRC technical division

After the 1 GeV Chasman-Green lattice was rejected, the Beam Dynamic Group (BDG) studied for almost one year two types of configuration, the FODO and the TBA (Triple Bend Achromat), showing that both of them were workable solutions. However, because the TBA is simpler and less compact, it had the preference of the BDG. Last April, after the decision made by the BOD to choose it, the BDG started to optimize this lattice

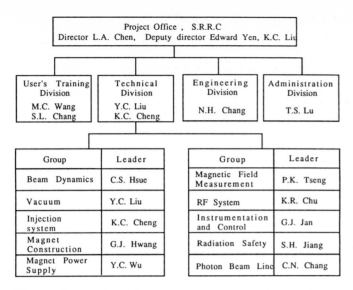

Fig.1 Organization chart for Technical
and User's Training Division

in order to meet the designed goals. In the meantime the BDG
worked on preliminary design of the transport line from the 1.3
GeV booster to the storage ring.

The storage ring is mainly composed of 6 superperiods. The
lattice is a TBA configuration that means that each superperiod
is composed of three bending magnets, several quadrupoles for
the focalization and one long straight section used to locate
the insertion devices (undulators and wigglers).

The 1.3 GeV full energy injection system is configured by a
low energy linac (may be 50 Mev) and a 1.3 GeV booster
synchrotron accelerator as well as a positron option. The
requirements and parameters of the injection system have been
studied by the Injection System group. Proposals for the
injection system will be requested from overseas companies and
the contract for construction will go to one of these. The final
decision on the injection system will be made at the workshop of
SRRC in February, 1988.

The vacuum group has done a lot of research and development
work for the ultrahigh vacuum (UHV) chambers using SUS 304
stainless steel and A6063-EX aluminum alloy materials. They have
investigated the outgassing rate of vacuum chamber made from
SUS 304 or A6063-EX materials after filling the chambers with
water [2,3]. Their results showed that the outgassing rate of
the Al alloy was greater than that of the stainless steel before
baking. However, after in situ vacuum baking, the outgassing
rate of the aluminum alloy becomes lower than that of the
stainless steel, even for the chamber which had been filled with
water. An aluminum alloy ultrahigh vacuum system has also been

studied. Pressures as low as 7.5×10^{-12} Torr were obtained in a large Al alloy vacuum chamber. The outgassing rate of 1×10^{-13} Torr·l/sec cm^2 for an Al 1050 aluminum alloy was measured after 10 hours following one day baking at $150^{\circ}C$. They investigated the pumping speeds for H_2, N_2, O_2 and Ar gases of a triode ion pump with A1100 Al alloy cathode. The results indicate that the Al alloy UHV system has good vacuum properties. However, the major material for the UHV system has not been decided yet. The bore radius and shape of the vacuum chamber is currently being studied.

The TBA lattice structure has been determined. The responsibility of the magnet manufacture group is to make acceptable quality magnets required for the complex. The number of magnets include 18 dipole, 48 quadrupole and 24 sextupole magnets. The material for the magnets uses the low carbon steel and lamination technique. These materials are provided by the domestic China Steel Company, but the electrical conductor of copper tubing is purchased from abroad. The design and construction of the magnet complex was carried out by the magnet group under professor Hwang's supervision. Three prototype dipole magnets have been made and will be measured by the magnet field measurement group.

The responsibility of the Magnet Power Supply Group (MPSG) is to design and implement the magnet current regulators for the dipoles, quadrupoles and sextupoles as well as the correcting or steering magnet. The prototype 75 kW bending magnet current regulator has been purchased from Bruker company in Germany. The programmable current control with standard interface and 10 ppm stability will be provided by the vendor. The 250 kW power station facility will be established at the Electronics and Electrical laboratory at National Taiwan University for initial use. The specifications and design phase of the TBA magnets are also being studied at the current stage. The construction of different magnet current regulators will be contracted to the domestic electric industry.

The Magnet Field Measurement Group (MFMG) was formed two years ago. The Hall probe, NMR probe and searching rotation coil as well as digital current integrator have been investigated and implemented during this period. The 1.7 m (W) X 3.5 m (L) X 2.0 m (H) high-precision X-Y-Z table has been constructed by the Machine Research Center (MRC) at the Institute of Technology and Research for Industry (ITRI). The short-term current accuracy is about 10^{-4}. The turn-key application software package has been developed and implemented by MFMG under an IBM PC/AT system. The stability of the Hall probe is less than 10^{-3}. The qualification and magnet field measurement on the TBA magnet will be carried out within the next quarter. Two full time staff and a part time group leader as well as 4 part time engineers are employed in this group.

The Radio Frequency group (RFG) was formed about one year ago. Six Doris cavities were purchased from DESY in Germany. The computer simulation and test/measurement facilities of the RF components were set up under professor Chu. Two full time staff and a group leader as well as 1 part time scientist are employed. The frequency of the RF cavity is 500 MHz. The test/measurement facility of the RF system has been purchased and the low-level electronic instrument will be built in the next semester. The temporary site of the RF group is located in the Department of Physics at National Tsing Hwa University.

The major responsibility of the instrumentation and control group is to design, develop and implement the computer control system and beam diagnostic instrument for TLS as well as to support the electronics and computer system. This group has been formed for 3 years and includes 5 full-time engineers, 1 associate research scientist, a group leader and 4 part-time graduate students. A 3-level hierarchy computer configuration, intelligent local controller and multi-task real-time operating software have been proposed. A Vax 8600 main frame, 4 sets of Micro VAX and multi-Microprocessor (68000 CPU) will be used in the computer control system. Detailed information about the TLS computer control system can be referred to in the literature [4].

The Radiation Safety Group (RSG) was formed several months ago. This group will take care of the radiation safety and radiation shielding considerations. The design and calculation of radiation shielding specification is in progress. One research scientist and one assistant engineer will be recruited and supervised by professor Jiang.

At the preliminary stage the Photon Beam Line Group (PBLG) is going to design and build up 3 beam lines for user. This group was also formed 3 years ago and has 2 full time assistant research scientists. They have designed and built up the ultra high vacuum facility to investigate the acoustic delay line. The results have been published and presented at the 1987 Synchrotron Radiation Instrumentation National meeting [5,6].

The site of SRRC has been selected in the Scientific -Based Industry Park at Hsin-Chu. The site has been ground-broken and the ground has been re-leveled for building construction. Phase one is going to construct the administrative and laboratory building, machine shop utility, and general facility. The construction engineering task has been contracted to China Engineering Company. The task will be finished within 1 and a half years. Phase two is going to construct the Storage Ring, Booster Synchrotron and Linac as well as utility house. The construction schedule proposed the other two years required. After that the different groups will be united at the SRRC site. Overall efficiency and performance should be improved at that time.

4. Discussion

Two major decisions are made by Taiwan Light Source during the last two years. One is the adoption of TBA lattice configuration and the other is the determination of ring parameters. Optimization of ring parameters and design of the beam transport line are being surveyed. The injection system with low energy Linac and 1.3 GeV Booster Synchrotron as well as positron option is also recommended by TRC and decided by BOD. There has been a considerable progress made by each group of the Technical Division since its formation.

The Engineering Division is located in Hsin-chu to monitor the phase one construction. The building construction should be finished within one and a half years. Taiwan Light Source is expected to provide synchrotron radiation light source for user's study in 1992.

Acknowledgement

I would like to thank each group in Technical Division for providing information and correcting the description in this paper. I wish to thank Mr. Christian Travier and Professor Hsue for their deep discussions on the subject of beam dynamics. I also wish to extend my thanks to SRRC director Dr. L.A. Chen and deputy director Professor Edward Yen for their support.

Reference

1. E. Yen : Brief report on the Present Status of the SRRC. In X-ray Microscopy Instrumentation & Biological Applications, eds. Cheng, P. C. and Jan, G. J., Springer Verlag, Berlin.(1986)
2. J. R. Chen, and Y. C. Liu, : Thermal Outgassing From Stainless Steel Vacuum Chambers. Chinese journal of Physics (Spring), 24: 29 - 36. (1986)
3. J. R. Chen, and Y. C. Liu : A comparison of outgassing rate of 304 stainless steel and A6063-EX aluminum alloy vacuum chamber after filling with water. Journal of Vacuum Science & Technology A, Vol.5, No.2, Mar/Apr 1987 : 262 - 264. (1987)
4. G. J. Jan, : Instrumentation and control system for Taiwan Synchrotron Radiation, Proceedings of 1986 International symposium of Mini. & Microcomputer, Feb. 5-7, 1986 Los Angles (USA).
5. Y. F. Song, C. I. Chen, C. N. Chang, Z. L.You, and F. K. Huang : Study of the transit time of pressure propagation in an acoustic delay line, Rev. of Scientific Instrument. Dec. 57(12), 3063 - 3065, (1986).
6. C. N. Chang, C. I. Chen, and Y. F. Song : Optical system for 1 meter Seya - Namioka monochromator in a SRRC beam line. Presented on SRI meeting, Wisconsin Madison. June, 21 - 25 (1987)

The Potential of Laser Plasma Sources in Scanning X-Ray Microscopy

A.G. Michette[1], *R.E. Burge*[1], *A.M. Rogoyski*[1], *F. O'Neill*[2], *and I.C.E. Turcu*[2]

[1]Physics Dept., King's College, The Strand,
London WC2R 2LS, UK
[2]Rutherford Appleton Laboratory, Didcot,
Oxon OX11 0QX, UK

1. Introduction

All published experiments on scanning x-ray microscopy have used synchrotrons [1], but the technique may only become routine with a more convenient source. Laser plasmas, used in x-ray lithography [2] and contact microscopy [3], are being studied by the authors and others [4]. Recent and projected advances in small lasers make their use in scanning microscopy feasible. The plasma soft x-ray emission has been measured using lasers with repetition rates of up to 20 Hz - a Nd:YAG laser giving 0.5 J of TEMoo radiation with wavelength 1064 nm in 8 ns pulses, and a KrF excimer laser giving ≈ 0.3 J of 248 nm radiation in 25 ns pulses. Since most x-ray microscopes use zone plates a system based on these is discussed; TRAIL and BYER [5] have considered a multilayer microscope.

2. Experimental Arrangements

For the Nd:YAG laser a glass lens focused the beam to a minimum spot diameter of ≈ 60 μm on 15 mm radius cylindrical targets, in a vacuum chamber to prevent atmospheric breakdown, giving a maximum irradiance $I_{max} \approx 2.2 \times 10^{16}$ W m^{-2}. The limit on the spot size was due to the beam divergence (≈ 0.4 mrad) and the low-cost non-optimized lens, which had to be frequently replaced due to coating by debris. The targets were remotely rotated and moved axially to allow fresh areas to be used whilst maintaining vacuum. Movement along the beam direction allowed focusing using a He/Ne laser with an offset to allow for the different focal length of the lens for Nd:YAG radiation. The target could be positioned to ± 30 μm, well within the lens focal depth.

A purpose-built vacuum photodiode [6] measured the flux. The conical cathode reduced space-charge effects due to electron liberation at its surface and gave high quantum efficiency since radiation was incident obliquely. A filter eliminated UV plasma radiation; electrons emitted by this were removed magnetically and by a negative bias on the cathode. The photodiode was positioned remotely at 40°-90° to the beam, the large focusing lens so far excluding smaller angles. A spatially resolving transmission grating spectrograph was also used. This has a 200 nm period grating made by contamination lithography [7] on 40 nm thick carbon film, gold shadowed at an angle to give 30 nm thick bars with 4:1 mark-to-space ratio. Other gratings, typically 25 μm × 25 μm with 110 nm periods, have been made on 50 nm thick boron nitride (Fig. 1).

The KrF experiments have so far been restricted to flux measurements at 45° to the beam. A fused-silica aspheric lens, protected by a quartz plate, focused the beam to a spot diameter

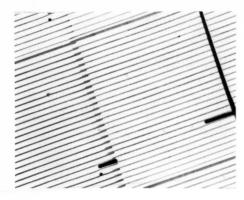

Fig. 1. An unshadowed 110 nm period grating made by contamination lithography

measured by obscuration as < 25 μm, giving $I_{max} \approx 1.7 \times 10^{16}$ W m^{-2} after attenuation by the optics. This spot size, larger than the diffraction limit, was chosen to ensure a low hard x-ray flux. The 250 μm diameter targets were positioned using two crossed He/Ne laser beams focused on 25 μm wires at their tops. The flux was measured with silicon PIN diodes; one, with a 0.5 μm V filter, was sensitive to soft (280-510 eV) and hard x rays. The other, with a 25 μm Be filter, indicated at most a few percent of hard x rays. Thus smaller spots could be used in future.

3. Soft X-Ray Spectra

Low-Z targets give mainly line spectra [8] with more continuum emission for higher Z. Since zone plates need monochromatic radiation the simplest system could use a low-Z target. A preliminary transmission grating spectrum of a carbon plasma is shown in Fig. 2. The lateral dimension of a line gives the size of the emitting region; that of He-like ions (cooler plasma) is larger than for H-like ions. For high-Z targets transmission grating monochromators could be used.

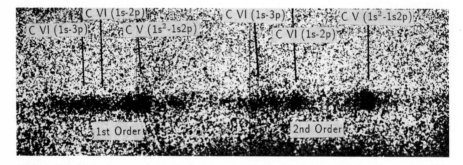

Fig. 2. A preliminary carbon target spectrum recorded with a 200 nm period grating

4. Soft X-Ray Fluxes

Using a large KrF laser the measured conversion efficiency ϵ of laser energy to soft x-ray energy is $\approx 6\%$ sr^{-1} for tungsten targets with $I \approx 10^{16}$ W m^{-2} [9]. For the same irradiances with smaller lasers smaller spot sizes must be used giving lower efficiencies. Preliminary results from the small KrF laser indicate $\epsilon \approx 1\%$ sr^{-1} for I greater than $\approx 1.5\times10^{16}$ W m^{-2} on tungsten.

The laser pulse length τ may affect the angular distribution of the emission. The approximately flat distribution between 45° and 90° for single pulses of the Nd:YAG laser at each target position is shown in Fig. 3. A $\cos\theta$ distribution has been reported [10] for $\tau = 0.5$ ns, while for $\tau = 50$ ns the emission goes as $\cos^{2.75}\theta$ for $\theta \leq 60°$ and is almost flat for larger angles [11]. This steep distribution may be due to a build-up of cold absorbing plasma.

Removal of target material by the beam forms a crater with two effects on the x-ray yield. An initial increase followed by a fall-off is shown in Fig. 4 for an Al target. This is thought to be due to the crater confining the plasma, increasing the emission time. As the crater deepens the beam is defocused and the emission eventually decreases. The second effect is peaking of the emission towards the beam axis (Fig. 3) due to obscuration by the crater at large angles.

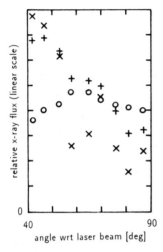

Fig. 3. The angular distribution of x rays from an aluminium target (o,+,x; 1st,4th,8th pulse)

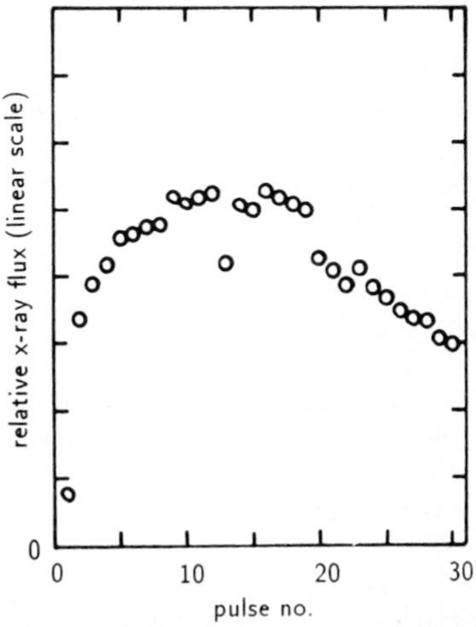

Fig. 4. X-ray signals for successive pulses at the same place on an aluminium target

A dip in x-ray emission as the target is scanned through the focus of the Nd:YAG laser is shown in Fig. 5. Similar effects have been observed previously [12], and are a matter of current investigation. It may be advantageous to slightly defocus the laser beam, but more detailed studies are required to determine the effect on the spectral quality of the emission.

5. Preliminary Specification of a Laser Plasma Source for Scanning X-Ray Microscopy

A high repetition rate source giving enough photons in a bandpass matched to the zone plate for fast imaging, ideally one pixel per pulse, is needed. The best hope is a KrF laser with better conversion to x rays than a longer wavelength laser and projected repetition rates of 1 kHz and pulse energies of 1 J. The small source, if necessary restricted, for coherent illumination of the zone plate is easier with a KrF laser. An estimate of the coherent flux with a 1 J laser follows.

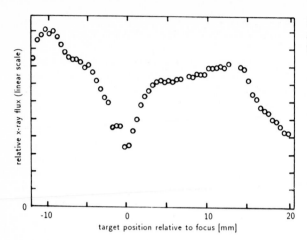

Fig. 5. Relative x-ray signals for an aluminium target scanned longitudinally through focus

The resolving power $E/\Delta E$ of a zone plate with N zones is $\approx 1.6N$ [13]; typically, with $N \approx 125$ [7], $E/\Delta E \approx 200$. The conversion efficiency for heavy targets at $E \approx 400$ eV into $\Delta E \approx 2$ eV can be $\approx 0.03\%$ sr^{-1} close to the beam axis. For a monochromator efficiency of 5% and a zone plate acceptance solid angle of 2×10^{-7} sr [14], corresponding to a source size of ≈ 10 μm and a target to zone plate distance sufficient to prevent damage by debris, $\approx 4 \times 10^4$ photons per pulse reach the zone plate, i.e. $\approx 4 \times 10^7$ s^{-1} at 1 kHz. A previous estimate for a carbon plasma, assuming no monochromating [14], gives a similar flux, taking into account the different laser systems considered. Such fluxes are too low for high-contrast one pixel per pulse imaging of most specimens but could possibly be increased by further source optimization. They give sufficient encouragement for continuing the study of laser plasma sources.

Acknowledgements

The work was supported in part by the Royal Society Paul Instrument Fund. We are grateful for the valuable help of technical staff at KCL and RAL. AMR was funded by an SERC studentship.

References

1. See the papers on scanning x-ray microscopy in this volume and in Proc SPIE 733, (1987)
2. D.J. Nagel et al: Proc SPIE 153, 46, (1978)
3. A.G. Michette et al: J. Phys D: Appl. Phys, 19, 363, (1986)
4. R. Popil et al: Phys. Rev. A, 35, 3874, (1987)
5. J.A. Trail and R.L. Byer: Proc SPIE 563, 90, (1985)
6. A.M. Rogoyski et al: to be published in Proc SPIE 831
7. C.J. Buckley: PhD Thesis, Chapter 2, London University, (1987) (unpublished)
8. F. O'Neill et al: Proc AIP 147, 354, (1986)
9. I.C.E. Turcu et al: accepted for publication in Appl. Phys. Lett., (1987)
10. M. Chaker et al: Proc SPIE 733, 58, (1987)
11. I.C.E. Turcu et al: to be published in Proc SPIE 831
12. J.M. Bridges, C.L. Cromer, and T.J. McIlrath: Appl. Opt., 25, 2208, (1986)
13. H.M. Rarback: PhD Thesis, SUNY, p51 (1983)
14. A.G. Michette et al: Proc SPIE 733, 28, (1987)

Future Plans for X-Ray Microscopy at the SRS

H.A. Padmore[1], P.J. Duke[1], R.E. Burge[2], and A.G. Michette[2]

[1]SERC Daresbury Laboratory, Warrington WA4 4AD, UK
[2]King's College, The Strand, London WC2R 2LS, UK

1. Introduction

The recent up-grade of the synchrotron radiation source (SRS) at the Daresbury Laboratory provides an improved low emittance photon source. In particular, the existence of a high brilliance source of x-radiation, and the need to provide diffraction limited optics for high resolution microscopy, has stimulated the re-examination of the problem of matching the x-ray source to the x-ray microscope.

2. Zone Plate Properties

A Fresnel zone plate is a circular diffraction grating in which the grating spacing dr_n of the nth grating element at radius r_n is given by $dr_n = mf\lambda/r_n$, where m is the diffractive order. The spatial resolution of the zone plate is limited by the size of the diffractive focal spot and is equal to $1.22\ dr_n/m$ by application of the Rayleigh criterion. The phase space area of the diffractive limited focal spot (ε) is the product of its transverse dimension and the angular aperture of the zone plate so that $\varepsilon = 1.22\lambda \times 10^{-3}$ mm.mrad (λ is measured in nm) which is independent of the diffractive order and the zone plate properties. By Liouville's theorem, this represents the maximum phase space area, transmitted by the optical system, which can be accepted by the zone plate (HOWELLS [1]) and defines the degree of spatial coherence of the x-ray source for optimal microscope performance. For example, assuming light of wavelength about 3 nm, this condition means that the acceptable phase space area, or photon beam emittance, is limited to about 4 μm.mrad. The emittance of the photon beam from the 10-period undulator installed in the SRS at the Daresbury Laboratory is about 200 μm.mrad in the vertical direction and is much larger in the horizontal direction so that considerable beam loss is inevitable and the optical system must be designed to minimise further losses.

The acceptable wavelength bandwidth ($\delta\lambda$) is given by $\lambda/\delta\lambda = 1.6$nm, which depends on the product of the number of zones and and the diffractive order. This is the temporal coherence condition and ensures that the chromatic blurring of the image is kept within the diffraction-limited size of the focal spot.

The table lists the properties of selected zone plates prepared recently at King's College (BURGE et al. [2]). In all cases a resolving power < 1000 is required and this does not make high demands on the diffractive power of a low aperture grating monochromator.

Existing monochromators, often designed for some form of XUV spectroscopy, have a higher resolving power than is required which leads to loss of useful photon flux when used to provide a monochromatic source for x-ray

Table: Properties of carbon zone plates

r_n (μm)	dr_n (nm)	f(mm) (λ=3 nm)	n	m	$\lambda/d\lambda$
25	100	1.6	126	1	210
23.5	75	1.13	158	1	260
21.5	75	1.03	145	1	240
22.5	50	0.71	230	1	377
28.5	30	0.55	476	1	780
18	30	0.35	303	1	500
18	18	0.21	498	1	816
25*	70	0.56	120	2	390

*This zone plate has every other absorbing zone missing to enhance the diffraction efficiency in the second order.

microscopy. Also, the compact design of some XUV monochromators leads to a very high demagnification and consequent high exit divergence. This is particularly true of the plane grating monochromator of the SX700 type [3] which is currently being used to form a demagnified image of the photon source provided by the SRS 10-period undulator. Although this photon source has made it possible to obtain the first scanned x-ray microscopy images at the SRS, its disadvantages have led us to look for an alternative.

3. Scanned X-ray Microscopy Beam Line

One of the simplest possible optical systems for a scanning transmission x-ray microscope (STXM) consists of a cylindrical mirror to provide hori- zontal focusing and a cylindrical diffraction grating to provide vertical dispersion and focusing of the undulator source. With the assumption that the included angles of reflection and diffraction are set by efficiency criteria and the entrance arm length is set by geometrical constraints, there is still a free choice of groove density and exit arm length. It can easily be shown that there are a range of solutions for the magnification from the limit where the demagnified image of the source is equal to the required pinhole size to the limit when the divergence is equal to the coherence angle defined by the pinhole size. Similar arguments apply to the range of allowed source slit width resolutions. The layout for such an arrangement is shown in fig. 1 as a branch line from the existing beam. The cylindrical premirror can be translated horizontally to allow radiation to proceed to another experiment. The grating can be prepared with regions of different groove density and moved through the beam to achieve optimum

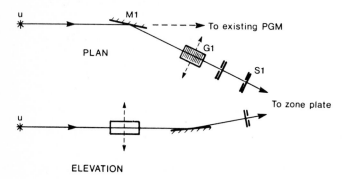

Fig. 1. Spherical grating monochromator on branch line U5.

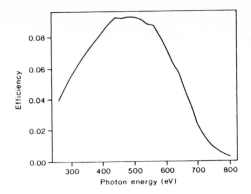

Fig. 2. Spherical grating monochromator efficiency as a function of photon energy.

performance over a range of wavelengths. The calculated efficiency as a function of wavelength is shown in fig. 2. It indicates that the overall efficiency, after allowing for the first order diffractive efficiency of the zone plate, should allow photon fluxes of 10^7 photons/sec to be obtained in a diffraction limited focal spot from a source whose brilliance is 3.6×10^{14} photons/(sec.mm^2.mrad^2) into a bandwidth of 0.33% when the stored electron beam current is 100 mA. It is planned to install a branch line with these properties during 1988.

4. Imaging X-ray Microscopy Beam Line

The possibility of constructing a second undulator at the SRS has led us to consider its use for full-field imaging microscopy. The high power density at the first optical element would preclude the use of a condenser zone plate as in the arrangement of NIEMANN et al. [4] used on a bending magnet beam line at BESSY. The best choice for the first element would be a multilayer mirror with a narrow band width matched, as described above, to the properties of the zone plate objective lens. The multilayer would need to be laterally graded to provide wavelength tunability.

This has led us to examine the possibility of replacing the condenser zone plate with a multilayer grating in an arrangement of two crossed cylinders as shown in fig. 3. Multilayer coatings have already been shown to be effective in increasing the diffraction efficiencies of gratings by JARK [5] in the soft x-ray region and by KESHI-KUHA [6] in the VUV. In addition, it has been shown by VIDAL et al. [7] that good efficiencies should be obtained at large grazing angles and with reasonable tuning ranges. Because the wavelength range of interest is lower than the carbon K-edge, we have restricted the possible grazing angle range to < 35° to avoid the practical problems of obtaining high reflectivities nearer to normal incidence.

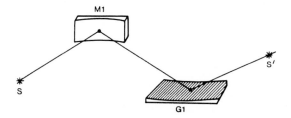

Fig. 3. Condenser stage for imaging microscope, s and s' are the source and condensed image.

The principal aberration which will affect the dispersive focus in this design will be the primary coma. The physical extent of this in the exit plane can be expressed (using the notation of HOWELLS [8]) as

$$dY = \frac{3 \gamma^2 r^2 r'}{4 \cos\beta \cos^2\alpha R^2} c_{30},$$

where r is the entrance arm length, r' is the exit arm length, R is the grating radius, γ is the full aperture and C_{30} is the coma coefficient. Fixing the demagnification and the source limited resolution as described above we are left with choices for r, the groove density, N, and the included angle, θ (α+β), which will give the desired spectral resolution. An example of one particular combination which minimizes the aberration as a function of wavelength is shown in fig. 4 and makes it clear that this arrangement can be used to give good tunability.

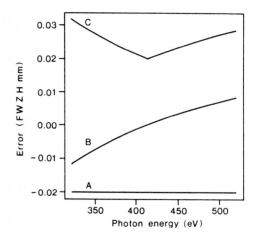

Fig. 4. Calculated aberrations for the condenser stage as a function of photon energy. Parameters are: dλ = 10^{-2} nm, s = 0.4 mm, θ = 110°, r = 15 m, r' = 741 mm and N = 1540 ℓ/mm. A: coma, B: defocus: C: total

These calculations have been performed using the parameters of the existing 10-period undulator at present installed on the SRS. This gives a vertical photon beam divergence of 0.5 mrad and a vertical source size, s, of 0.4 mm. Because the principal aberration scales as the square of the vertical aperture, multilayer condensers will be capable of much better performance on lower emittance sources such as the Advanced Light Source at Berkeley and the proposed European Synchrotron Radiation Facility.

5. Conclusion

Optical systems for x-ray microscopy, matched to the properties of the 10-period SRS undulator, have been presented. These should provide high photon flux for rapid collection of scanned image data and full-field imaging.

Acknowledgements

REB and AGM wish to acknowledge financial support from the SERC. The support and encouragement of the Director and staff of the Daresbury Laboratory is gratefully acknowledged.

References

1. M. Howells, J. Kirz, S. Krinsky: BNL 32519, Dec. (1982)
2. R.E. Burge, A.G. Michette, P.J. Duke: Scanning Electron Microscopy (1987) to be published
3. G.R. Morrison et al: these proceedings
4. B. Neimann: these proceedings
5. W. Jark: Opt. Commun. 60, 201 (1986)
6. R.A.M. Keshi-Kuha: Appl. Optics 23, 3534 (1984)
7. B. Vidal, P. Vincent, P. Dhez, M. Nevière: SPIE 563, 142 (1985)
8. M.R. Howells: Nucl. Instrum. Meth. 172, 123 (1980)

Part II

X-Ray Optics and Components

Theoretical Investigations
of Imaging Properties of Zone Plates
Using Diffraction Theory

J. Thieme

Universität Göttingen, Forschungsgruppe Röntgenmikroskopie,
Geiststraße 11, D-3400 Göttingen, Fed. Rep. of Germany

1 Introduction

Based on Kirchhoff's diffraction theory, the modulation transfer function MTF has been calculated as a criterion for the imaging quality of a micro zone plate. The micro zone plate will be used for imaging with high magnification in x-ray microscopes. The optical arrangement of an x-ray microscope is sketched in Fig.1 [1].

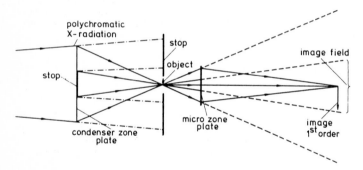

Fig.1 Optical arrangement of an x-ray microscope

A condenser zone plate images the radiation from the source into the object plane. The object in this plane is close behind a pinhole. Due to the wavelength dependence of a zone plate, the condenser zone plate and the pinhole work together as a linear monochromator [2]. The high resolution x-ray lens, the micro zone plate, images the object with high magnification. In the condenser zone plate is a central stop. It causes the zero-order radiation of the micro zone plate to form a hollow cone. Therefore in the image plane mainly radiation of the first order of diffraction is present and the zero-order radiation does not play a role. The quality of the image of the object is strongly dependent on the imaging properties of the used micro zone plate. These properties have been theoretically investigated for the micro zone plate MZP5. The parameters of this zone plate are listed in Table 1.

Table 1 Parameters of the micro zone plate MZP5

radius of the innermost zone r_1 [μm]	0.624
zone plate diameter $D=2r_n$ [μm]	21.6
zone number n	300
width of the outermost zone dr_n [μm]	0.018
focal length $f_{\lambda=2.48nm,1^{st} order}$ [μm]	157.0
D/f, λ = 2.4 nm	1:7

MZP5 will be used at a magnification of V = 1000, which is valid for all the results shown here.

MZP5 will e.g. be manufactured as a sputtered sliced zone plate [3]. That means, on a thin wire layers of absorbing material will be deposited alternating with material as transparent as possible. The thickness of the layers will be made according to the zone plate law, which determines the zone widths of a zone plate. After sputtering, a slice of the wire represents a zone plate. Due to this process, the micro zone plate MZP5 will have a central stop. In the following calculations this central stop is taken to have a diameter of 14 μm. This stands for a reduced zone number of n = 176.

2 The Fresnel-Kirchhoff Diffraction Formula

The basis for the investigations of the imaging quality of MZP5 is the diffraction theory of Kirchhoff [4]. It describes the complex disturbance at a point P, when monochromatic light is emitted from a point source Q and diffracted by an aperture dS (see Fig.2). This can be written as

$$U(P) = -\frac{i \cdot A}{2 \cdot \lambda} \cdot \iint_{dS} \frac{e^{ik(r_q+r_p)}}{r_q \cdot r_p} \cdot \left[\cos(\vec{n},\vec{r}_p)-\cos(\vec{n},\vec{r}_q)\right] dS, \quad (1)$$

which is known as the Fresnel-Kirchhoff diffraction formula. Here $\vec{n}$ is the normal to the aperture dS and A is an amplitude factor, which is set to unity hereafter. The conditions for the validity of this equation are that the optical paths r_q and r_p are large compared to the wavelength λ and that the size of the diffracting aperture, which means a zone width, is small compared to the object and image distances z_q and z_p but large compared to the wavelength λ. The exponential function $\exp\{ik(r_q+r_p)\}$ in the integrand describes the phase at the point P of a wave travelling along the optical path r_q+r_p. The term $1/(r_q \cdot r_p)$ determines the height of the amplitude at P.

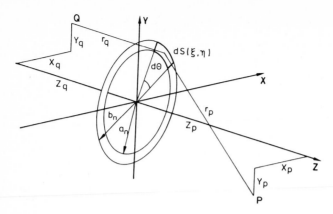

Fig.2   Application   of   the   Fresnel-Kirchhoff   diffraction
formula to a zone plate

This formula may be transformed in the case that the micro
zone plate MZP5 is the diffracting aperture. First, since a
zone plate has a rotational symmetry, polar coordinates can be
introduced in the plane of the zone plate. The element of area
dS of the diffracting zone is described in that way. Second,
the difference of the two cosine functions in (1) deviates
little from the value of 2 even for the micro zone plate MZP5
with a high aperture ratio. The deviation can be neglected in
the following calculations and the difference is set to the
fixed value of 2. Third, the term $1/(r_q \cdot r_p)$ can be changed to
$1/(z_q \cdot z_p)$ and can be moved outside the integral. With these
changes the diffraction formula is

$$U(P) = - \frac{i}{\lambda \cdot z_q \cdot z_p} \cdot \int_0^{2\pi} \int_{a_n}^{b_n} \rho \cdot e^{ik(r_q+r_p)} \, d\rho d\theta \ . \qquad (2)$$

The integration limits a_n and b_n are the inner and the outer
limits of an open zone. This integral is valid for one zone.
To get the amplitude at P coming from the whole zone plate,
the amplitudes from each open zone must be summed. The square
of the resulting amplitude is the intensity at the point P.
Scanning the image plane by changing the coordinates of P
leads to an intensity distribution.

3 The Modulation Transfer Function MTF

It is of interest to see how structures in the object are
transferred to the image. Any object can be analysed as a
Fourier series. A structure in an object is then represented
as a spectrum of sine and cosine functions. Therefore a
standard object for the following calculations, also taken in

expert literature, is a sine grating [5]. It describes one particular spatial frequency out of a complete spectrum. Such a grating is a line object with a sinusoidal intensity distribution perpendicular to the orientation of the lines. The minimum of intensity is zero and the maximum is for all the lines the same and is normalised to unity: $I_{min} = 0$ and $I_{max} = 1$. Imaging this object with a zone plate yields an intensity distribution in the image plane that is similar to that of the object. It will again be a grating. The grating constant is the constant of the object multiplied by the magnification. The diffraction of the light ensures that there is light in the minima of the grating at the image. This yields a reduction in the difference between maximum and minimum of intensity, compared to that of the object. The contrast C is defined as the difference between maximum and minimum intensity divided by the sum of the two:

$$C = \frac{I_{max} - I_{min}}{I_{max} + I_{min}} . \tag{3}$$

C is equal to unity in the object, but it is less than unity in the image. The contrast in the image of a sine grating determines therefore the loss on transfer of one spatial frequency. Changing the grating constant in the object and repeating the procedure yields as a result the description of a curve contrast=f(spatial frequency). This curve represents a modulation transfer function, abbreviated MTF [6]. It tells with what contrast great or small structures, i.e. small or great spatial frequencies, in an object will be transferred to the image. It is characteristic of the optics used.

The object field chosen here is very small and has an extension of 0.2 by 0.3 μm. The zero-order radiation of that field has no practical influence on the image. Thus the MTFs which will be shown here illustrate the imaging properties in the first order of diffraction.

4 The calculation method

The calculations were performed for the condition that every point of the object radiates independently of the others. The grating is then described by a two-dimensional scan of point sources that illuminate the zone plate incoherently. That means, the diffraction image of point sources must be calculated. The intensity distribution in the image plane must be calculated with a computer.

To determine the phase of a wave at an image point P the optical path $r_q + r_p$ in the exponential function must be developed in a power series. This power series is a function of the coordinates of the diffracting aperture. The summation has to be stopped after an appointed member of the infinite series to get a result. This introduces an approximation in determining the phase. The assumptions under which analytic solutions of the diffraction formula can be achieved are not valid in the case of using MZP5 [7]. The approximation of Fresnel diffraction is to take only linear and quadratic terms of the coordinates of the aperture into account. This is also invalid for MZP5. To calculate the correct phase of the wave at the image point P the fourth order of the power series has also to be taken into account.

To verify this, the MTF obtained in the case of Fresnel diffraction is compared to a MTF where the fourth order is included in the calculations, shown in Fig. 3. A theoretical MZP5 without a central stop is used with monochromatic radiation. The resolution limit of MZP5 according to the Rayleigh criterion [8] corresponds to a spatial frequency of 45 lines per micrometer. The differences are plain to see. Including even higher orders of the power series in the calculations yields no change in the results: the fourth-order approximation is sufficient for the investigation of MZP5.

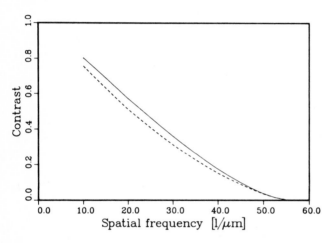

Fig.3 Comparison of a MTF obtained in the case of Fresnel diffraction (dashed line) and a MTF where terms up to the fourth order are taken into account for the calculation of the phase (solid line)

5 The influence of the bandwidth of radiation on the imaging properties

The condenser of an x-ray microscope limits the bandwidth $\Delta\lambda$ of the radiation incident on the micro zone plate. It has to fulfil the demands of the micro zone plate, which has to be used with a certain monochromaticity to image in good quality [9]. Designing a zone plate system for an x-ray microscope therefore raises the question: What is the relation between the bandwidth of the radiation and the imaging properties of the micro zone plate?

74

The modulation transfer functions are helpful here. They are calculated in the following way: The object is a sine grating. It is assumed that in the bandwidths which are mentioned in the figures the synchrotron radiation has the same intensity for any wavelength. The spectrum is scanned with a specific bandwidth. For each wavelength the intensity distribution in the image plane is calculated and added to the other ones. The resulting distribution describes the image of an object which radiates with that bandwidth. The contrast can be determined and repeating that, as described before, yields the MTFs.

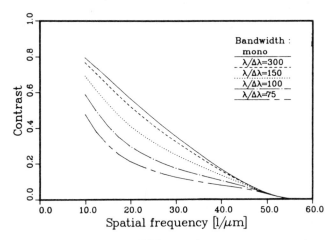

Fig.4 MTFs of MZP5 without a central stop for several reciprocal relative bandwidths

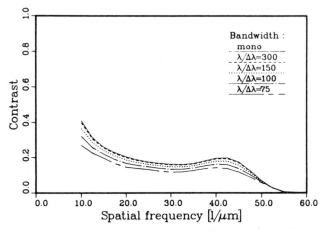

Fig.5 MTFs of MZP5 with a central stop of 14 μm diameter for several reciprocal relative bandwidths

Plotted in Fig.4 are the MTFs imaging with the theoretical MZP5 without a central stop using monochromatic radiation and radiation with four different reciprocal relative bandwidths $\lambda/\Delta\lambda$. Figure 5 shows the same MTFs for MZP5 with a central stop of 14 μm diameter.

Comparing the two figures, the well-known fact is seen that for small structures a zone plate with a central stop yields a better contrast than a complete one, but this has to be paid for with a loss in contrast in the larger structures. Another thing to notice is that a broader bandwidth yields less contrast. However, switching from a reciprocal relative bandwidth of $\lambda/\Delta\lambda = 300$, which equals the zone number, to $\lambda/\Delta\lambda = 75$, 30 % of the contrast at the spatial frequency of 45 l/μm is lost but there is a gain of four times more light. This gain is true both with or without a central stop.

6 The influence of the ratio of the width of the opaque and the transparent zones on the imaging properties

If the width of the open zones is enlarged and that of the absorbing ones is reduced, or vice versa, strictly retaining the correct position of the zones, the efficiency of the zone plate in the first order of diffraction decreases. The spatial resolution, however, still remains the same, which is shown in Fig.6. There $\Delta = 20$ % means that the width of the open zones is enlarged by 20 %. The MTFs are calculated for monochromatic radiation. The five MTFs lie on the same curve. Naturally, this result holds true not only for MZP5 but for all zone plates.

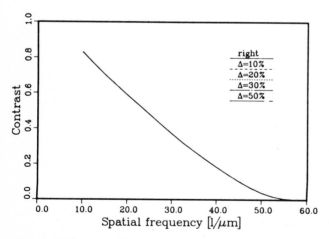

Fig.6 MTFs of MZP5 without a central stop for several ratios of the widths of the opaque and the transparent zones

7 The influence of statistical errors in the manufacture of the zone plate on the imaging properties

MZP5 will e.g. be manufactured as a sputtered sliced zone plate as mentioned above. The sputtering process introduces statistical errors to the zone widths of the resulting zone plate, according to the limited accuracy of the sputtering control. These statistical errors are simulated with a computer in the following way: A statistical and Gaussian distributed displacement is added to the ideal zone radii. In Fig.7 $\Delta r = 10$ nm means that the HWHM of the Gaussian distribution is 10 nm. The result of the calculation is that $\Delta r = 10$ nm, which is half the width of the outermost zone, does not lead to a loss in contrast. Figure 7 shows all MTFs on one curve. The MTFs are calculated for monochromatic radiation. The reason for this good behaviour is the statistical nature of the displacements. The sign of the displacements is also distributed statistically. That means the limits of the open zones vary up and down around their ideal position, but the position of the center of the zone does not vary with the same magnitude.

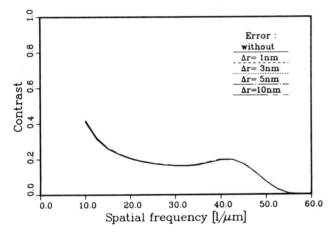

Fig.7 MTFs of MZP5 with a central stop of 14 μm diameter for several statistical errors in the manufacturing

8 The influence of systematical errors in the manufacture of the zone plate on the imaging properties

Considering that the sputter rate is not controlled and the diameter of the wire is not measured during the sputtering process, it is possible that each zone is a little bit too thick or too thin. This introduces a systematical error in relation to the zone plate law. In these calculations it is assumed that the difference between the ideal and the actual

77

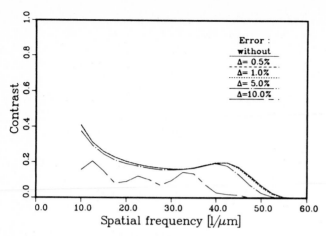

Fig.8 MTFs of MZP5 with a central stop of 14 μm diameter for several systematical errors in the manufacturing

sputter rate is constant. An error of Δ = 0.5 % means that the width of every zone is 0.5 % broader than in the ideal case. That leads to a displacement of each zone, which will increase with the zone number, because every zone gets all errors of the inner ones in addition to its own error. The zone sequence can be approximated by another zone plate law, if the systematical error is small. Another zone plate law leads to another focal length and the object distance and the image distance have then to be readjusted, if the magnification of V = 1000 is still to be used. If the error is too big, another zone plate law cannot be fitted to that sequence. Then an object plane with a corresponding image plane in which the conditions for good imaging are fulfilled cannot be found. The MTFs for MZP5 illuminated with monochromatic radiation are shown in Fig.8. The curves show that up to a systematical error of Δ = 1 % the MTFs still lie on one curve. Larger errors, however, cause loss in contrast.

References

1. G. Schmahl, D. Rudolph, B. Niemann and O. Christ: "Zone Plate X-Ray Microscopy", Quarterly Reviews of Biophysics 13, 297-315 (1980)

2. B. Niemann, D. Rudolph and G.Schmahl: "Soft X-Ray Imaging Zone Plates with Large Zone Numbers for Microscopic and Spectroscopic Applications", Optics Communications 12, 160-163 (1974)

3. D. Rudolph, B. Niemann, G.Schmahl: "Status of the Sputtered Sliced Zone Plates for X-Ray Microscopy", SPIE Proceedings 316, 103-105 (1982)

4. M. Born, E.Wolf: <u>Principles of Optics</u>, 6th ed., (Pergamon Press, Oxford 1980)
5. L. Bergmann, C. Schaefer: <u>Lehrbuch der Experimental-physik</u>, <u>Band III</u>, <u>Optik</u>, 6th ed. (de Gruyter, Berlin 1974)
6. H.H. Fink: <u>Untersuchungen zum Punktbild und zur Modulationsübertragungsfunktion Fresnelscher Zonen-platten im optischen Spektralbereich</u>, Dissertation, Universität Tübingen 1975.
7. A.G. Michette: <u>Optical Systems for Soft X-Rays</u>, (Plenum Press, New York 1986)
8. D. Rudolph and G. Schmahl: "High Power Zone Plates for a Soft X-Ray Microscope", Ann. NY Acad. Sci. <u>342</u>, 94-104 (1980)
9. G. Schmahl, D. Rudolph, P. Guttmann and O. Christ: "Zone Plates for X-Ray Microscopy" in <u>X-Ray Microscopy</u>, ed. by G. Schmahl, D. Rudolph, Springer Series in Optical Sciences <u>43</u>, 63-74 (Springer, Heidelberg 1984)

Microzone Plate Fabrication
by 100 keV Electron Beam Lithography

V. Bögli[1], P. Unger[1], H. Beneking[1], B.Greinke[2], P. Guttmann[2],
B. Niemann[2], D. Rudolph[2], and G. Schmahl[2]

[1]Institute of Semiconductor Electronics,
  Aachen Technical University, Sommerfeldstraße,
  D-5100 Aachen, Fed. Rep. of Germany
[2]Universität Göttingen, Forschungsgruppe Röntgenmikroskopie,
  Geismarlandstraße 11, D-3400 Göttingen, Fed. Rep. of Germany

Introduction

Fresnel zone plates which are used as imaging lenses for soft x-rays are
required to have high resolution, high efficiency, and low distortions. In
terms of practical realisation, this means smallest possible widths of the
outer zones, high aspect ratios of the absorber rings, and a positional
accuracy of the structures to a fraction of the outer zone width. Additionally, the absorptance of the membranes carrying the zone plate has to be
minimized.

To achieve high resolution and accurate edge definition of the fabricated
structures, we use a 100 keV electron beam lithography system. The deflection field distortions that always inhere in such systems are corrected by
an alignment procedure using an externally-generated registration mark
array. Deflection field alignment has turned out to be an important condition to get optimum efficiency and image resolution of the fabricated zone
plates. The exposed resist image is transferred to a metal absorber structure by electroplating into a polyimide mould using a suitable trilevel
resist system.

This paper presents zone plates which have been used as imaging objective
lenses in the Göttingen x-ray microscope /1/. Images of artificial and
biological specimens have been taken at the electron storage ring BESSY in
Berlin. Efficiencies of electroplated and formerly fabricated ion milled
zone plates have been evaluated and compared with the theoretically achievable values.

Lithography

The e-beam lithography system used for nanostructure fabrication is basically a HB501 scanning transmission electron microscope (STEM) which has
been converted into a lithography machine. The system features as well as
the method of fast zone plate exposure have been described in /2-4/. In
this paper, we therefore will just confine to the deflection field calibration procedure. This calibration step has drastically improved the
accuracy and imaging quality of our zone plates.

The linear deflection errors and their influence on the exposed zone
plates are summarized in Fig. 1. While the shift and rotational errors in

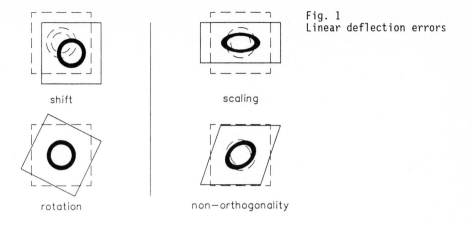

Fig. 1
Linear deflection errors

shift

scaling

rotation

non—orthogonality

our case do not affect the imaging properties of the rotational symmetrical zone plate, mis-scaling and non-orthogonality of the cartesian deflection axes will lead to astigmatism and worsen the image resolution. This
is pointed out on the right hand side of Fig. 1, where the elliptical
distortions of a ring resulting from the deflection errors can be seen. In
our lithography machine, we are able to correct these linear deflection
errors with the aid of an automated mark alignment system. Registration of
small fiducial marks is performed using the transmitted electron detector
signals of the STEM. Atomic number contrast as well as topographic contrast of appropriately generated calibration marks on thin Si_3N_4 membranes
are utilized together with the darkfield and brightfield detectors mounted
in our system. Details of the registration procedure and the achievable
calibration accuracy can be found in /5,6/.

As our lithography system does not provide a laser interferometer controlled specimen stage, we are restricted to an externally-generated
calibration standard. For correction of the linear deflection errors, a

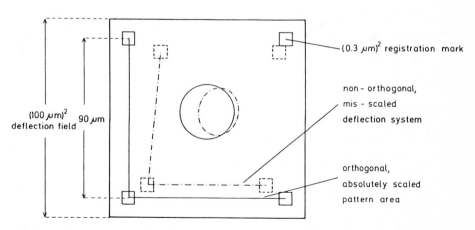

$(0.3 \mu m)^2$ registration mark

non - orthogonal,
mis - scaled
deflection system

$(100 \mu m)^2$
deflection field   90 μm

orthogonal,
absolutely scaled
pattern area

Fig. 2 Deflection system alignment on externally-generated reference marks

well suitable reference is an array of four registration marks which are situated near the corners of the deflection field. Fig. 2 schematically shows the method of deflection system alignment. The four registration marks have been exposed in an EBMFII e-beam machine, which is calibrated with a laser interferometer. This system, however, has inferior resolution and linewidth control and cannot be used to fabricate the microzone plates by itself. After exposure into a bilevel resist system (PMMA/MA:OEBR1010) and development, the marks are generated by evaporation of a 50-nm-thick Au layer and lift-off.

Due to the low resolution of the EBMFII compared to the high standards of nanolithography, the marks of only $(0.3\ \mu m)^2$ size are not ideal concerning shape and edge definition. They are, however, well suitable for calibration of the HB501 with our special registration method of total area scan and first momentum analysis to determine their center positions. The only requirement is that the relative distance of the center positions must be exact, whatever the absolute geometry of the marks looks like. This demand is easier to fulfill, as shape distortions due to beam astigmatism or beam blanking delay in the EBMFII will be the same for all exposed shapes.

Inserting this sample into our nanolithography system, the reference marks are scanned and their center positions are determined as described. From the positions found, the deviations are calculated and the electronic deflection field correction unit is set appropriately. This scanning and error calculating correction procedure is repeated until the delineations are in the order of the detection accuracy (1 LSB in our system, corresponding to a 6-nm step). Of course, the achieved calibration accuracy is limited by the generated calibration standard itself. Field stitching experiments have shown that with the described method, an error of 40 nm at the edges of a $(100\ \mu m)^2$ deflection field will remain /6/. As the deflection system errors are assumed to be linear mainly, the distortions of the exposed zone plates will decrease with decreasing diameter towards the center. For the fabricated zone plates with a diameter of about 50 μm, the maximum positional error is estimated to about 20 nm. Higher order distortions can generally be corrected by electronic methods, but are more difficult to determine and require greater efforts /7,8/.

Pattern transfer

After e-beam exposure and development, the resist patterns have to be transferred into an absorber layer of sufficient thickness. We recently have described an improved ion milling process with a thin titanium mask layer to generate gold absorber rings /9/. The intermediate mask layer thereby improves the steepness of the absorber ring sidewalls, compared to the method of using the PMMA structures directly as an ion milling mask. The achievable aspect ratio, however, is still limited, because the maximum obtainable sidewall angle is in the range of about 80 degrees. A method of generating zone plates by electroplating into the PMMA resist mould has been reported in /10/. To get absorber rings with a height of 150 nm or more at a width of 50 nm and below, we developed a process using a trilevel resist system followed by an electroplating step. The schematic of this process is shown in Fig. 3.

The silicon nitride carrier membrane is covered by a 10-nm-thick Ni layer, serving both as an adhesion promotor of the following polymer layer and later as the electroplating base. It is followed by a thick layer of polyimide, into which the structures are to be transferred. A thin Ti layer is evaporated onto the polymer, and finally the e-beam sensitive

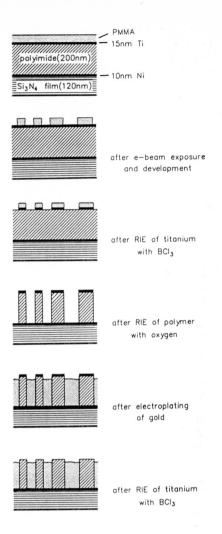

PMMA
15nm Ti
polyimide(200nm)
10nm Ni
Si$_3$N$_4$ film(120nm)

after e-beam exposure
and development

after RIE of titanium
with BCl$_3$

after RIE of polymer
with oxygen

after electroplating
of gold

after RIE of titanium
with BCl$_3$

after removal of the polymer
in an oxygen plasma

Fig. 3 Schematic of the electroplating process using a trilevel
resist system

resist of approximately 35 nm thickness is spun on top. After e-beam exposure and development of the zone plate structure, the PMMA resist pattern is transferred into the thin Ti mask layer by reactive ion etching (RIE) with BCl$_3$. In the following oxygen RIE step of the polyimide, the titanium is oxidized and forms a stable mask. The obtained polymer structures show almost vertical sidewalls and can be etched to a high depth. In the next step, these structures are used as a mould to electroplate the metal absorber rings. The gold absorbers have been grown onto the Ni plating base without problems. This is due to the low sputter attack of Ni during the oxygen RIE. By careful control of growth conditions and rate, the metal structures can be grown to almost the total height of the polymer mould. Finally, the mould is removed in an oxygen plasma.

With this method, a microzone plate with an outermost ring width of 55 nm and a height of the absorber rings of more than 150 nm has been generated. An SEM micrograph of this zone plate with 250 zones is shown in Fig. 4. From the outermost section view, it can be seen that the edge definition is excellent up to the smallest features.

X-ray microscope images

To evaluate the properties of our fabricated zone plates, imaging experiments have been performed in the Göttingen x-ray microscope at BESSY in Berlin. At first, the influence and importance of the deflection field correction shall be demonstrated. For this purpose, another zone plate serves as an object. The topography of this zone plate is known from SEM and STEM pictures. Fig. 5 shows a comparison of the images taken with

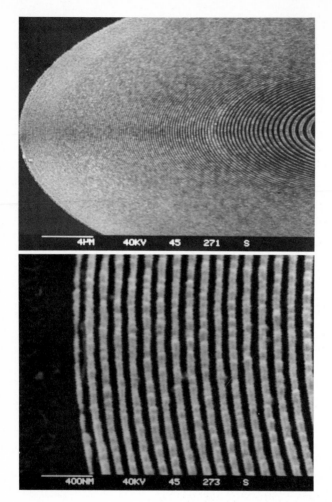

Fig. 4
Electroplated gold
zone plate with a
maximum aspect ra-
tio of 3:1

different objective lenses /11/. The first picture (5a) has been taken
with a zone plate that was e-beam exposed without deflection field correction. Resolution of this image is poor, contrast decreases to the outer
region, and a strong astigmatism has been observed. An image of the same
object, but made with the holographically fabricated MZP3 from Göttingen
/12/, is shown in Fig. 5b for comparison. The properties of this objective
lens are better, and details up to the 50-nm-wide outermost rings are
resolved. Fig. 5c finally presents an image obtained with an e-beam generated zone plate after deflection field calibration. The bright zero order
ring results from the optical setup, but does not affect the imaging
properties of the zone plate. All details up to the finest rings are
clearly resolved, with almost no decrease in contrast.

a) objective lens:
   e-beam generated zone plate
   100 zones, 50 nm min. zone width
   <u>uncorrected</u> deflection field

b) objective lens:
   holographically-generated MZP3
   for comparison
   250 zones, 55 nm min. zone width

c) objective lens:
   e-beam generated zone plate
   100 zones, 50 nm min. zone width
   <u>corrected</u> deflection field

Fig. 5
X-ray microscope images obtained with
different objective zone plates.
object: slightly distorted 50-nm
Au zone plate on Si_3N_4 membrane

1μm

Fig. 6
X-ray microscope
image of a human
fibroblast

With an e-beam generated copy of the MZP3 (250 zones, 55 nm minimum zone width), still fabricated with the improved ion milling process, pictures of biological specimens have been taken. In Fig. 6, we present an X-ray microscope image of a human fibroblast, which shows details in the 50-nm-range.

Efficiency measurements

To get more information about the properties of the different fabricated microzone plates, and to see where further improvements will be necessary, exact efficiency measurements have been performed using an x-ray experimental chamber at the BESSY storage ring. The results are presented in Table 1. We have compared the ion milled zone plate with 80 nm absorber height to the electroplated structure having 150 nm thick absorber rings at wavelenghts of 4.5 nm and 2.4 nm, respectively. To eliminate the influence of the nitride carrier membrane, the absorption in this membrane has been separately measured and taken into account, resulting in the groove efficiency. The theoretical values have been calculated in /13/.

Table 1. Efficiency comparison of ion milled and electroplated zone plates

method of pattern transfer	Au absorber height	membrane	efficiency (%)		groove efficiency (%)		theoretical groove efficiency (%)	
			2.4nm	4.5nm	2.4nm	4.5nm	2.4nm	4.5nm
sputter etched	80 nm	110nm Si_3N_4	not measured	2.7±0.3	not measured	5.7±0.6	7.5	7.0
electroplated	150 nm	110nm Si_3N_4 +10nm Ni	4.9±0.5	2.8±0.3	8.6±1.0	7.3±0.5	11.5	10.2

A comparison of the absolute efficiency with the groove efficiency values shows a strong influence of the Si_3N_4 membranes. It is therefore desirable to make these carrier films thinner to gain a further increase of efficiency. However, a lower limitation is thereby given by the demand of stability of the fragile membranes.

Conclusion

We have shown that it is possible to generate high quality imaging microzone plates by electron beam lithography. Correction of the deflection field distortions thereby plays an important role to achieve accurate lenses with minimum imaging errors. By using an electroplating step combined with a trilevel resist system for pattern transfer, absorber structures with a width of 55 nm at an aspect ratio of 3 : 1 can be fabricated. X-ray microscope images of artificial and biological objects have proven the high obtained imaging quality. From the results of efficiency measurements on zone plates generated with different methods of pattern transfer, it can be said that with the electroplated structures (showing an efficiency of 4.9 % at 2.4 nm wavelength) we have moved one step closer towards the achievable maximum.

Fig. 7 finally shows that the capability of the lithography system is not exhausted with the 50-nm structures. It is a STEM micrograph of a zone plate with a minimum zone width of 20 nm etched into a 15-nm-thick titanium mask layer. Although there are still some stability problems to overcome to transfer this mask into a sufficient thick polymer, it seems to be likely that zone plates with these dimensions will be available in the near future.

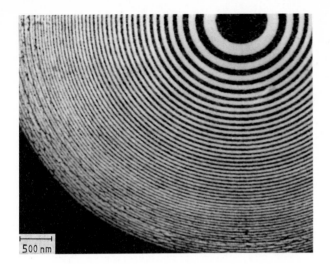

Fig. 7
STEM micrograph of a
20-nm zone plate in
titanium

Acknowledgements

We wish to thank P.C. Cheng for preparing the human fibroblast specimen,
H.J. Geelen for exposing the calibration mark array, and A. Kloidt for
investigations on the electroplating process. This work has been funded by
the Stiftung Volkswagenwerk, and by the Federal Minister for Research and
Technology (BMFT) under the contract number 05 320 DAB.

References

/1/ D. Rudolph et al., in X-Ray Microscopy, edited by G. Schmahl and
     D. Rudolph (Springer-Verlag, Berlin 1984), p. 192.
/2/ E. Kratschmer et al., in Proc. Microcircuit Engineering 1983, edited
     by H. Ahmed, J.R.A. Cleaver, and G.A.C. Jones (Academic Press, London
     1983), p. 15.
/3/ D. Stephani et al., J. Vac. Sci. Technol. B 1(4), 1011 (1983).
/4/ V. Bögli et al., in Soft X-Ray Optics and Technology (Berlin 1986),
     edited by E.E. Koch and G. Schmahl, Proc. SPIE 733, p. 449 (1987).
/5/ E. Kratschmer et al., in Proc. Microcircuit Engineering 1984, edited
     by A. Heuberger and H. Beneking (Academic Press, London 1984) p.15.
/6/ V. Bögli and H. Beneking, Microelectronic Engineering 3, 117 (1985).
/7/ M. Fujinami et al., Proc. Intern. Electron Device Meeting, Washington
     1981, p. 566.
/8/ K. Takamoto et al., J. Vac. Sci. Technol. B 4, 675 (1986).
/9/ P. Unger et al., Microelectronic Engineering 5, 279 (1986).
/10/ Y. Vladimirsky et al., to be published in J. Vac. Sci. Technol. B 6(1)
     (1988).
/11/ P. Unger et al., to be published in J. Vac. Sci. Technol. B 6(1) (1988).
/12/ G. Schmahl et al., in X-Ray Microscopy, edited by G. Schmahl and
     D. Rudolph (Springer-Verlag, Berlin 1984), p. 63.
/13/ J. Kirz, J. Opt. Soc. Am. 64, 301 (1974).

Zone Plates for Scanning X-Ray Microscopy: Contamination Writing and Efficiency Enhancement

C.J. Buckley, M.T. Browne, R.E. Burge, P. Charalambous, K. Ogawa, and T. Takeyoshi

Department of Physics Research, King's College,
London University, The Strand, London WC2R 2LS, UK

1 Introduction

The resolution performance of a soft x-ray microscope which uses a zone plate as its focusing element is primarily dependent on the scale and accuracy with which the zone plate can be made. In order to be capable of attaining spatial resolution in the 10 - 100 nm region it is necessary to fabricate zone plates which have finest rings of this scale. Also the placement of the zones must be accurate to a fraction of the finest zone width across the zone plate diameter if aberrations are not to limit resolution.

At present zone plates with sub - 100 nm zone widths are being fabricated by holographic [1] and electron beam lithographies [2,3,4] using conventional resists. These methods have produced successful zone plates, however the finest zone widths of these zone plates have been limited to about 50 nm. In the case of conventional electron beam lithography, the finest zones which can be fabricated with equal mark and space are limited by the extent of the scattering of the electron beam as it penetrates the resist. This problem has been largely overcome in an alternative process known as electron beam contamination writing [5]. This technique can produce carbon lines which are 20 nm wide and 100 nm high with a mark to space ratio of 1:1. A 100 nm thickness of carbon absorbs about 50% of the radiation at a wavelength of 3.5 nm and thus in principle zone plates can be made by contamination writing which are capable of focusing soft x rays down to a spot size of 20 nm.

The method by which the contamination process defines lines on a thin film substrate has been described in previous publications [6,7]. Here the latest procedures by which the contamination process is employed to make more accurate zone plates are presented, and a summary of zone contrast enhancement is given.

2 Zone Plate Design Considerations

For a zone plate which has a finest zone width dr to be a useful focusing element in a scanning soft x-ray microscope, its first order focal length (f_1) must be sufficiently large that an order selecting aperture [8] can be placed and manoeuvred between the zone plate and the specimen. At a wavelength λ, the first order focal length in terms of the zone plate radius (r_{zp}) is

$$f_1 = \frac{2r_{zp}dr}{\lambda} \qquad (1)$$

Practical interpretation of this relationship leads to a minimum zone plate radius of 25 μm when dr is a few tens of nm. To be able to fabricate an accurate zone plate (with such dimensions) by electron beam lithography the beam position distortions introduced by abberations in the electron optic system must be compensated for. For displacements of the electron beam of only a few microns the beam placement distortions are small but increase rapidly with distance from the electron optical axis. For example, the distortion of the positioning of the electron beam in an HB5 STEM has been measured [8] and for a displacement of the electron beam from the scan origin of 4.5 μm the difference between the applied deflection and that obtained is about 2 nm, while at a beam displacement of 10 μm the difference is in excess of 10 nm. Correcting for these distortions over several tens of microns is difficult as they can be complex in nature.

To take advantage of the low level of beam position distortion at relatively small beam displacements, a zone plate can be formed by patching several small fields together by moving the substrate. The field stitching accuracy required for this process is a few nanometres which would be extremely difficult to achieve by stage movements alone. However, having drawn a line by the contamination technique, it is possible to locate it with a high degree of precision, using the electron beam, without further contaminating the line to any significant degree. This process, known as registration, can be used to stitch adjacent fields of 4.5 μm in size together to an accuracy of $\pm$ 1 nm. Thus by a combination of stage movement and electron beam registration it is possible to fabricate zone plates which have zones with widths down to a few tens of nm and which are accurately positioned over a diameter of several tens of microns.

3 Registration Marks and Field stitching

In order to patch the drawing fields accurately together it is necessary to create a mark which can be used as a reference. A suitable mark is a simple 0.5 μm diameter cross. Having drawn such a cross-shaped registration mark, the substrate can be moved a few microns by the stage, and the position of the cross can be precisely determined by a registration scan on the cross. The technique is further used to monitor drift of the substrate, which is corrected for.

The registration process described above can be used to patch a number of drawing areas together to form a zone plate of some tens of microns in diameter. To create a set of reference marks over the full diameter of the zone plate drawing field, a central reference circle (diameter about 1.5 μm), is drawn and registration crosses are drawn radially from this origin. The pattern definition process is simplified by a substrate rotation mechanism which allows adjacent registration marks to be laid down by repetitious electron beam deflections and stage displacements. All important measurements are thereby reduced to one dimension. The pattern of registration marks is referred to as a skeleton, an example of which is shown in fig.1 (here the registration marks are connected by radial spokes). The registration marks define a series of radial markers which can be used as references from which to place zone arc segments.

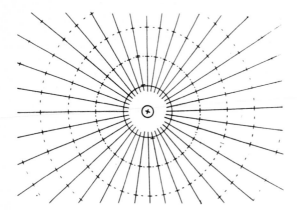

Fig. 1 A pattern of carbon contamination registration marks connected by radial spokes. The marks are used to accurately position zone arcs to form a zone plate pattern. The central circle has a radius of 1.5 μm.

Assuming that no significant errors are present in the positioning of the registration marks, the skeleton could be used to accurately define the position of the zones which could be drawn in 4.5 μm patches using the registration technique in conjunction with the stage rotation/translation mechanism. Thus the entire zone plate pattern can be formed by a patchwork of zone segments. The placements of the registration marks are subject to errors however. The origins and correction of these errors are discussed in the following section.

4 Errors in Registration mark and Zone Positioning

Errors in the positioning of registration marks and zones arise from the positional distortion of the electron beam scanning system, height changes of the substrate, electrical instabilities (which result in magnification changes) and drift. Also each time one registration mark is used as the datum to define an adjacent one, there is an error associated with the measured position of the reference mark.

The process discussed above can produce registration marks at constant radial separations to an accuracy of ± 1 nm in a particular sector. However the separation of the registration marks can vary from one sector to another. This is because the surface on which the substrate stage rotates is not perfect and differences of a few microns can be introduced in the focus which lead to changes in the x and y scan amplitudes of the electron beam. This can be corrected to a certain extent by monitoring the focus change and applying an increment or decrement to the placement of the registration marks. However, some error still remains.

The residual errors can be accounted for by using an inter-sector registration technique. The registration process makes it possible to determine the relative position of a registration mark in one sector with respect to the corresponding registration mark in the adjacent sector. Having defined the registration marks for a sector, a part of a single zone arc is drawn adjacent to the registration mark in each field. After the rotation and the registration marks for the next sector have been defined, the position of the arcs from the previous sector are determined at the field overlap and their displacements from the expected positions are recorded. The values of these displacements are stored and it should be possible to use this information to determine the relative positions of the registration marks from the correct circular pattern. However due to the non orthogonality of the x and y electron beam scanning system there is a constant element in all the displacements which can be determined and removed.

5 Magnification Checks

Immediately before the zone arcs are drawn the separation between registration marks is measured and this measurement is used to correct for changes in magnification caused by electrical instabilities. If there is a difference between the current magnification and that recorded when the skeleton was defined, its magnitude can be assessed and the zone pattern for the field can be stretched or compressed accordingly.

6 Arc Writing

Having defined a precise set of radial markers, the zones can be written as follows. The data for an arc representing the central radius of a particular zone is calculated and this arc is drawn in 20 nm wide contamination lines a sufficient number of times at small displacements to form the solid zone segment. The arc data is held in a table and consists of a series of fine horizontal displacements from a vertical scan in the y direction. The displacements are from a datum, which is calculated to be a set distance from the registration mark thus providing the correct radius from the pattern centre Having written the zone arcs for each field in a sector, the substrate cartridge is rotated and the fields for the next sector are filled. This process is repeated until the entire zone plate is drawn.

7 Error Correction

The registration marks define a series of radii which are used as references when drawing the zones. These marks can be considered as lying on concentric reference circles. However not all the marks lie exactly on circles, as small displacements are incurred during the drawing of the skeleton. The inter-sector registration process mentioned above provides a measure of the relative displacement of each registration mark from the mean circle. Thus when the arcs are drawn with reference to the registration marks, the displacements are added or subtracted to the position of every arc in the field so that the arcs (which together make up a complete zone) are all positioned at the same radius.

If no correction was made for the non-orthogonality of the x and y scan of the electron beam then only the centre of each arc would lie at the correct radius from the centre of the zone plate. The resulting pattern would have the appearance of a series of concentric ratcheted wheels rather than smooth circles. However the magnitude of the nonorthogonality may be obtained from the inter-sector registration process and the ratchet effect can be removed by tilting the arc patterns by adding a graded displacement to the data used to define the arc.

Small magnification changes can be brought about by electrical and mechanical instabilities in the time between the definition of the skeleton and the drawing of the zones. These can be corrected for by measuring the distances between the registration marks in a field and comparing the values obtained to those recorded when the skeleton was defined. If differences arise then the zone pattern can be either expanded or contracted such that the arc pattern fills the field correctly.

Thus the application of the corrections described reduces the distortion of the zone plate pattern, and allows zone plates to be made which have finest zone widths of a few tens of nanometres which result in practical first-order focal lengths with soft x rays. Fig. 2 shows a scanning electron micrograph of contamination written zone plate which has a finest zone width of 35 nm. The zone plate was fabricated on a 50 nm thick boron nitride membrane.

Evidence that the correction schemes result in good matching of zone arc segments can be obtained by viewing the moire fringe pattern obtained when

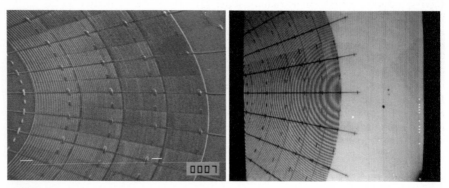

Fig. 2   Fig. 3

Fig. 2 Scanning electron micrograph of a carbon contamination zone plate. The zone plate has a finest zone width of 35 nm.

Fig. 3 Scanning transmission electron micrograph of a carbon contamination zone plate which has a finest zone width of 35 nm. The linear raster creates a moire fringe pattern which indicates correct matching of zone segments.

a contamination written zone plate is imaged in bright field by a linear raster in STEM. The moire fringe pattern which results from this type of imaging is another zone plate pattern. Misplacement of one field of zone segments with respect to its nearest neighbour by a fraction of a zone width in either direction shows up as a discontinuity in the moire pattern. Fig. 3 shows a carbon contamination zone plate imaged by the method described above. The zone plate shown has a finest zone width of 35 nm, and shows no fringe displacement greater than one tenth of a fringe. This corresponds to field matching accuracies of 7 nm which are within the required tolerance for this zone plate.

8 Contrast Enhancement

The carbon contamination material of which the zone plates are comprised does not attenuate a high fraction of the incident x-ray amplitude. This results in a diffraction efficiency which is considerably lower than that obtained from a perfect absorber. The efficiency has been considerably improved by replicating the carbon zone plates in gold by proximity printing x-ray lithography. Fig. 4 shows a gold replica which has a finest zone width of 75 nm [9]. At present faithful replicas have not been made which have a finer zone width than this. However the most recent experiments have produced replica zone plates which have sub-50 nm zone widths.

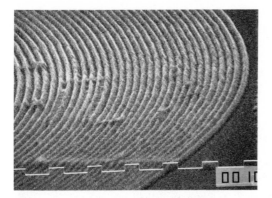

Fig. 4 A scanning electron micrograph of a gold replica zone plate which has a finest zone width of 75 nm. The zone plate was fabricated from a carbon master using x-ray lithography.

9 Summary

Improvements to the carbon contamination lithographic scheme have resulted in the fabrication zone plates which have a finest zone width of 35 nm and which are accurate to the tolerance required over a diameter of more than 50 μm. The technique is under refinement, and it is hoped that finer zone plates will be made with sufficient accuracy in the near future. The contrast enhancement by x-ray lithography has produced gold replica zone plates which have a first order efficiency which is close to a factor of 10 higher than the carbon originals. These replica zone plates have been used to produce images with near theoretical resolution in the STXM at the Daresbury SRS and at BESSY in Berlin. The x-ray replication technique has recently been used to produce zone plates with finer zone widths, and the performance of these zone plates will be assessed at Daresbury in the near future.

10 References

1 Schmahl G., et al, Zone plates for x-ray microscopy, X-ray Microscopy, Springer Series in Optical Sciences 43, 63-74, (1984).

2 Bögli V., Unger P. and Beneking H. Electron-beam lithography and nanometre structures: fabrication of micro zone plates. Soft X-Ray Optics and Technology, Proc. SPIE 733, in press, (1986).

3 Kern D., et al, Electron Beam Fabrication and Characterization Studies of Fresnel Zone Plates. Advances in Soft X-ray Science and Technology, Proc. SPIE, (1984).

4 Aritome H., Nagata K. and Namba S. Fabrication of zone plates with a minimum zone width smaller than 100 nm by EBL. Microelectronic Engineering 3, Elsevier Science (1985).

5 Browne M.T., Charalambous P., Burge R.E., Duke P.J., Michette A.G. and Simpson M.J. A new lithographic technique for the manufacture of high resolution zone plates for soft x rays. Journal de Physique C2, 45, 89-92, (1984).

6 Buckley C.J., Browne M.T. and Charalambous P., Contamination lithography for the fabrication of Zone Plate X-ray lenses. E-Beam, X-Ray & Ion Beam Techniques for Submicrometre Lithography IV, Proc SPIE 537, 213-217, (1985).

7 Buckley C.J., Feder R. and Browne M.T. Soft X-Ray Nano Lithography of Semitransparent Masks for the Generation of High-Resolution High-Contrast Zone Plates. Short Wavelength Coherent Radiation: Generation & Applications, Am.Inst.Phys.Conf.Proc. 147, 368-381, (1986).

8 Buckley C.J. The Fabrication of Gold Zone Plates and their use in Scanning X-ray Microscopy. Doctor of Philosophy Thesis, Kings College, London University, p156, p32 (1987).

9 Buckley C.J., Ogawa K., Browne M.T., Charalambous P., Kenney J.M., Rosser R., Ade H., Mcnulty I., Feder R. and Cheng P.C. Zone Plate Replication by Contact X-ray Lithography. X-ray Microscopy, Springer-Verlag, 247-253, (1987).

Phase Zone Plates
for the Göttingen X-Ray Microscopes

R. Hilkenbach, J. Thieme, P. Guttmann, and B. Niemann

Universität Göttingen, Forschungsgruppe Röntgenmikroskopie,
Geiststraße 11, D-3400 Göttingen, Fed. Rep. of Germany

1 Introduction

For biological and medical research it is important to be able
to look at living and therefore wet specimens. As was pointed
out by H. WOLTER [1] in 1952, an X-ray microscope operating in
the wavelength region 2.36 nm $\leq \lambda \leq$ 4.5 nm would be suitable
for this purpose. In this "water window" cell organelles
absorb radiation more strongly than water. Therefore specimens
in a wet environment can be imaged with sufficient contrast.

In this wavelength region the complex index of refraction
is close to unity. Therefore it is practically impossible to
build refractive lens optics. Diffraction optics can be
employed instead. Circular gratings called zone plates can be
used to image X-rays. As a result of the theory explained
below, zone plates made of gold yield a maximum efficiency of
only 12 %, not including the absorption of radiation in the
support foil. It is desirable to improve this because smaller
losses of photons result in increased contrast, due to the
reduction of photon noise, or shorter exposure times and
therefore smaller irradiation doses for the specimen. Also
smaller, cheaper and thus less intense X-ray sources can be
employed for future X-ray microscopes.

2 The Theory of Phase Zone Plates

The complex index of refraction of materials is

$$n = 1 - \delta - i\beta \; .\tag{1}$$

A material of thickness t will attenuate the amplitude of
incident radiation by a factor $\exp(-2\pi\beta t/\lambda)$. Compared to a
wavefront travelling through vacuum, the phase of the
radiation penetrating the material will be retarded by
$\Phi = 2\pi\delta t/\lambda$. Taking the ratio $\eta = \beta/\delta$ the theoretical
efficiency of a real zone plate can be shown to be [2]

$$I/I_0 = \frac{1}{\pi^2} \cdot (1 + e^{-2\eta\Phi} - 2 \cdot e^{-\eta\Phi} \cdot \cos\Phi) \; .\tag{2}$$

This formula is derived under the assumption that the zones
have a box shaped profile so that the walls of the filled

zones are perpendicular to the plane of the zone plate. It is assumed also that the areas of the filled zones equal those of the empty ones.

If one neglects the phase shift and considers only the absorption of radiation in the material, one obtains $\delta = 0$ and therefore $\eta = \infty$. Then both exponential terms vanish and the theoretical efficiency of a so-called "amplitude zone plate" turns out to be 10.1 %.

The two parts β and δ of the index of refraction n can be expressed by the atomic scattering factors f_1 and f_2 :

$$\beta = r_0 \lambda^2 \ xf_2 \ / \ 2\pi \tag{3}$$

$$\delta = r_0 \lambda^2 \ xf_1 \ / \ 2\pi \ . \tag{4}$$

where r_0 means the classical electron radius and x the number of atoms per unit volume. The ratio η becomes $\eta = f_2/f_1$. These equations are valid only in the case of pure elements. The groove efficiency of a germanium zone plate for three wavelengths as a function of the thickness of the material is shown in Fig.1.

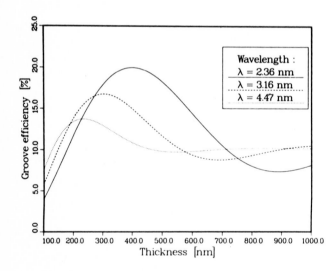

Fig. 1 The theoretical efficiency of a germanium zone plate calculated for three wavelengths and plotted versus the thickness of the germanium layer.

If one considers a compound of elements, the values of the different components must be summed up according to their relative abundance.

$$\overline{xf}_1 = \sum_q x_q f_{1q} \tag{5}$$

$$\overline{xf}_2 = \sum_q x_q f_{2q} \tag{6}$$

Here x_q denotes the number of atoms per unit volume of type q and f_{1q} and f_{2q} are the atomic scattering factors f_1 and f_2 of the element q. In the energy region of E = 100 - 2000 eV the values of f_1 , f_2 are measured and calculated for the elements Z = 1 - 94 and are given in tabular form by HENKE et al. [3]. The theoretical efficiencies of some materials at a wavelength of λ = 2.36 nm are listed in Table 1.

Table 1 The maximum theoretical efficiency of zone plates made of different materials and the corresponding thicknesses for the wavelength λ = 2.36 nm.

Material	Efficiency [%]	Thickness [μm]
Beryllium	30.4	0.84
Magnesium	27.6	0.92
Chromium	24.9	0.50
Nickel	24.7	0.28
Silicon	22.9	0.64
Germanium	20.0	0.40
Gold	12.0	0.18

3 The Parameters of Actual Zone Plates

The Göttingen X-ray Microscopes require zone plates with the parameters listed in Table 2 [4].

Table 2 The Parameters of the Condenser Zone Plates (KZP) and the Micro Zone Plate (MZP)

	KZP 3	KZP 4	KZP 5	KZP 6	MZP 3
Diameter $2r_n$ [μm]	9000	2500	2500	9000	55.6
Radius of innermost zone r_1 [μm]	37	50	36	23	1.8
Number of zones n	15000	630	1200	38000	251
Width of outermost zone dr_n [μm]	0.15	0.99	0.46	0.06	0.055
Focal length [mm] $f_{\lambda=4.5nm,1^{st}\ order}$	304	551	290	119	0.69
Focal length [mm] $f_{\lambda=2.36nm,1^{st}\ order}$	570	1033	544	222	1.30

The condenser zone plate KZP 4 was chosen as the first zone plate to be built as a phase zone plate for λ = 2.36 nm because of its rather coarse outermost zone width of 1 μm. This zone plate is used in the scanning X-ray microscope developed by B.NIEMANN [5].

4 The Manufacture of the Germanium Condenser Zone Plate KZP4

The semiconductor material germanium has been chosen for the construction of phase zone plates because of its phase shifting properties and its ease of handling and processing. The required thickness of the layer of 404 nm is deposited upon a polyimide (PIQ) foil of 0.3 μm thickness using an electron beam evaporator. As can be seen in Fig.1 an uncertainty of about 100 nm in the thickness of the layer is tolerable since the theoretical efficiency drops from 20% to 18%. The foil is supported from the back for all processing steps by a glass carrier. The photoresist AZ 1350 is spun on these substrates and prebaked. Then the zone plate pattern is exposed holographically using the optical arrangement described by J. THIEME [6]. After developing and postbaking this pattern is transferred into the germanium layer using a reactive ion etch (RIE) process. $CBrF_3$ is used as an etchant revealing an etch rate of 70 nm/min. For this process a pressure of 30 mtorr, a flow of 5.0 sccm of etching gas and 0.3 W/cm^2 of high frequency power is applied. The remaining resist is stripped off and the germanium layer round the zone plate is removed in a second RIE process to reduce the mechanical tensions of the layer. After the mounting of a support ring the PIQ foil with the zone plate is separated from the glass carrier and the zone plate is ready for use in the X-ray microscope.

5 Measurements of the Efficiency

The wavelengths of λ = 2.36 nm and λ = 4.47 nm were obtained from the electron storage ring BESSY using a grazing incidence grating monochromator. The contributions of the wavelengths caused by the higher orders of this grating were diminished by a chromium filter of 1 μm thickness.

The zone plate diffracts radiation into different orders. One specific order of diffraction can be selected by placing a pinhole in the image plane corresponding to that order. The image of the synchrotron source in this order is smaller than the diameter of the pinhole of 0.4 mm. This means the radiation of that order can be measured without loss. The contribution of the higher orders to the radiation in that pinhole is very small. It depends on the efficiency of the zone plate in these orders and on their spot sizes in the plane of the pinhole. The result is that the radiation of the higher orders can be neglected. A wire with a diameter of 0.6 mm at the center of the zone plate totally absorbs the radiation of the undiffracted zero-order beam. For the measurements a semiconductor diode just behind the pinhole has been used.

The radiation focussed in one specific order of diffraction by the zone plate was measured, as well as the radiation without the zone plate. The ratio of these signals, normalized on the effective areas collecting the radiation, is the absolute efficiency of the zone plate in that order. This figure contains the absorption of the supporting PIQ foil.

For comparison with theoretical calculations the absorption of X-rays in the PIQ foil has to be taken into account. This is done by dividing the absolute efficiency of the zone plate by the transmission of the supporting foil. The resulting figure is the groove efficiency.

6 Results

The results of the measurements can be seen in Tables 3, 4 and 5. In Table 3 the transmission of PIQ foils is listed. Table 4 lists the absolute efficiency of germanium and gold zone plates at the wavelengths λ = 2.36 nm and λ = 4.5 nm. In Table 5 the derived groove efficiencies are shown.

Table 3 The transmission of PIQ foils

Thickness [μm]	Transmission [%]	
	λ = 2.36 nm	λ = 4.47 nm
0.3 ± 0.05	71.9 ± 0.8	87.7 ± 0.9

Table 4 The measured absolute efficiencies for the first order of diffraction for the zone plates

Material	Efficiencies [%]	
	λ = 2.36 nm	λ = 4.47 nm
Gold	4.0 ± 1	4.4 ± 1
Germanium	10.8 ± 1	7.7 ± 1

Table 5 The groove efficiencies at λ = 2.36 nm and λ = 4.47 nm

Material	Efficiencies [%]	
	λ = 2.36 nm	λ = 4.47 nm
Gold	5.6 ± 1	5.1 ± 1
Germanium	15.0 ± 1	8.8 ± 1

An increase in efficiency by a factor of three for the germanium zone plates with respect to the gold zone plates is shown. Since the groove efficiency of the germanium zone plate at λ = 2.36 nm clearly exceeds 10.1% the phase effect is demonstrated.

At λ = 4.5 nm both types work mainly as amplitude zone plates and should have the same efficiency. In contrast to this prediction the germanium zone plate reveals a groove efficiency improved by a factor of 1.5. This result demonstrates that the transfer of the pattern into germanium works better. In addition the gold zone plates were overexposed or overetched to a small extent so the gaps are broader than the lines.

The groove efficiency of the germanium zone plate at $\lambda = 2.36$ nm was measured to be 15%. As mentioned above, the theoretical figure for the groove efficiency is 20.0%. The difference can be explained by the variation of the exposure across the zone plate diameter due to an inhomogeneous distribution of intensity in the exposing zone plate pattern. The inner zones are overexposed, thus leading to a decrease in efficiency. Using a filter with radially increasing transmission, this can be corrected. When one takes into consideration the worse structures of the gold zone plates, an increase in efficiency of a factor of two still remains when using germanium instead of gold zone plates, as predicted by the theory.

Another result of these measurements is that the atomic scattering factors of germanium as published by B.L. HENKE [3] are confirmed by experiment. However, the accuracy of the measurements is insufficient to obtain improved figures for the data tables.

7 Outlook

As the next step, micro zone plates made of germanium are being constructed. The much smaller structures of these zone plates require improved methods for the pattern transfer process due to the higher ratio between the thickness of the layer and the width of the structures.

Experiments with other materials, especially nickel, seem promising as well. A nickel zone plate KZP 4 has been constructed successfully but has not been measured up to now.

8 Acknowledgements

This work has been supported by the German Federal Ministry for Research and Technology (BMFT) under the contract number 13 N 5328.

The measurements have been done at the "Berliner Elektronen-speicherring Gesellschaft für Synchrotronstrahlung m.b.H." (BESSY) in Berlin.

9 References

1. H. Wolter, "Spiegelsysteme streifenden Einfalls als abbildende Optiken für Röntgenstrahlen", Ann. d. Physik 6. Folge, 10, 94-114 (1952)
2. Kirz, J., "Phase zone plates for x-rays and the extreme uv", J. Opt. Soc. Am., 64, 301-309 (1974)

3. Henke, B.L., Lee, P., Tanaka, T.J. ,Shimabukuro, R.L. and B.K. Fujikawa, "Low-energy interaction coefficients: Photoabsorption, scattering and reflection", Atomic Data and Nuclear Data Tables, $\underline{27}$, 1-144 (1982)
4. Schmahl, G., Rudolph, D., Guttmann, P. and O. Christ, "Zone Plates for X-Ray Microscopy", in: X-Ray Microscopy, G. Schmahl, D. Rudolph, eds., Springer Series in Optical Sciences $\underline{43}$, 63-74 (Springer, Heidelberg 1984)
5. B.Niemann, "The Göttingen Scanning X-Ray Microscope", in: X-Ray Microscopy, G. Schmahl, D. Rudolph, eds., Springer Series in Optical Sciences $\underline{43}$, 217-225 (Springer, Heidelberg 1984)
6. Thieme, J., "Construction of Condenser Zone Plates for a Scanning Microscope", in: X-Ray Microscopy, G. Schmahl, D. Rudolph, eds. , Springer Series in Optical Sciences $\underline{43}$, 91-96 (Springer, Heidelberg 1984)

Axisymmetric Grazing Incidence Optics for an X-Ray Microscope and Microprobe

S. Aoki

Institute of Applied Physics, University of Tsukuba,
Ibaraki 305, Japan

1. Introduction

There are several approaches to form point-to-point x-ray images with grazing incidence optics. A single ellipsoidal mirror and the Kirkpatrick-Baez microscope are used routinely at many synchrotron radiation facilities and laser fusion laboratories. These devices, however, suffer inevitable aberrations such as coma and astigmatism at grazing incidence. In order to overcome this problem the tandem toroidal mirror [1] and the Wolter-type mirror have been developed [2, 3]. Recent advances in the fabrication technique have made it possible to use these mirrors in the short-wavelength x-ray region.

In this paper we describe recent advances in Wolter's optical system for several purposes. Our main programs are :
1) soft x-ray microscope for biological specimens,
2) x-ray microscope for laser fusion diagnostics,
3) x-ray fluorescence microprobe for microanalysis.
Design criteria and some experimental results are presented.

2. Design criteria and Fabrication

The surface of Wolter-Type-I mirror is a combination of a hyperboloid and an ellipsoid, as shown in Fig.1. This mirror can be used for either the imaging microscope or the scanning microscope. The geometrical features of the mirror system are determined by the following three parameters :

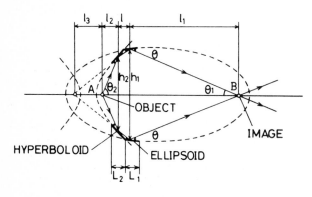

Fig.1. Concept of the Wolter type-I mirror, and its design parameters, θ is the average grazing angle of incidence.

1) the bore diameter of the mirror,
2) the magnification or demagnification,
3) the grazing angle of incidence.

For a given grazing angle and the magnification, a field of view is linearly proportional to the size of the mirror, that is, the diameter and the length of the mirror [4].

The average grazing angle of incidence (θ) is determined by the wavelength to be used. An approximate expression of the critical grazing angle of incidence θ_c is given by [5]

$$\theta_c = 1.6 \times 10^5 \, \lambda \sqrt{\rho}$$

where λ is the wavelength in centimeters and ρ is the density of the reflection surface in g/cm^3. The magnification (M) of the mirror system is given by the angular aperture ratio θ_2/θ_1. It is easily shown that $\theta_1 + \theta_2 = 4\theta$. Other parameters are determined from the requirements for the object distance and the object size. Of course, the length of the mirror is limited by the fabrication technique.

We show how to determine the paramaters for three applications.

(1) Soft x-ray microscope (SXM) for biological specimen

It is well known that the x-ray wavelengths between the absorption edges of oxygen and carbon are useful for imaging biological specimens in a living state. Therefore the shortest wavelength to be used is about 20A. In this scheme the resolving power of the microscope must be better than 0.2μm to make up for the optical microscope. The length of the optical system should be as small as possible to keep the system stable. The larger magnification is favorable for the real-time observation with the low resolution detectors.

(2) X-ray microscope for laser fusion diagnostics (LFD)

Space-resolved observations of the x-ray emission from laser-produced plasmas give us important information on the uniformity of plasma heating and the stability of plasma compression. Although the spatial resolution is modest (-1μm), the field of view must be larger than several hundreds micron in diameter. Most of the x-ray energies which we want to observe are below 5 keV. The proper distance between the target and the mirror is required to avoid possible contamination of the mirror by debris from the target.

(3) Microbeam x-ray fluorescence analysis (XRF)

X-ray fluorescence (XRF) has been utilized for trace element analysis in diverse fields. However, owing to the lack of the efficient optical device, spatially resolved XRF analysis have not been actively studied. XRF analysis needs relatively short wavelengths to excite characterstic x-ray of various elements. We determine the average grazing angle of incidence to be 7mrad by taking account of the exciting energy (10keV), the size of the mirrors and the length of the optical system. The optical system using synchrotron radiation is shown in Fig.2. The first Wolter mirror is a

103

pair of long mirrors which is designed to collect x-rays as many as possible, and the second is a relatively short one designed to make a small x-ray spot.

Design parameters of three optical systems which have been so far examined are shown in Table 1.

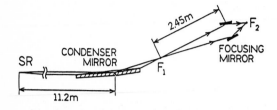

Fig.2. Relayed optics for the synchrotron radiation microprobe.

Table 1   Parameters of three optical systems

Program	SXM	LFD	XRF
Mirror No.	M_S	M_L	M_{x1}, M_{x2}
Wavelength region (A)	10-50	2-50	1-10
Grazing angle θ(mrad)	18.5 (glass)	17.5 (Au coat)	7 (Pt coat)
Number of Wolter mirror	1	1	2
Magnification	20	10	1/20 x 1/13
Mirror diameter,h_1(mm)	10	32	10, 32
Mirror length,L_1+L_2,(mm)	60	70	70,820(320+500)
Optical system,l_1+l_2+l(m)	1.4	3	11.5 + 2.5

3. Experimental Results

Small Wolter mirrors (M_S, M_L, M_{x1}) were produced by a glass replica technique. The detailed fabrication process was described in our previous paper [6]. The long Wolter mirror (M_{x2}) which consists of two separate mirrors was produced by a conventional optical polishing. Each fabrication process is as follows :

    Condenser Mirror :
        1. Optical design
        2. Grinding with the shaped cast iron mandrel
        3. Lapping with the pitch polisher
        4. Coating with metals

    Object Mirror :
        1. Optical design
        2. Grinding the master mandrel
        3. Lapping the master mandrel
        4. Replication with the Pyrex glass
        5. Coating with metals.

Every small mirror was tested by using a conventional micro-focus x-ray generator. The testing arrangement is shown in Fig.3. The test mirror was placed in a vacuum chamber. A photographic film or a plate was set either in a vacuum chamber or outside the chamber depending on the wavelength.

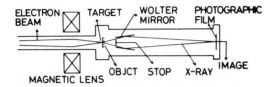

Fig.3. Arrangement for testing Wolter mirrors

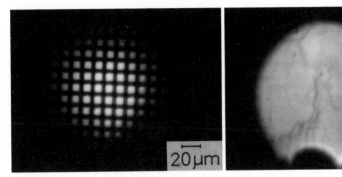

Fig.4

Fig.5

Fig.6

Fig.7

Fig.4. An x-ray micrograph of a copper mesh #2000 obtained with the mirror M_S.

Fig.5. An x-ray micrograph of a copper mesh #1000, showing a larger field of view.

Fig.6. An x-ray micrograph of a stainless steel mesh #400 imaged with the mirror M_{X1}. $\lambda = 1.54A$ (CuKα).

Fig.7. A micro-focused x-ray spot obtained by using a synchrotron radiation source. $\lambda = 1.5A$.

The highest resolution was obtained by using the mirror M_S with CKα(λ=44.8A). An x-ray micrograph of the copper mesh #2000 is shown in Fig.4. The resolution is better than a half micron.

A large field of view was obtained by using the mirror M_{x1} with CKα. The best focus was obtained over an area having a diameter of more than 2mm (See Fig.5).

The highest x-ray energy ever imaged with the mirror M_{x1} was CuKα (λ=1.54A). An x-ray micrograph of a stainless steel mesh #400 is shown in Fig.6. The last mirror M_{x1} was used as a microfocusing mirror combined with the long mirror M_{x2} for XRF analysis [7]. A focused x-ray image at the second focal plane is shown in Fig.7. In this case a 60μm-dia. tantalum pinhole was put at the first focal plane to define the second x-ray source.

4. Discussion

The reason why the resolution is not so good as the theoretical one is mainly due to the imperfection of the surface figure in the soft x-ray region and due to the surface roughness in the hard x-ray region. Assuming that the grazing angle of incidence at the mirror surface is θ, peak-to-valley surface roughness must be less than $\lambda/8si\theta$, which is easily derived from the Rayleigh's criterion, that the path difference must be less than a quarter of the wavelength. In our case the surface roughness is about 40A and the figuring irregularity is about 0.5μm. The experimental data coincide with this value qualitatively.

Summary

Several kinds of Wolter mirror have been fabricated and tested. A wide range of x-ray wavelengths can be imaged by changing the design parameter of the grazing angle of incidence. A further improvement of the resolution should be achieved with the development of the polishing technique.

Acknowledgements

We would like to acknowledge Prof. Y. Gohshi of the University of Tokyo and Prof. T. Yamanaka of Osaka University for their stimulating discussions and encouragement. We also wish to thank the staffs of Nikon Co. Ltd. for producing Wolter mirrors.

References

1. Y. Sakayanagi and S. Aoki : Appl. Opt. 17 (1978) 601
2. H. Wolter : Ann. Phys. 10 (1952) 94
3. M. J. Boyle and H. G. Ahlstrom : Rev. Sci. Instrum. 49 (1978) 746
4. R. C. Chase : Appl. Opt. 15 (1976) 3094
5. E. F. Kaelble : Handbook of X-rays (McGraw-Hill, New York 1969)

6. S. Aoki, M. Shiozawa, K. Kamigaki, H. Hashimoto,
   M. Kokaji, Y. Setuhara, H. Azechi, M. Yamanaka,
   T. Yamanaka, Y. Izawa, and C. Yamanaka :
   Jpn. J. Appl. Phys. 26 (1987) 952
7. Y. Gohshi, S. Aoki, A. Iida, S. Hayakawa, H. Yamaji
   and K. Sakurai : Jpn. J. Appl. Phys. 26 (1987) L1260

Bragg-Fresnel Optics:
Principles and Prospects of Applications

V.V. Aristov

Institute of Problems of Microelectronics Technology and
Superpure Materials, USSR Academy of Sciences,
142432 Chernogolovka, Moscow District, USSR

Possibilities of control and focusing of X-rays have been discussed repeatedly since the discovery made by Röntgen. It turned out, however, that applications of the principles of optical element fabrication adopted in
visible light, infrared radiation and other wavelength ranges are rather
limited due to negligible difference of the refractive index from unity, a
relatively large absorption coefficient and the necessity to fabricate optical elements with the accuracy compared to a radiation wavelength. In the
last years the methods of X-ray optics acquire further extensive development
because of the advances in microstructuring technology, namely, structure
fabrication with element sizes up to one hundred angstroms, sputtering
technology and that of the growth of thin films of different materials, and
due to the advances in the investigation of X-ray diffraction. In our
opinion, at present there exists a possibility to set and solve the task of
fabricating effective focusing X-ray elements with the structure of three-
dimensional Fresnel zones, that is, Bragg Fresnel optics. These elements
can be made on the basis of multilayer interference mirrors for the nanometer
wavelength range ($0.5nm \leq \lambda \leq 10nm$) and semiconductor perfect crystals with
heterostructures for the wavelength range of $0.1Å \leq \lambda \leq 5Å$. The principal pe-
culiarity of Bragg-Fresnel X-ray elements lies in the fact that coherent
Bragg scattering by separate layers is used in them. As it will be shown,
this phenomenon permits increasing their diffraction efficiency, widening
the spectrum range and angular aperture and gives a possibility to realize
amplitude and phase modulations of radiation, to switch X-ray elements by
an electrical signal, ultrasound and light signals.

The principles of Bragg-Fresnel optics suggested by us in 1986 are given
below. The prospects of its application for image formation, information
recording, transmission and processing are discussed. The review is compiled
chiefly on the investigation data performed in the Institute of Microelectronics Technology and Superpure Materials of the USSR Academy of Sciences
in 1984-1986.

1. PLANE AND QUASI-PLANE X-RAY DIFFRACTION OPTICS

The most typical feature of X-rays lies in the fact that the refractive index
of a substance strongly depends on the radiation wavelength even in the regions far from element absorption edges:

$$n \approx 1 - \frac{e^2}{mc^2} \frac{\lambda^2}{2\pi} F_H, \tag{1}$$

where e is the electronic charge, m is the electronic mass, c is the velocity of light, F_H is the structure amplitude of radiation scattering $\vec{K}_o$ in the direction $\vec{K}_H$, $\vec{K}_o$ and $\vec{K}_H$ are wave vectors of incident and scattered waves. At $|\vec{K}_o| = |\vec{K}_H|$, the value F_H is equal to the number of electrons in volume unit of the scattered substance. It follows from (1) that any X-ray lens should reveal a curvature (surface curvature radii r_1 and r_2 should be of an order $f(n-1)$, where f is the focal distance) and chromatic aberration proportional to λ^2. In this view, the suggestion expressed by Baez in 1952 [1] to use a Fresnel zone plate as a focusing element was an innovation and permitted constructing a zone plate with chromatic aberration proportional to λ and a resolution up to 0.1–0.5 μm for X-ray radiation in the nanometer wavelength range. Certain progress in the construction of Fresnel zone plates and their application in X-ray microscopy could not eliminate such disadvantages of this optics as a three-dimensional volume of a zone plate (it has up to 10^3–10^4 wavelengths in its depth) as a result of an insignificant difference between the refractive index and unity, and the near-comparability of radiation wavelength with interatomic distances in a scattering substance. It can be readily shown that the plane optics approximation used in calculations of a zone plate is valid only under the conditions of small scattering angles $\varphi_{max} = (1 - n)^{\frac{1}{2}} \ll 1$ that, in its turn, restricts the resolution to the value $\delta = 2\lambda / \sqrt{1 - n}$. Moreover, any inhomogeneity (defect) generated on fabricating a plate causes the occurrence of phase inhomogeneities and deteriorates image quality. Thus, plane Fresnel optics can be effectively used only for radiation with the wavelength $\lambda \gtrsim 1$nm. It has a resolution limit of the order of 50 nm and a maximum angular aperture of the order 10^{-1}–10^{-2}. That is why fabrication of a phase and, in particular, a blazed phase lens presents some problem.

As it was mentioned above, for construction of X-ray optical elements one should give up the old concepts conventionally employed in visible light optics and examine the possibilities for X-ray optical elements to be developed with a structure of three-dimensional Fresnel zones. It means we should turn our attention from plane zones to three-dimensional Fresnel zones [2,3]. Therewith, as shown below, X-ray optical elements can be produced for a wide wavelength range from 0.1Å to 100Å with a possibility in principle of achieving a resolution of 1–10nm and diffraction efficiency of 100%. A perfect Bragg-Fresnel lens is free of chromatic aberrations (Table 1).

2. PRINCIPLES OF THREE-DIMENSIONAL BRAGG-FRESNEL OPTICS

2.1 Bragg-Fresnel Lens

A perfect Bragg-Fresnel lens is a three-dimensional system of isophase surfaces reflecting a spherical wave from point A_1 to point A_2 (Table 1). The

Table 1. Summary of properties of focussing systems

	Optics	Fresnel optics, 1952	Fresnel-Bragg optics, 1986
Basic equations	$\dfrac{1}{f} = (n-1)\left(\dfrac{1}{r_1} + \dfrac{1}{r_2}\right)$ $n-1 = -\dfrac{e^2}{mc^2}\dfrac{F}{2\pi}\lambda^2 \ll 1$	$f = \dfrac{r_1^2}{\lambda}; r_N^2 = N\lambda f + \dfrac{N^2\lambda^2}{4}$	$N-1 \ll \lambda\dfrac{[\vec{K_1}\times\vec{r_c}]^2}{r_1} + \lambda\dfrac{[\vec{K_2}\times\vec{r_c}]^2}{r_2} \leq N$
Chromatic aberrations	λ^2	λ	is absent, an image is formed in a reflected wavelength range
Limiting resolution δ max	—	$\dfrac{500-1000 \text{ Å}}{\dfrac{2\lambda}{\sqrt{1-n}}}$	is determined by technological possibilities and is likely to be 10–100 Å
Maximal angular aperture φ max	—	$\sqrt{1-n}$	is evidently determined by technological possibilities
Wavelength range λ	—	≥ 10 Å	$0.1 \text{ Å} \leq \lambda \leq 1000 \text{ Å}$
Possibility of amplitude and phase modulations	—	none	possible
Possible fabrication of hybrid schemes	—	none	switching of the X-ray signal by electrical, optical and sound signals is possible
Possible fabrication of various X-ray optical elements	—	fabrication of lenses and gratings is possible	fabrication of lenses, prisms, interferometers, modulators and mirrors is possible

distance between the surfaces is determined such that the path difference for radiation of wavelength λ reflected to point A_2 is a multiple of λ. This perfect lens presents itself a three-dimensional hologram, the properties of which are dictated by the properties of three-dimensional diffraction [4]:

- the resolution and shape of a diffraction maximum are determined by the Laue function L, that is, by the "lens" aperture x,y and its thickness z

$$L = \frac{\sin \frac{2\pi}{\lambda} x}{\frac{2\pi}{\lambda} x} \; \frac{\sin \frac{2\pi}{\lambda} y}{\frac{2\pi}{\lambda} y} \; \frac{\sin \frac{2\pi}{\lambda} z}{\frac{2\pi}{\lambda} z} \; ; \tag{2}$$

- each surface reflects an incident wave to the focal point A_2 independently of the radiation wavelength λ. Hence, a perfect Bragg-Fresnel lens is free of chromatic aberrations;

- the lens vision field is small and controlled by its aperture. The distance between the radiation source and lens along the axis OA_1 can be changed over a wide range. Therewith, the distance OA_2 varies as in the case of conventional zone plate optics;

- the displacement of reflecting planes towards each other by a value Δa in adjacent regions of the lens leads to phase changes of waves scattered by these regions by a value proportional to $2\pi\Delta a/a$, where a is the lattice parameter. It enables one to control reflected beams;

- in most experiments Bragg-Fresnel lens reflection is unlikely to be described in the first Born approximation and should be estimated considering the effects of multiple scattering.

Experimental realization of a Bragg-Fresnel lens depends on the progress in the technology of modulated superlattice growth. In this case three-dimensional modulation of a structure is required. A Bragg-Fresnel lens can be also fabricated on the basis of common three-dimensional lattice (multi-layer glasses or perfect crystals) by modulating lattice reflectivity according to the structure of three-dimensional Fresnel zones, the position of which is determined by superimposing the Bragg system of reflecting planes on the system of perfect lens isophase surfaces. It can be shown, that Fresnel zone boundaries, determined in this position through the shift of imposed lattices by a half of the interplanar distance, are given by the expression

$$N - 1 \leq \lambda \frac{[\vec{K}_1 \times \vec{r}_C]^2}{r_A} + \lambda \frac{[\vec{K}_2 \times \vec{r}_C]^2}{r_B} \leq N \; . \tag{3}$$

Here N is a zone number, $|\vec{K}_1| = |\vec{K}_2| = 1/\lambda$; r_A and r_B are the distances source-point A to lattice, lattice to focal-point B; r_C is a point coordinate on the lattice; $|\vec{K}_1|$ and $|\vec{K}_2|$ are the wave vectors of the incident and diffracted waves directed along the optical axis coinciding with the exact

Bragg direction. Equation (3) describes the system of ellipsoids (Fig. 1), the form and position of their axes depend on the relations $\vec{K}_1$, $\vec{K}_2$, r_A and r_B [5-7]. Figure 2(a) gives a photograph of a one-dimensional Bragg-Fresnel lens made out of a Si single crystal. Figure 2(b) gives a focused image of the source. During the experiments on the lenses out of Ni-C multilayer mirrors and Si single crystals we managed to obtain focusing with a resolution up to 7 μm and diffraction efficiency up to 20% [8].

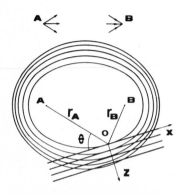

Fig. 1. Interference pattern from two spherical waves (equal phase lines)

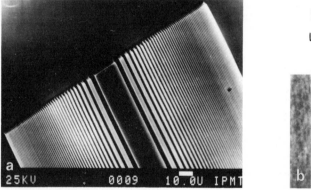

Fig. 2. (a) Bragg-Fresnel lens, (b) focused source image

2.2 Bragg-Fresnel Prism

The principle of three-dimensional diffraction can be used for fabricating not only a lens but also analogues of other elements, namely, a Bragg-Fresnel prism [9]. Spherical wave diffraction by a three-dimensional lattice is described by the expression

$$\psi_H(\hat{X}) = A_0 \int_{-\infty}^{+\infty} \text{sinc}\, \eta \, \exp\left[2\pi i (\eta \varphi_H - \frac{\eta^2 \sigma^2}{2t^2}) \right] d\eta ; \qquad (4)$$

here $\hat{X}$ is counted off from point O in the observation plane (Fig. 1). Point O satisfies the exact fulfilment of the Bragg condition at the crystal depth of $t/2$, A_0 is the amplitude of a diffracted wave at the point $\hat{X} = 0$ from the plate with the thickness $\sigma = \sqrt{R\lambda}\,\gamma_0/\sin 2\theta$, $R = R_1(\gamma_H/\gamma_0)^2 + R_2$, $\gamma_0 = (\vec{S}_0 \cdot \vec{n}_0)$, $\gamma_H = (\vec{S}_H \cdot \vec{n}_0)$. An optical analogue of the problem considered is scattering of the plane wave $A_0 \exp(2\pi i(\vec{S}_H \cdot \vec{R})/\lambda)$ by the slit with the size $L = t \sin 2\theta/\gamma_0$. It means that, as in the slit L, the crystal thickness can be divided into a system of layers parallel to the crystal irradiation surface

$$t_{N-1} \le 2 \mid \vec{n}_0(\vec{R} - \vec{R}_0)\mid < t_N. \tag{5}$$

Here $t_N = 2\sigma\sqrt{N}$. The characteristic features of structure (5) are similar to those of one-dimensional Fresnel layers. The foregoing states that a wave field structure can be changed by amplitude or phase changes in crystal reflectivity in depth. Fig. 3(a) presents calculated intensity distribution in spherical wave diffraction by a perfect crystal plate (dotted curve) ($t = 19\mu m$) and by a plate with an amorphous layer $t_S = 13\mu m$ thick, located at the depth of the second phase layer $r_S = 5.8\mu m$ (solid curve) [10]. Insertion of amorphous interlayers at the sites of even Fresnel zones produces a pattern similar to plane wave diffraction by an amplitude zone plate of thickness

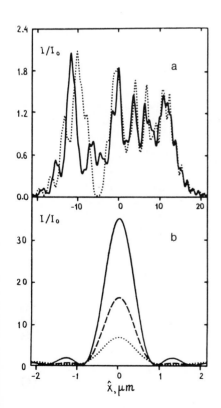

Fig. 3. Intensity distribution upon spherical wave diffraction: (a) on a perfect wafer (dotted curve) and a wafer with an amorphous layer (solid curve); (b) on heterostructures with amplitude (dotted curve), phase (dashed curve) and kinoform (solid curve) modulation

L.   A wave field structure can be changed by both the amplitude and phase modulations, that is, by the shift of reflecting planes.   A shift by a/2 alters a reflected wave phase by π [11].   A scattering wave front can be effectively transformed by changing the lattice parameters of the structure itself.   It is easy to show, that if a heterostructure has a linear change of the lattice period $a(z) - a_0(1 + Kz)$, the size of the first layer varies in Bragg symmetrical diffraction and is given by the formula

$$\sigma' = \sigma\ell/(\sigma^2 \pm \ell^2)^{\frac{1}{2}}. \tag{6}$$

Here $\ell = (d_H/|K|)^{\frac{1}{2}}$, where d_H is the interplanar distance, ℓ is the size of the first phase layer in plane wave diffraction [7], the "+" sign corresponds to $K > 0$, the "-" sign to $K < 0$. The inclusion of multiple scattering makes the function of a prism more complicated due to the fact that in a dynamic approximation of the diffraction theory the refraction index in the vicinity of Bragg scattering depends on deviation of an angle of radiation incident on the lattice from the exact Bragg value.   At $K < 0 \sigma = \ell$, $\sigma' \to \infty$ and the whole crystal bulk, just as a blazed lens, focuses radiation into a point.   In case of (555) Si reflection, $\ell = \sigma$ at $K = 6m^{-1}$, here $\Delta a/a = KZ(t < 20\mu m)$ and does not exceed 10^{-4} and can be easily made in heterostructures.   Fig. 3(b) presents intensity distribution in the focus of a "lens" at amplitude (dotted curve), phase (dashed curve) and kinoform modulations by the plate of 19 μm thick.   It is obvious that a crystal with modulated reflectivity is a

diffraction prism which transfers the image from point $\vec{R} = 0$ to point

$$\vec{R} = R_1 \cdot \vec{S}_0 + R_2 \cdot \vec{S}_H$$

$$R_1(\gamma_H/\gamma_0)^2 + R_2 = F. \tag{7}$$

An increasing coefficient M is independent of the distance $\tilde{R}$ and equals $|\gamma_H|/\gamma_0$. Prism dispersion $\Delta\lambda/d\theta = \lambda \cdot ctg\,\theta$, spectral resolution $\lambda/\Delta\lambda$ is equal to $F\,tg\,\theta/\Delta f$, where the size of a focal spot $\Delta f = F\gamma_0 d_H/(t \cdot \cos\,\theta)$ and in the example considered constitutes $3 \cdot 10^5$.

2.3 Diffraction Lattices and Interferometers

The conventional X-ray triple-crystal interferometer [12] is cut out of a perfect single crystal and represents, in principle, a two-beam device in which Bragg scattering is used for divergence of a beam into two parts, for beam convergence on the third crystal-analyzer and for the formation of an interference pattern in parallel beams.   The fabrication technology of different superlattices in perfect crystal enables producing multilayer microinterferometers and complex gratings.   At present, only pioneer experiments have been performed to study the properties of Bragg diffraction superlattices and the contrast of diffraction patterns on microinterferometers [13, 14].

2.4 X-Ray Radiation Modulators

For X-rays the absorption and refractive indices of a substance negligibly differ from unity and slightly vary under the influence of different physical effects even when using radiation with a wavelength near the edge of element absorption. It means that the principles of optical radiation modulation are not valid for X-ray wavelength range. For X-ray radiation the modulators can be fabricated only by using coherent Bragg scattering. Therewith, the following physical phenomena can form their basis.

a) For amplitude modulation one should withdraw the crystal from a reflecting position. Here the required deformation value $\Delta a/a$ is of an order $10^{-4} - 10^{-5}$. The penetration depth of radiation with $\lambda = 1\text{Å}$ in the crystal is of an order 10 μm. Hence, for generating such a modulation the film should be contracted by the value of an order of several angströms. It is achieved by transmitting ultrasound waves (both volume and surface ones) through the crystal;

b) For phase modulation, the lattice should be shifted by a half of the interplanar distance. For high order reflections this value constitutes tenth fractions of an angström. In principle, phase modulation is also possible when introducing deformation, since a diffraction wave phase changes on deviation from the exact Bragg position. This phenomenon is used in the method of standing waves and permits the lattice deformation value required for the modulation discussed under a) to be decreased by one or two orders;

c) The deformation lattice value required for obtaining amplitude and phase modulations is likely to be additionally decreased when using multiwave diffraction, structural forbidden reflections, strong excitation. Therewith, under radiation with a wavelength near the absorption edge of one of the elements forming a lattice (a crystal or an interference mirror), radiation modulation is likely to be obtained without any change in lattice parameters only by shifting the absorption edge in response to an external effect.

It is obvious that a coherent arrangement of reflecting layers permits constructing different X-ray optical elements, namely, mirrors, lenses, prisms, lattices and modulators.

3. POSSIBLE APPLICATIONS OF X-RAY OPTICAL ELEMENTS

The use of X-ray radiation for image formation, transmission, recording and processing of information is one of the most important areas where X-ray elements may be employed. In this area the following results can be obtained:

a) Development of X-ray microscopy, interferometry and holography.

An X-ray microscope with a spatial resolution up to 100 Å and a holographic microscope for investigating biological objects and thin films are to be constructed. Construction of X-ray microinterferometer will make possible a local analysis of crystal structures with sensitivity to lattice parameter changes up to $10^{-7}-10^{-8}$.

b) Development in X-ray spectrometry.

Diffraction lattices can be constructed including echelons with resolution up to 10^6 ($R = \lambda/\Delta\lambda$).

c) Development of surface X-ray local spectroscopy.

It is envisaged to develop an X-ray microscope with resolution up to 100 Å with detection of secondary radiations. This device is likely to be solely intended for a local nondestructive control over surface chemical composition. This microscope, when used in the wavelength range 0.1Å–5Å and combined with the method of standing waves, permits local structural study of a perfect crystal surface.

d) Development of technology of submicron X-ray lithography.

The development of this technology meets some difficulties, but even today, in some cases, X-ray lithography permits fabricating structures with ultra small sizes. Highly focused beams may be employed for gene engineering, that is, for directed local changes in the structure of molecules.

e) Construction of X-ray electron systems of information recording, processing and transmission.

The possibility exists of constructing X-ray electron elements performing the convolution operation, Fourier transformation, etc. X-ray electron elements can be also used for stroboscopic measurements required in microelectronics.

One can assume that X-ray electron systems will provide means for obtaining recording density up to 10^{12} bit/cm^2 at the information transmission rate up to 10^{12}–10^{14} bit/sec. It is of interest to discuss the possibility of using X-ray optical modulators and optical elements for transmission, since a beam with divergence of 10^{-6}–10^{-8} rad and a high rate of information transmission are obtainable.

Advances in development of the above trends are related to the intensity of investigations in the fields of physical fundamentals of X-ray optics, X-ray electron modulators and radiation detectors, construction of coherent, synchrotron and other sources of X-ray radiation as well as to development of fast solid state electronics.

Research in the field of X-ray optics calls for progress in the nanometric microstructuring technology (with resolution at the level of 50–100Å), the technology of thick epitaxial films growth with a pre-set structure and chemical composition and the construction of materials answering special requirements.

REFERENCES

1. A.V. Baez: J. Opt. Soc. Am. <u>42</u>, 756 (1952).
2. V.V. Aristov, E.V. Shulakov: Kristallographiya (1987) (in press).
3. E.V. Shulakov, V.V. Aristov: The Theory of Diffraction Fresnel Topography of Perfect and Defect Crystals, J. Appl. Cryst. (1987) (in press).
4. V.V. Aristov, V.Sh. Shekhtman: Uspekhi fizicheskih nauk <u>104</u>, 1, 51 (1971).
5. V.V. Aristov, A.A. Snigirev, Yu.A. Basov, A.Yu. Nikulin: AIP Conf. Proc. <u>147</u>, 253 (1986).
6. V.V. Aristov, Yu.A. Basov, A.A. Snigirev: Pisma v ZhETF <u>13</u>, 2, 144 (1987).
7. V.V. Aristov, Yu.A. Basov, S.V. Redkin, A.A. Snigirev, V.A. Yunkin: Nucl. Instr. Methods (1987) (in press).
8. V.V. Aristov, S.V. Gaponov, V.M. Genkin, Yu.A. Gorbatov, A.I. Erko, V.V. Martynov, Z.A. Matveeva, N.N. Salashchenko, A.A. Frayerman: Pisma v ZhETF <u>44</u>, 4, 207 (1986).
9. V.V. Aristov, E.V. Shulakov: X-Ray Optical Diffraction Elements on the Basis of Heterostructures, Optics Comm. (1987) (in press).
10. Here we consider Si(555) reflection, $\lambda = 0.0709 nm(M_oK_\alpha)$, $\tilde{R} = 40$ cm, $\sigma = 3.24\mu m$, extinction depth $\Lambda = \lambda \sin\theta / |\gamma_H| = 89\mu m$.
11. The index of refraction for X-rays $\lambda \sim 0.1nm$ differs from unity by the magnitude of an order $10^{-5}-10^{-6}$, so that a common change in a transmitted wave is possible by a material of about 10 μm thick.
12. U. Bonze, M. Hart: Appl. Phys. Letters <u>6</u>, 155 (1965).
13. V.V. Aristov, A.I. Erko, A.Yu. Nikulin, A.A. Snigirev: Opt. Comm. <u>58</u>, 5, 300 (1986).
14. V.V. Aristov, A.Yu. Nikulin, A.A. Snigirev, P. Zaumseil: Phys. stat. sol. (a) <u>95</u>, 81 (1986).

High Resolution Image Storage in Polymers

D.M. Shinozaki

Department of Materials Engineering,
The University of Western Ontario, London, Ontario,
N6A 5B9, Canada

1. Introduction

The recent advances in soft X-ray microscopy have resulted in examples of high
resolution images of biological materials which are significantly improved
over earlier ones. The method with the highest resolution is the simplest one,
namely microradiography. While directly magnifying X-ray microscopes or
scanning X-ray microscopes have an inherent appeal to all microscopists,
being exactly analogous in operation and application to other established
microscopies, the contact imaging method has itself certain advantages which
are already being exploited. For example, short exposure time images taken
with pulsed sources will be necessary to freeze the mobile structures in wet
or living biological cells. The serial recording method of the scanning
microscope obviously will preclude its use in this application. The direct
imaging microscope (using zone plates) utilizes only about 10% of the
photons incident on the specimen to form the image. It is not clear that the
extremely high dose rates necessary to image wet cells will not distort or
destroy the imaging optical components such as the zone plate itself. In
many instances of contact microradiography using pulsed sources, the specimen
itself has been destroyed. The relative advantages of contact methods for
this application are clear.

The use of polymer resists to record the image and the subsequent
examination of the topographic image using electron microscopy in the contact
method is inherently less attractive to the microscopist because it has no
direct optical analog. In addition, the recording is nonlinear. In the
absence of detailed calibration data, the contact images so far reported in
the literature have been largely non-quantitative. However, they still
constitute a large fraction of the highest quality images produced by soft
X-rays. It is of central importance in microradiography to quantify the
image formation processes; exposure, development and examination of the
resist. This paper is an attempt to examine some of the basic parameters
which can limit the resolution of the microradiographic method.

2. Image Formation in Resists

The exposure of the resist to soft X-rays produces radiation damage in the
polymer. For positive resists this manifests itself as a decrease in
molecular weight as result of bond breakage. If a chemical bond in the
backbone of the long chain is involved, the length of the chain decreases,
and the molecule goes into solution more readily.

For most applications related to X-ray lithography, in which replication
of images with scales of the order of 0.1 microns is of great interest, the

radiation damage process and dissolution of the resist may be accurately modelled from the viewpoint of averages over large numbers of molecules and large numbers of photons absorbed in the resist. In this case, the resolution is limited by the photoelectron path length in the resist, which is of the order of tens of angstroms (depending on the resist chemistry and the incident photon energy). The image elements of interest in this case (0.1 x 0.1 micron) contain a large number of resolution elements (400 pixels 50 x 50 A), and the visibility of a 0.1 x 0.1 micron element depends on averaging the discrete irradiation-induced chain scission events over neighbouring 400 pixel elements. The detailed argument follows that of Spiller and Feder in their estimate of the performance of an ideal resist used for X-ray lithography [1].

Over large areas of resist, the incident irradiation dose is uniform and is characterized by the average intensity of the incident X-ray beam multiplied by the time of exposure. However the intensity is a flux of photons, which on a microscopic spatial scale are not uniformly distributed. Assuming the photons arrive at the resist surface independently, the dose incident on neighbouring elemental areas will be different. For spatially random arrivals of photons, the impact locations on the resist surface will show as a random distribution and not as a uniform distribution. Random distributions in the plane of the resist surface will show a certain amount of clustering, especially at relatively low doses. That is, if the average incident dose is such that the average dose per resolution pixel is statistically small, then large differences in dose will be observed among the pixels. In the case described by Spiller and Feder, the probability for recording a defect in the lithographic image was calculated. The conclusion was that to increase the visibility of small structures, the contrast of the mask and the dose absorbed in the resist should be increased.

3. Soft X-Ray Induced Surface Roughness

The wet development of PMMA resist after exposure to soft X-rays results in a noise which is observed as a surface roughness. A typical observation is shown in Figure 1. The resist has been exposed to soft X-rays through a TEM

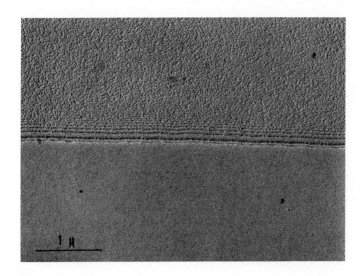

Figure 1

grid, so the regions under the grid bars have not been irradiated. The resist was then developed in the usual solvents for a time typical of that used for images of biological material. A more detailed description of the experimental observations will be given in another paper. The boundary between the exposed and unexposed regions is clearly seen in the micrograph as a line separating regions of very different surface roughness.

The surface noise shown is not visible in PMMA which is examined directly in the electron microscope because the contrast of the resist image is very low unless a metal shadowed carbon replica is used. The direct examination of the PMMA surface using SEM or TEM furthermore results in severe radiation damage in the electron beam sensitive polymer. Experiments show that the mass loss suffered by the resist under normal operating conditions in an electron microscope is as much as 50% or more [2]. The small surface features of an exposed and developed image in PMMA thus would be inherently difficult to observe because of the low contrast, and would be drastically altered by the rapid and massive mass loss during examination in the electron microscope.

4. Exposure Statistics

One source of the noise in the resist image arises from the statistics of photons incident on the surface of the PMMA; essentially an extension of the model used by Spiller and Feder to describe the optimum dose for masks of different contrast in lithography. To estimate the number of photons incident on an area of resist for the typical high and low dose exposures, the source output was calibrated. The Canadian Synchrotron Radiation Facility at the University of Wisconsin 1 GeV storage ring was used in these experiments. At 800 Mev (with a ring current of approximately 200 ma after injection), the output from the "Grasshopper" monochromator was contaminated with significant amounts of second order radiation over the longer wavelengths. To reduce this, a wavelength of 20 A was selected. Only the uniform part of the beam was selected by using a 75 micron diameter aperture after the monochromator. The photoelectron current from an aged Au surface was measured, and compared to the current stored in the ring. As expected there was linear relationship, and thus the incident dose during any contact imaging experiment could be monitored by recording the ring current and the time of exposure (ma-min). Knowing the energy of the incident photons, and using the published data on the quantum efficiency of Au surfaces in the soft X-ray region, the exact incident dose (both in numbers of 20 A photons and in average energy per unit area) could be calculated.

The development curves were then measured and compared to the results published earlier by Spiller and Feder (Figure 2). The difference may be ascribed to the differences in development conditions (temperature, flow rate, etc.) and to the absence of measured incident doses in the earlier work.

The range of exposures used (100 ma-min) to (1000 ma-min) correspond to the typical experimental exposures used to obtain low and high dose images in PMMA. Knowing the numbers of photons incident on the resist per unit area for these exposures, a simple model can be constructed to estimate the point to point doses across the resist surface (Figure 3).

Considering a strip of resist 5 nm wide and 1 micron long, divided into 50 A x 50 A pixels, the average exposure for a high dose image in PMMA would involve 1.8×10^5 photons with wavelength of 20 A. Assuming

Figure 2

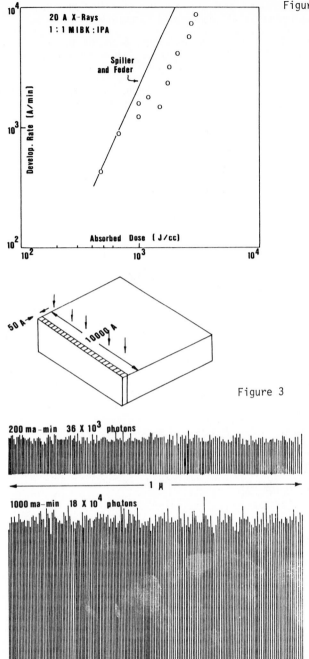

Figure 3

Figure 4

the photons are randomly incident on this strip. The distribution of incident photons is inhomogeneous along this strip (Figure 4).

The peak to valley dose range is of the order of 15 to 30% in the low dose image (100 ma-min) which corresponds to a range of development rates of 360 to 500 A/minute. At the high dose (1000 ma-min) exposure, the development rates vary from 8000 to 10000 A/minute from pixel to pixel along this strip. The spatial distribution is of the order of several hundred angstrom. An examination of the noise in the micrograph (Figure 1) reveals a similar spatial distribution.

5. Molecular Conformation

At the scales of interest in contact imaging, there is an additional factor which will affect the visibility of the image in the resist. Amorphous polymers such as PMMA consist of long chain molecules with

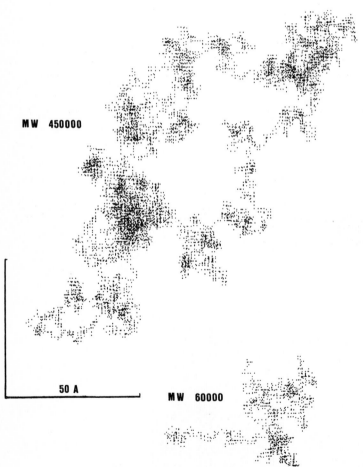

MW 450000

50 A

MW 60000

Figure 5

122

large molecular weights. The conformation of such long chains is apparently similar to that in a theta solvent. Each molecule therefore assumes shape in which the freely jointed chain is in a randomly coiled conformation. The shape and average size of such molecules can be estimated. An example is shown in Figure 5 for a freely jointed chain of PMMA of molecular weight 450,000. A molecule of molecular weight 60,000 is also shown.

Real polymer resists are made of a distribution of molecular weights which can be quite broad. For example, a high molecular weight PMMA (M_w = 450,000) contains a significant mass fraction of molecules of mass 1×10^6 and of mass 6×10^4. In a typical sample of PMMA resist, therefore, there is a distribution of molecular weights which results in a spatial distribution of low molecular weight molecules. With irradiation, the molecular weight will vary from point to point in the resist. Greeneich has shown that the dissolution rate changes very sharply at molecular weights of the order of 10^4 of less [3]. Hence the dissolution rate is a function of position on the surface of the resist, with regions with an average molecular weight greater than 10^4 being effectively insoluble.

6. Conclusions

The development of the resist surface varies from point to point. Two parameters which may contribute to the inhomogeneous development are: (1) the locally inhomogeneous distribution of absorbed energy and (2) the local variation in molecular size which exists in the resist before irradiation, and will persist after exposure. The noise in the resist surface which results can be observed using replica techniques, and is not necessarily visible by direct examination of the resist.

References

1. E. Spiller, R. Feder: In X-Ray Optics, ed. by H.J. Queisser, Springer Ser. App. Phys., Vol.22 (Springer, Berlin, Heidelberg 1977), p.35
2. D.M. Shinozaki, B.W. Robertson: In X-Ray Microscopy: Instrumentation and Biological Applications, ed. by P.C. Cheng, G.J. Jan (Springer, Berlin, Heidelberg 1987), p.65
3. J.S. Greeneich: J. Electrochem. Soc. 122(7), 970 (1975)

Application of Charge Coupled Detectors in X-Ray Microscopy

W. Meyer-Ilse

Universität Göttingen, Forschungsgruppe Röntgenmikroskopie,
Geiststraße 11, D-3400 Göttingen, Fed. Rep. of Germany

Abstract

Camera systems which utilize CCD (charge coupled device) sensors are successfully used for x-ray detection. The required performance of a CCD camera system in x-ray microscopy is discussed. A CCD camera system was constructed and first results are reported.

1 Introduction

In the x-ray microscope (XM) up to now the images are taken using photographic emulsions, in the main on the basis of photographic films. These films have the advantage of comparably high detective quantum efficiency (DQE) [1] and a good spatial resolution across a large area. The last point means, there are a very great number of distinguishable image elements (pixels).

The main lack of photographic detection is the necessity of developing the image. The developing process is time consuming and it prevents interactive work at the microscope. Solid state image detectors like CCD's have the advantage of negligible time for processing, and the images are already digitized and ready for image processing. That is why we are developing a solid state camera system on the base of a CCD to be used for soft x-ray image detection.

2 Requirements of the camera system

The requirements of a soft x-ray camera system to be used in x-ray microscopy are listed below in the order of their importance:

a. High detective quantum efficiency (DQE) of the system.

b. Large number of pixels
(in the order of 40 000, more preferred).

c. High dynamic range of each pixel (more than 100:1).

The spatial resolution and the energy resolution are of minor attention in x-ray microscopy. However a high spatial resolution allows lower x-ray magnifications. To achieve a high DQE is the main difficulty. A large number of pixels and a sufficient dynamic range are no problems with modern CCD's. X-ray microscopy utilizing amplitude contrast mechanism of biological samples uses photon energies between 280 eV and 570 eV. A CCD to be used in the XM must be able to work in this range. In phase contrast microscopy this wavelength range can be extended from 280 eV up to photon energies of about 2 keV [2,3]. The extension to higher energies makes it easier to use a CCD in the XM. On the other hand, phase contrast microscopy has advantages at 280 eV and therefore demands a CCD which can be operated at this wavelength. First experiments during the construction of a camera system are reported in [4] and [5]. This paper presents the first practical images taken with a solid state camera in the XM.

2.1 Detective Quantum Efficiency (DQE)

The detective quantum efficiency (DQE) of the system can be separated in two parts: The conversion efficiency from x-rays to electrons in the sensor, and the efficiency of reading these electrons with an amplifier. The conversion efficiency limits the entire DQE and therefore it must be as high as possible.

2.1.1 Direct method to increase the conversion efficiency

The main problem of direct x-ray detection with an CCD sensor without x-ray to visible converter is the absorption of photons in insensitive surface layers (dead layers) of the CCD. The CCD NXA 1011 (Valvo, Philips) has 26% of its front surface free of any electrical contacts, only the thin insulating layer and some scratch protection absorbs photons before they reach the sensitive regions of the CCD. The direct efficiency of this CCD is insufficient. Radiation damage of the CCD occurs after several seconds of exposure with 280 eV photons in the x-ray microscope.

Special CCD's, which allow backside illumination or utilize thinned dead layers may overcome the problems depicted in the previous paragraph, but up to now they are not tested in our group.

2.1.2 Indirect method to increase the conversion efficiency

A CCD sensor with a x-ray to visible converter provides a good efficiency for x-ray energies below 1 keV. In addition radiation damage is drastically reduced.

2.1.3 Amplifier noise

Low noise readout of the CCD is essential to reach a high DQE. However the acceptable noise level depends on the application of the camera. CCD camera systems designed for good energy resolution must obtain extremely low noise. Noise levels of less than 8 electrons RMS are achieved [6,7]. Such a camera is usually operated in single photon mode (less than one photon per pixel average) to allow spectroscopy. Therefore the noise in electrons RMS must be less than the number of electrons produced by one photon. In order to take images in the XM the CCD must be operated in a multiple photon mode. In this mode the acceptable noise level is restricted by the photon noise.

The excellent low noise of the camera described in [6,7] needs comparably low frequency readout. The higher acceptable noise level in a multiple photon camera allows higher readout frequencies.

3 System design

The CCD camera system for soft x-rays utilizes a frame transfer sensor NXA 1011 from Valvo (Philips). This CCD consists of 604 pixels horizontal 10 μm apart and 288 pixels vertical 15.6 μm apart. To prevent thermal noise the CCD is operated at 77 K.

The CCD is directly covered with an x-ray to visible converter $Gd_2O_2S:Tb$ with 10μm grains. The conversion efficiency varies over the whole CCD less than 20 %, however this efficiency distribution is constant and corrected using image processing.

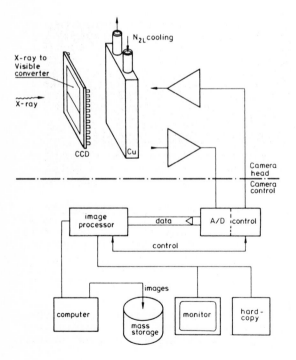

Fig.1 Configuration of the CCD camera system for soft x-ray microscopy.

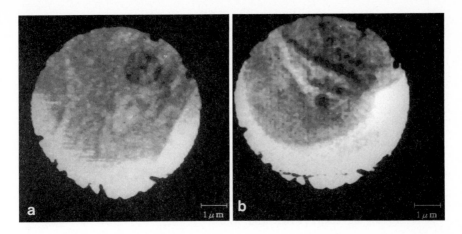

Figure 2: a) Unstained dried human fibroblasts imaged at
λ = 4.5 nm using the Göttingen x-ray microscope at BESSY
equipped with a CCD. Exposure time was 14 sec with 210 mA
storage ring current. The x-ray magnification was 532 times.
b) Two spores *Dawsonia superba* imaged at λ = 4.5 nm using the
Göttingen x-ray microscope at BESSY equipped with a CCD.
Exposure time was 10 sec with 450 mA storage ring current. The
x-ray magnification was 532 times.

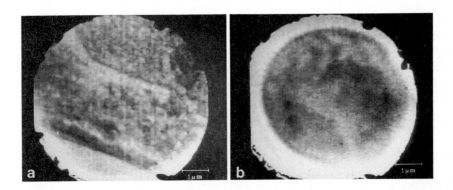

Figure 3: a) Unstained dried human fibroblasts imaged at
λ = 4.5 nm using the Göttingen x-ray microscope at BESSY
equipped with a CCD. Exposure time was 22 sec with 156 mA
storage ring current. The x-ray magnification was 780 times.
b) Spore *Dawsonia superba* imaged at λ = 4.5 nm using the
Göttingen x-ray microscope at BESSY equipped with a CCD.
Exposure time was 30 sec with 280 mA storage ring current. The
x-ray magnification was 780 times.

The output of the CCD comes through three output gates. These gates are read out each with 3.75 MHz. The output signals are amplified in three amplifiers before they are connected to the A/D converters. There are three A/D converter each running with 11.25 MHz to allow double sampling of each pixel. The offset of the first amplifiers is digitally controlled in a closed loop using the A/D output during black reference pixels. Finally the three data channels are multiplexed into one channel with a data rate of 11.25 MHz and fed into the image processor. This design results in a 8 bit A/D conversion of a complete image in 20 ms with reasonable low noise.

Specifications:

Total conversion efficiency at 280 eV (4.5 nm): 1.7 e^-/Photon
CCD output:.................................. 3.5 $\mu V/e^-$
Readout noise of amplifier and A/D:........... 534 e^- RMS
DQE:.. 2.7   %
Improved value of DQE expected:.......... over 32   %.

The DQE of 2.7 % was measured with a reduced gain of the amplifier. An advanced version of the amplifier with a higher gain is under construction. This version increases the signal to noise ratio. A substantially increased DQE of more than 32 % and a readout noise below 100 e^- RMS is expected.

4 Images taken with the CCD in the x-ray microscope

The following images are taken with the CCD with x-ray to visible converter as described above. Each image is median filtered.

Acknowledgements

The images shown in this paper were taken with the Göttingen x-ray microscope at BESSY, Berlin. The author wishes to thank the members of the Göttingen x-ray microscopy group for discussions and cooperation with the experiments. Individual thanks are addressed to P.C. Cheng who prepared the fibroblasts and V.Sarafis who collected the spores. This work has been funded by the German Federal Minister for Research and Technology (BMFT) under the contract number 05 266SL I.

References

1. L. Jochum: "Detective quantum efficiency of film emulsions", in: Soft X-Ray Optics and Technology, ed. by E. E. Koch, G. Schmahl, SPIE Proc. 733, 492-495, (1986)

2. G. Schmahl, D. Rudolph: "Proposal for a Phase Contrast X-Ray Microscope", in: X-Ray Microscopy - Instrumentation and biological applications, ed. by P. C. Cheng, G. J. Jan, (Springer 1987)

3. G. Schmahl, D. Rudolph, P. Guttmann: "Phase Contrast X-Ray Microscopy - Experiments at the BESSY Storage Ring", this Volume

4. R. Germer, W. Meyer-Ilse: "X-ray TV camera at 4.5 nm", Rev. Sci. Instrum. 57, 426-427 (1986)

5. W. Meyer-Ilse: "Soft X-Ray Imaging using CCD Sensors", in: Soft X-Ray Optics and Technology, ed. by E. E. Koch, G. Schmahl, SPIE Proc. 733, 515-518 (1986)

6. J. P. Doty, G. A. Luppino, G. R. Ricker: "X-Ray CCD Cameras: I. Design of Low Noise, High Performance Systems", in: Multilayer Structures and Laboratory X-Ray Laser Research, ed. by N. M. Ceglio, P. Dhez, SPIE Proc. 688, 216-221 (1986)

7. G. A. Luppino, N. M. Ceglio, J. P. Doty, G. R. Ricker, J. V. Vallerga: "X-Ray CCD Cameras II: A Versatile Laboratory System", ibid. 222-233

Advances in X-Ray Optics – '87*

N.M. Ceglio

University of California, Lawrence Livermore National Laboratory,
Livermore, CA 94550, USA

This contribution is the third in a series of review articles,
Ceglio [1,2], on recent developments in x-ray optics by the Advanced
X-ray Optics Group at Lawrence Livermore National Laboratory and its
collaborators. [3] In the past three years significant progress has been
made in the development of novel x-ray sources, x-ray optical components,
and x-ray detectors.

1. X-RAY SOURCES:

A new x-ray source operating in the energy range from 1keV to 5keV
is schematically illustrated in Figure 1 (shown for Titanium anode
characteristic emission at 4.5keV). The source operates as an optically
driven diode in which x-rays are generated by the illumination (with
visible light) of a red-sensitive multialkali photocathode (S-20) to
produce electrons that are accelerated onto a thin film transmission
x-ray anode. The characteristic x-ray spectrum can be controlled by
varying the atomic number of the active anode material. Since the
photocathode and anode are in close proximity (3mm gap), the x-ray
emission is a spatial and temporal replica of the input light signal,

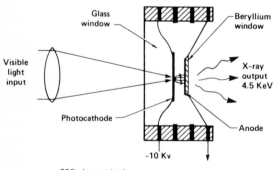

- S20 photocathode
- Anode is 4000 Å titanium on 5 mil beryllium window
- 3 mm gap

Figure 1: Schematic illustration of x-ray diode source. Visible light is focused onto the S20 photocathode producing photoelectrons that are accelerated across the 3mm gap into the thin film anode, which is coated on the interior surface of a thin Be window. X-rays produced emerge from the Be window and are a temporal and spatial replica of the photocathode illumination.

---

*This work was performed under the auspices of the U.S. Department of
Energy by Lawrence Livermore National Laboratory under Contract
No. W-7405-Eng-48.

providing great versatility as a laboratory source. The x-ray diode operates at brightness levels up to 10^{12} photons/sec $-cm^2$ $-str$. Its spot size may be varied from 30µm to 1cm, and it may be operated in a CW or pulsed mode. In the pulsed mode it is capable of producing x-ray bursts as short as 1-10 psec duration (using a picosecond laser driver), Stearns [4].

Laboratory x-ray lasers operating at soft x-ray wavelengths were first demonstrated at LLNL in 1984, D.L. Matthews [5]. This early work exploited $2p^53p \rightarrow 2p^53s$ transitions in Neon-like Selenium ions produced by high intensity optical laser illumination of a thin foil. The dominant emissions were the J = 2→1 transitions at 206.3Å and 209.6Å. Since then significant work has been done to extend the laser phenomena to shorter wavelengths, MacGowan [6,7], using Neon-like and Ni-like transitions in higher atomic number materials. Table I summarizes recent progress in producing high power, short wavelength sources of amplified spontaneous emission (ASE).

Table I: Recent progress toward short wavelength lasers. Shown are experimentally verified characteristics of short wavelength amplifying media, MacGowan [6,7].

Wavelength	Measured Gain	Lasing Medium	Gain Length Demonstrated	Output Power Demonstrated
206.3Å; 209.6Å	6 cm^{-1}	Ne-like Se	$\alpha L = 16$	10^6 watt
106.4Å	2 cm^{-1}	Ne-like Mo	$\alpha L = 4$	10^2 watt
131Å; 132.7Å	4 cm^{-1}		$\alpha L = 7$	10^3 watt
71Å	1 cm^{-1}	Ni-like Eu	$\alpha L = 4$	10^2 watt
50.3Å	1 cm^{-1}	Ni-like Yb	$\alpha L = 2$	10^2 watt

2. X-RAY OPTICAL COMPONENTS:

There is a broad interest in the x-ray community for efficient normal incidence x-ray optical components. Potential applications for such optics include high resolution x-ray spectroscopy, x-ray interferometry, and x-ray laser cavities. Thin film multilayer deposition techniques have been used to great advantage in producing normal incidence mirrors and beamsplitters at soft x-ray wavelengths. We have previously reported the performance of normal incidence mirrors and beamsplitters around 210Å, N.M. Ceglio [8], D.G. Stearns [9], M. Kuhne [10]. Figures 2 and 3 illustrate the performance of normal incidence mirrors and beamsplitters, respectively, around 130Å. As the data indicate , these Molybdenum/Silicon multilayer mirrors can have a normal incidence reflectivity in excess of 50% at 130Å, and the Silicon Nitride supported beamsplitters can achieve a normal incidence transmission approaching 50% and reflectivity greater than 10% at 130Å.

Normal incidence multilayer optical components have been used to produce multipass cavity amplification of x-ray laser emission increasing

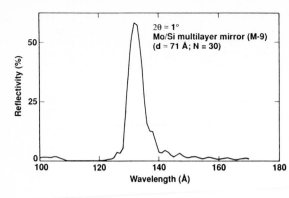

Figure 2: Measured performance of a multilayer mirror at θ = 0.5° off normal incidence; d is the multilayer period, N is the number of periods comprising the mirror.

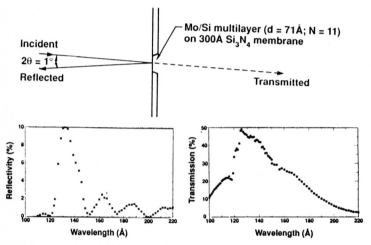

Figure 3: Measured performance of a multilayer beamsplitter at θ = 0.5° off normal incidence. Reflectivity is reduced from mirror values because of the smaller number of layer pairs (i.e., N = 11).

the output power by more than an order of magnitude, Ceglio [11,12,13]. This is illustrated in Figure 4 in which the triple pass output (at 206.3Å) of a pulsed x-ray amplifier medium is plotted. The cavity geometry is schematically illustrated. The double pass emission (second peak) is a factor of twenty greater than the single pass output. The modest intensity of the third pass emission is due to the rapid decrease of the gain at late time in this pulsed x-ray amplifier. In this experiment mirror reflectivity was 20%. It had a 12cm radius of curvature, and was 4.5cm away from the center of the 2cm long gain medium. The beamsplitter had a reflectivity of 15%, a transmission of 5%, and was placed 3cm from the center of the gain medium. Cavity experiments were also successfully conducted at λ=131Å, and will be reported elsewhere, Ceglio [14].

A wholly new x-ray optical component, the Highly Dispersive X-ray Mirror (HDXM), is illustrated in Figure 5. It is a normal incidence multilayer mirror into (or onto) which a high resolution (i.e., submicron

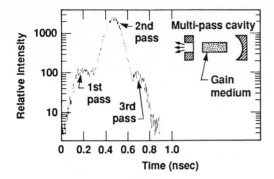

Figure 4: Measured triple pass output of an x-ray laser cavity. Laser wavelength is 206.3Å.

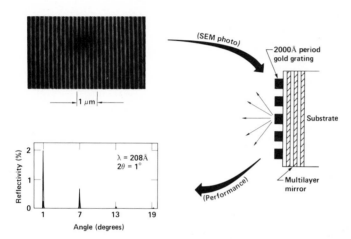

Figure 5: Highly Dispersive X-Ray Mirror. A submicron grating pattern is deposited on top of a multilayer mirror. The measured dispersion of a near normal incidence beam is shown.

linewidth), two-dimensional diffraction structure is patterned. The diffraction structure can be a simple linear diffraction grating (to disperse an incident beam into a number of output beams, as in Figure 5), or a Fresnel zone plate (to focus incident x-rays to a spot), or a complex holographic pattern (to diffract an incident coherent beam into any desired pattern). The HDXM is in essence a synthetically generated x-ray reflection hologram, with a wide range of potential applications for control and manipulation of coherent x-ray beams.

The practical value of guided transmission of EM radiation has long been appreciated. We have developed synthetic waveguide structures and demonstrated guided wave phenomena at soft x-ray wavelengths, Ceglio [15]. Figure 6 illustrates the experimental set up for the guided wave demonstration. A polarized, monochromatic (λ=208Å), synchrotron beam is incident on the waveguide structure, which is a free standing transmission grating with period (p=3000Å), thickness (t=4700Å), and spacewidth (s=2000Å) as shown. The pattern of diffracted radiation emerging from the waveguide channels was measured as a function of

133

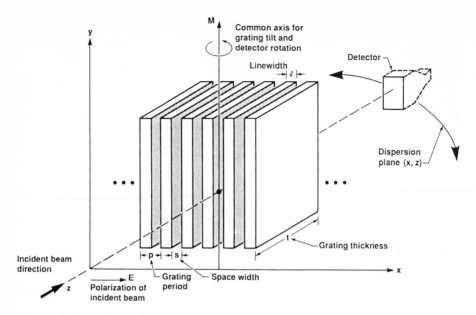

Figure 6: Experimental set-up for the measurement of a diffracted synchrotron beam passing through a thick waveguide structure.

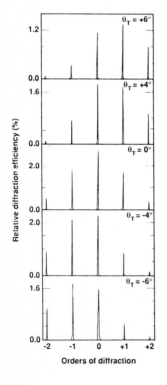

Figure 7: Measured diffraction patterns at λ = 208Å emerging from a waveguide structure having (1 = 1000Å; s = 2000Å; p = 3000Å; t = 4700Å) at a variety of tilt angles (θ_T). The angular separation between diffracted orders is $\Delta\theta = \lambda/p$. The asymmetry of the diffraction pattern is evidence of guided wave propagation down the narrow channels.

waveguide tilt about the axis, M. As shown in Figure 7, the asymmetry of
the diffraction pattern is very sensitive to grating tilt. This allows a
degree of directional control over the diffracted x-ray energy, somewhat
analogous to the directional control of a radar using a phased array of
sources. In this case the array of sources are the channels of the
transmission grating and the amplitude and phase distribution in the
array are determined by the propagation characteristics of the
waveguide. Further experiments have demonstrated propagation of soft
x-rays through waveguide channels at extreme angles, i.e., for which the
beam has no geometric line of sight through the channels.

3. X-RAY DETECTION:

A balanced development program in x-ray optics must not overlook the
significant advantages that accrue from improvements in x-ray detection.
In this regard we continue to pursue charge coupled devices (CCD's) for
direct x-ray imaging and non-dispensive spectroscopy, Luppino [16], Doty
[17]. The joint LLNL/MIT x-ray CCD camera uses a TI 4849 virtual phase
CCD. It has a 390 x 584 pixel array. Each pixel is 22μm square, and
the overall detector area is 8.6mm x 12.8mm. The active detector is
coated with an insulating oxide layer less than 800Å thick. This
allows x-ray detection down to energies below the C_k absorption edge
($\approx$ 283eV). The CCD camera may be used as a non-dispersive spectrometer
providing an energy resolution better than 140eV, see Figure 8. The
detector has an operating temperature of -30°C. Camera readout noise is
less than $8e^-$ (rms), serial and parallel transfer efficiencies are
$\approx$.99999, and dark current is less than 10^{-11} amp/cm^2 (at room
temperature). The camera is used as a combination imager/spectrometer
capable of single photon detection over a spectral range from 270eV to
15keV.

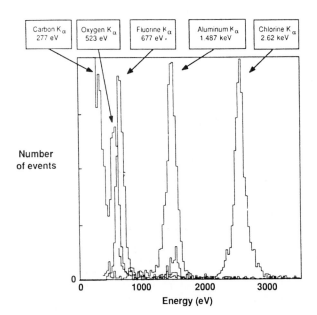

Figure 8: Non-dispen-
sive x-ray spectrum
measured with the x-ray
CCD camera. A variety
of characteristic x-ray
source spectra are
overlaid.

135

CONCLUSION:

There exists an exciting variety of emerging capabilities in x-ray sources, optical components and detection. It is very likely these will play an important role in a mature technology for x-ray microscopy. The precise role of these new capabilities is not yet clear. What is clear is that they are much too important to ignore.

ACKNOWLEDGEMENTS:

In addition to the contributions of our collaborators listed in reference [3], the work discussed above represents the significant efforts of members of the x-ray optics group: A.M. Hawryluk, D.G. Stearns, D.P. Gaines, G. Howe, W. Cook, S. Hill, H. Petersen, E. Utterback, and R. Hicks.

REFERENCES:

1. N.N. Ceglio: In Low Energy X-Ray Diagnostics, ed. by D.T. Attwood and B.L. Henke (American Institute of Physics, 1981) pg. 210.

2. N.M. Ceglio: X-Ray Microscopy, ed. by G. Schmahl and D. Rudolph, Vol. 43 (Springer - Verlag, 1984) pg. 97.

3. The work herein reported has involved LLNL personnel as well as the collaborative efforts of researchers from other institutions: MIT's Center for Space Research (G.R. Ricker, J. Doty, and G. Luppino); LLNL's Laboratory X-Ray Laser Group (D. Matthews, B. MacGowan, S. Brown, J. Trebes, and C. Keane); VUV Radiometric Laboratory of the PTB at the Berlin Storage Ring, BESSY, (M. Kuhne, P. Muller, and B. Wende); MIT's Submicron Structures Laboratory (H. Smith, E. Anderson, M. Schattenberg).

4. D.G. Stearns: Nucl. Inst. and Meth. A242, 364 (1986).

5. D.L. Matthews et al.: Phys. Rev. Letter 54, 110 (1985).

6. B.J. MacGowan et al.: J. App. Phys. 61, 5243 (1987).

7. B.J. MacGowan et al.: "Demonstration of Soft X-Ray Amplification in Nickel-Like Ions", submitted to Phys. Rev. A.

8. N.M. Ceglio et al.: Journal de Physique C6, 277 (1986).

9. D.G. Stearns et al.: Proceedings SPIE 688, 91 (1986).

10. M. Kuhne et al.: Proceedings SPIE 688, 76 (1986).

11. N.M. Ceglio et al.: Proceedings SPIE 688, 44 (1986).

12. N.M. Ceglio et al.: "Time Resolved Measurement of Double Pass Amplification of Soft X-Rays", submitted to Phys. Rev. Letter.

13. N.M. Ceglio et al.: "Multipass Amplification of Soft X-Rays in a Laser Cavity", submitted to Optics Letters.

14. N.M. Ceglio et al.: To be submitted to Applied Physics Letters.

15. N.M. Ceglio et al.: "Demonstration of Waveguide Phenomena at Soft X-Ray Wavelengths", submitted to Optics Letters.

16. G.A. Luppino et al.: Proceedings SPIE 688, 222 (1986).

17. J.P. Doty et al.: Proceedings SPIE 688, 216 (1986).

Zone Plates for the Nanometer Wavelength Range

V.V. Aristov, A.I. Erko, L.A. Panchencko, V.V. Martynov, S.V. Redkin, and G.D. Sazonova

Institute of Problems of Microelectronics Technology and
Superpure Materials, USSR Academy of Sciences,
142432 Chernogolovka, Moscow District, USSR

The development of efficient optical elements for focusing and image transmission at soft X-ray wavelengths (0.5 nm - 10 nm) opens the way for realization of new methods in X-ray microscopy, local photoelectron analysis, plasma physics, astronomy, etc. In fact, only the last decade has witnessed the appearance of technological possibilities for fabricating high resolution X-ray diffractive elements thanks to advanced methods in microelectronics technology.

At present, Fresnel zone plates with an absorbing gold pattern are widely used. They are of low efficiency, theoretically not exceeding 10%. Yet, a number of materials are known to be suitable for fabrication of transparent, phase, X-ray optical elements at wavelengths 0.5 nm - 10 nm /1/. In this case, energy efficiency increases up to 40%. For blazed phase elements it increases up to 70% /2/.

This paper presents experimental data on fabrication of phase X-ray diffraction optics elements from a Si single crystal and characteristics calculated for an X-ray diffraction lens at wavelengths 0.4 nm - 2 nm. For the wavelength band 2.0 - 40 nm Bragg-Fresnel optical elements on the basis of multilayer structures are described.

The calculations of an X-ray wave phase shift in layers were performed on the basis of tabulated data on atomic scattering factors /3/. Considering the fact that the scattering factor has real and imaginary components responsible for a wave phase shift and radiation absorption

$$f = f' + if'',$$

we can write the equation of the complex optical constant

$$n = 1 - \delta - i\beta = 1 - \frac{Nr_e\lambda^2}{2\pi}(f' + if''),$$

where δ = refractive index of a material; β = absorption coefficient; $N = N_a\rho/W$ = concentration of atoms of a substance; ρ = substance density; W = atomic weight; N_a = Avogadro number; r_e = classical electron radius; λ = radiation wavelength. The phase shift of the wave transmitted through a material layer t thick

$$\Delta\varphi = \frac{2\pi}{\lambda} t\delta.$$

Hence, the thickness of a layer shifting the phase by π required for a maximum contrast of a zone plate is

$$t_\pi = \frac{\pi W}{N_a \rho r_e \lambda f'} .$$

For a number of substances the value t_π does not exceed several microns. Account must be taken of the layer absorption when evaluating the efficiency of Fresnel zone plate radiation focusing. Here, the radiation fraction diffracted in the first order does not exceed /2/

$$\epsilon_1 = \left[\frac{1 + \exp(-\pi f''/f')}{\pi} \right]^2 .$$

According to the evaluations performed for a number of materials with sufficiently good X-ray optical properties, silicon, carbon, and carbon containing polymer thermally stable films are most suitable from the technological point of view.

The calculated data of the efficiency of zone optics elements from silicon are shown in Fig. 1. It is seen that silicon is practically an ideal material at wavelengths 0.4 - 1 nm where diffraction efficiency is close to theoretical one for purely phase objects.

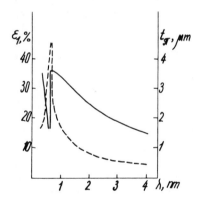

Fig. 1. Efficiency of the first order diffraction (solid curve) and thickness of silicon material (dashed curve) as a function of radiation wavelength for zone optics

The optics for high resolution image transmission must be free of wave front phase distortions caused by surface roughness and fluctuations of the refractive index inside the medium. As it is known /4/, the coefficient of surface roughness is of great importance for fabrication of multilayer structures and grazing incidence X-ray optical elements. In this case, the value of roughnesses should not exceed $\lambda/10 - \lambda/20$ being angström units at X-ray wavelengths. Requirements in surface quality are less for transparent optics and found to be $\Delta t \approx 0.1\, t_\pi$ for the same values of wave front distortion. Actually, silicon surface distortions can achieve 0.1 μm having no marked influence on the image quality. Significantly greater influence is likely to be made by the structure of a phase-shifting layer. Since material inhomogeneities, namely macroscopic grain boundaries, scatter an X-ray wave, the peculiarities of zone plate fabrication technology from gold and silver associated with inevitable growth of grains at electroplating do not permit obtaining effective focusing close to theoretical limits.

Figure 2 shows a microphotograph of a zone plate based on a Si single crystal. Precise relief formation was provided using electron beam lithography and reactive ion etching.

The parameters of the zone plate fabricated are as follows:

Thickness of a Si layer	3	μm
Number of zones	80	
Diameter of the first zone	12.6	μm
Aperture	114	μm
Minimum zone size	0.35	μm
Focal distance for $\lambda = 0.8$ nm	5	cm
First order diffraction efficiency	36%.	

Since a silicon structure is formed on a polyimide membrane, it can be used as a transparent X-ray lens at wavelengths 0.4 - 2.0 nm.

Figure 3 shows a microphotograph of a diffraction lattice formed from a Si single crystal on a transparent polyimide membrane.

All the possibilities of X-ray radiation focusing in submicron spots do not reach their limit when using optical diffraction elements formed either as freely suspended structures or on transparent membranes. Newly developed Bragg-Fresnel optics elements employing diffraction properties of profiled multilayer mirrors seem highly promising /5/. Phase profile formation on a multilayer structure surface permits fabricating focusing elements with very high stability to heat load and mechanical deformation. An important feature of these elements is their achromatism. Figure 4 presents an intensity distribution curve in the image plane on transmitting a slit image with the dimension of 50 μm reduced by 10 times. Radiation of an X-ray tube with a Fe anode was collimated by two slits S_1 and S_2 and then focused by a profiled multilayer mirror. A phase profile was fabricated by sputtering a gold layer.

Fig. 2. Silicon zone plate for X-ray radiation focusing in the wavelength range 0.4 - 2.0 nm

Fig. 3. Zone diffraction lattice based on Si for the wavelength range 0.4-2 nm

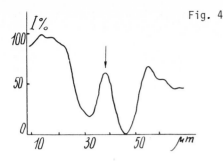 Fig. 4

 Fig. 5

Fig. 4. Intensity distribution curve in the image plane of a focusing Bragg-Fresnel element

Fig. 5. Bragg-Fresnel lens on the basis of a multilayer mirror with ellipsoidal zones

Thus, the paper demonstrates the possibilities of fabricating phase diffraction optics of a nanometric range on the basis of a Si single crystal and multilayer mirrors.

Multilayer mirrors were fabricated in Applied Physics Institute Academy of Sciences by S.V. Gaponov's group.

Figure 5 shows the structure of an aberration – corrected lens with ellipsoidal zones on the basis of a multilayer mirror. The minimum zone width is 250 nm. A reflecting lens was specially designed for an X-ray microscope, which is mounted on the beam of the "Siberia-1" synchrotron radiation source.

Literature

1. J. Kirz: J. Opt. Soc. Amer. _64_, 301 (1974).
2. N.M. Ceglio: AIP Conference Proceedings, _75_, 210 (1981).
3. B.L. Henke, D. Lee, I.J. Tanaka, R.L. Shimabukuro, B.K. Fujikawa: Atomic Data and Nuclear Data Tables _27_, 1 (1982).
4. S.V. Gaponov, S.A. Gusev, Yu. Ya. Platonov, N.N Salashchenko: Zh. Tehn. Fis. (USSR), _54_, 747 (1984).
5. V.V. Aristov, S.V. Gaponov, V.M. Genkin, Yu.A. Gorbatov, A.I. Erko, V.V. Martynov, L.A. Matveeva, N.N. Salaschenko, A.A. Fraerman: Pisma Zh. Eksp. Teor. Fis. (USSR), _44_, 207 (1986).

Editors' note: According to additional information kindly supplied by the authors, the silicon zone plate in Fig. 2 is supported on a 0.5 micron window of boron-doped silicon.

Sputtered-Sliced Linear Zone Plates for 8 keV X-Rays

R.M. Bionta, A.F. Jankowski, and D.M. Makowiecki

Lawrence Livermore National Laboratories, L-278, P.O. Box 5503, Livermore, CA 94550, USA

We describe the fabrication and testing of sputtered-sliced linear zone plates, designed to focus 8 keV ($\lambda = 1.54$ Å) x-ray photons from conventional Cu x-ray sources.

1. Introduction

Zone plates have been successfully used in soft x-ray microscopy systems. Similar zone plates operating at 5 to 20 keV would vastly improve industrial x-ray instrumentation. Unfortunately, such zone plates cannot be fabricated by conventional electron beam and holographic lithography techniques because of the need for small feature sizes and large aspect ratios. We are investigating an alternative fabrication technique[1], using alternating layers of transparent and opaque materials sputtered onto a wire core, which is then sliced perpendicular to the core axis. This sputter-slice technique should work well at high energies, where sputtered materials are more transparent. Large aspect ratios simply mean thicker slices.

2. Zone Plate Fabrication

For simplicity we have chosen to dispense with the wire core and instead produce linear structures by sputtering alternating layers onto optical flats. Coated substrates are encapsulated in epoxy, then sliced into wafers and polished to final thicknesses in the range of 10 to 100 μm. Figure 1 shows a sputtered sliced linear zone plate made of Al and Ta with a focal length of 40 cm at 8 keV. This linear zone plate focuses x rays in one dimension, forming line images of point sources much as a cylindrical lens does in ordinary optics.

Fig. 1 SEM of a sputtered-sliced linear zone plate mounted across a 1 mm diameter hole in a brass disk. The light lines are Ta; the dark lines, Al. The lens is 112 μm across and 30 μm thick

3. Testing

We test our zone plates by using them to image the x rays produced by a Cu microfocus x ray source (Figure 2). Lenses of focal length f are tested by placing them a distance $2i$ from the x-ray source so that they form an image of the x-ray source at a distance o beyond the lens according to the thin lens law $1/f = 1/i + 1/o$. The images are recorded at a series of pixels by stepping a 30 μm dia pinhole across the image. Since zone plate focal lengths depend on photon energy, we record the entire pulse height spectrum from the Si(Li) detector at each pixel position.

Figure 3 shows a pulse height spectrum recorded at a typical pixel. The clean and distinct Cu Kα and Cu Kβ lines allow us to obtain monochromatic x-ray images at 8 keV by plotting the number of photons in the Kα peak as a function of pinhole position.

Prior to testing zone plates, we evaluate the microfocus x-ray source spot by imaging it with a 10 μm wide slit. Figure 4 shows a typical image. We find that the average spot size is $\sim$ 37 μm FWHM, and that the x ray spot is slightly asymmetric.

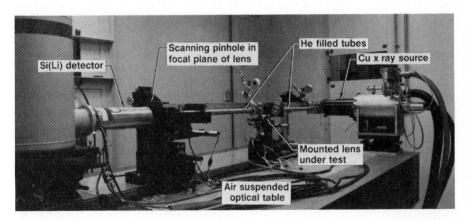

Fig. 2 X-Ray lens test bed. The lens forms an image of the Cu x-ray source in the plane of the scanning pinhole which is measured by stepping the pinhole across the Si(Li) detector and recording the energies of the photons at each pinhole position

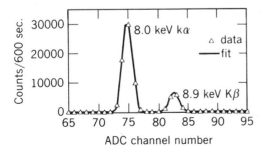

Fig. 3 Si(Li) detector pulse height spectrum, as recorded at each pinhole position. The characteristic Cu, Kα and Kβ lines at 8.0 and 8.9 keV have FWHM of 207 eV (Kα) and 219 eV (Kβ) with < 4% background

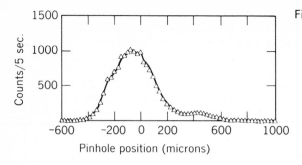

4. Zone Plate Results

We tested the 40 cm focal length linear zone plate by placing it 80 cm from the x-ray source, so that it formed an image of the source 80 cm behind the plate at a magnification of 1X. Spectra were collected across the image for 60 sec at each of 240 pinhole positions separated by 2.5 μm. Figure 5 shows the resulting image at 8 keV. The zone plate is 100 μm across, so it casts a shadow that widens to ~ 200 μm at the scanning pinhole. This shadow is the dip in intensity seen at pinhole positions between -0.12 to 0.12 mm. The small peak within the shadow is the focused image of the x-ray source which has the same small tail on the right as seen in the slit image of the source in Fig. 4. The flat background under the image is the zero-order radiation which passes through the zone plate undeflected.

The solid line in Figure 5 is the expected performance of this lens from a numerical simulation, based on the Huygens principle (see [1]). In this simulation the x-ray wave front transmitted through the lens is represented by an array of x-ray-radiating antennas located in the plane of the lens and separated by 200 Å. The amplitudes and phases of the antennas are set as if the x rays passed through 30 μm of Al or Ta (depending on the antenna position along the lens) according to the optical constants of Al or Ta.[2] In order to calculate the electric field amplitude at an array of points in the image plane separated by 1 μm, we first obtain the electric field intensity by squaring the calculated

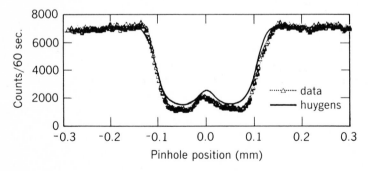

Fig. 5 One-dimensional image at 8 keV taken with the zone plate in Fig. 1

amplitude at each point. To obtain the solid line in Figure 5, we convolute the resulting intensity with a Gaussian shape similar in size to the x-ray emitting region of the anode. We then multiply the result by a constant so that it matches the measured x-ray intensity outside the shadow of the zone plate.

The predicted distribution is very similar to the measured data; the biggest difference is in the amounts of zero-order-intensity x rays. We measure 16.7 $\pm$ 0.8% into zero-order, whereas the model predicts 21.2% into zero-order. The agreement is much better in the amount of x rays focused in the first-order peaks, which is often referred to as the efficiency of the lens. The measured efficiency for this lens (defined here as the area under the first-order peak divided by the total area in the shadow) is 2.50 $\pm$ 0.05% as compared to a predicted efficiency of 2.64%.

The excellent agreement between prediction and measurement means that we can confidently pursue higher efficiency designs that utilize phase modulation instead of amplitude modulation.

5. Summary

We have demonstrated that transmissive optics can be fabricated for 8 keV radiation by sputtering alternating layers of two materials and then slicing the coatings into thin slabs. This technique shows promise for the fabrication of highly efficient phased focusing devices for these energies.

Acknowledgements

The authors are indebted to the technical staff of "O" Division's X-Ray Microscopy Project who made this work possible, including E. Ables, K. Cook, L. R. Mendonca, K. Miller, H. Olson, R. Tilley, R. Vital, T. Viada, P. Gabriele, and L. Kennedy, as well as the useful advice given by T. Barbee, H. S. Park, and L. Wood.

Work performed under the auspices of the U.S. Department of Energy by the Lawrence Livermore National Laboratory under Contract No. W-7405-ENG-48.

References

1. Richard M. Bionta, Appl. Phys. Lett., <u>51</u>, 725, (1987).

2. B.L. Henke, in <u>X-Ray Data Booklet</u> ed. by D. Vaugham, (Lawrence Berkeley Laboratory, Technical Information Department, Berkeley, 1985), p2-28.

Measurement of Resolution in Zone Plate X-Ray Microscopy

C.J. Buckley

Department of Physics Research, King's College,
London University, The Strand, London WC2R 2LS, UK

1 Introduction

Image formation in scanning x-ray microscopy is achieved by rastering a semi transparent specimen across a small x-ray probe and counting the number of photons transmitted by the specimen at position of the scan. The finest soft x-ray probes generated so far have been produced by modified fresnel zone plates [1,2]. In scanning x-ray microscopy a zone plate is used to demagnify a source of x rays. The dimensions of the probe are primarily determined by the width of the finest zone (dr) of the zone plate. If the zone plate has been fabricated with sufficient accuracy and is illuminated with x-rays which are both spatially and temporally coherent then the probe shape will resemble an airy pattern which has a full width at half maximum (FWHM) which is very nearly equal to the finest zone.

In practice the zone plate is never illuminated with totally coherent radiation. However if the temporal coherence of the radiation accepted by the zone plate obeys

$$\frac{\lambda}{\delta\lambda} \geq \frac{r_{zp}}{2\,dr} \tag{1}$$

and the spatial coherence is such that

$$S\Theta_{zp} = \lambda, \tag{2}$$

where λ is the wavelength of the radiation, r_{zp} is the radius of the zone plate, S is the size of the x-ray source and Θ_{zp} is the angle subtended by the zone plate to the source, then a good compromise is obtained between the physical size of the probe and the photon flux contained within it. Under these conditions and with the unwanted orders removed by an aperture and a central stop equal to 40% of the zone plate radius, the FWHM of the probe will be equal to 1.1 dr. The resolution capability of scanning microscopes is often discussed in terms of the probe size, or the contrast versus spatial frequency performance. Here two simple methods which can be used to measure these quantities are described and the influence that practical test objects have on these measurements is discussed.

2 Probe Size Measurement by Knife Edge Method

A measure of the FWHM of the probe can be obtained in a scanning microscope by interpretation of the image of a knife edge specimen [3]. Imagine a specimen half of which is made from a material which is opaque to x rays. The specimen has a well-defined edge and is large relative to the scale of the probe but is thinner than its depth of focus. When this specimen is scanned across the probe at focus, the photon flux transmitted by the specimen is progressively reduced until the entire extent of the probe is obstructed by the opaque half of the knife edge specimen. Plot a of fig. 1 shows the calculated reduction in transmitted flux as this process is performed. The reduction in transmitted flux (ΔF) as an ideal knife edge specimen is scanned a distance equivalent to the theoretical FWHM across the centre of the probe, is 59% [4]. The distance scanned to cover the FWHM of the probe is represented by the distance c to d on fig. 1a, and the respective reduction in transmitted flux (ΔF) is represented by points e to f (the 79.5% to 20.5% levels). Applying this philosophy to data recorded from an experiment, the scan distance required to reduce the transmitted flux from 79.5% to 20.5% of maximum can be used as a measure of the FWHM of the experimental probe.

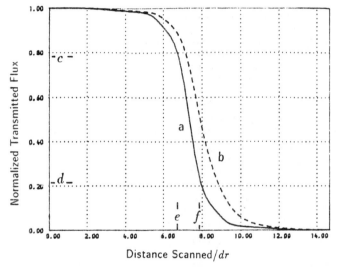

Fig. 1 The reduction in transmitted x-ray flux as a) an ideal knife edge and b) a non-ideal knife edge is scanned across the probe produced by a zone plate.

2.1 Knife Edge Method: Practical Considerations.

The above method was discussed in terms of an ideal knife edge specimen. In practice all edges will have a finite distance over which their absorption goes from zero to maximum. The effect of this on the knife edge experiment is to increase the distance that the knife edge must be scanned in order to reduce the transmitted flux from 79.5% to 20.5% of maximum, thereby giving a falsely large measure of probe size. Plot b of fig. 1 shows the calculated change in transmitted flux as a copper knife edge which has a wedge profile is scanned across a probe produced by a zone plate (dr=100 nm) is illuminated by 3.5 nm radiation. Here the wedge has an included angle of 60°. If both the shape of knife edge and it's absorption characteristics are well known, the effect can be taken into account, and an appropriate ΔF may be derived which gives the correct FWHM of the probe when used on data. Unfortunately there is no general rule for the choice of ΔF as it is dependent on the edge profile, the edge material and the scale of the probe. Fig.

Probe FWHM	Wedge Angle	ΔF
100 nm	90°	59%
100 nm	80°	56%
100 nm	60°	54%
100 nm	45°	52%
50 nm	80°	56%
50 nm	60°	50%
50 nm	45°	41%

Probe FWHM	Wedge Angle	ΔF
100 nm	90°	59%
100 nm	80°	54%
100 nm	60°	49%
100 nm	45°	36%
50 nm	80°	52%
50 nm	60°	33%
50 nm	45°	24%

Fig. 2a Flux reductions ΔF used to measure probe size for gold edges

Fig 2b Flux reductions ΔF used to measure probe size for copper edges

2 tabulates ΔF for a range of wedge angles and probe sizes as an example of two different knife edge materials which have wedge profiles.

An alternative method for the evaluation of resolution employs a thin grating as a test object. Here the contrast versus spatial frequency performance is assessed. This method and its practical interpretation is discussed below.

3 Grating Test Object: Line Width and Contrast.

Periodic objects such as gratings are used as calibration and resolution test objects in many optical devices [5]. Consider the case of a grating which has line widths which decrease in a regular way which is imaged in a scanning x-ray microscope. Visual inspection of the image gives an immediate assessment of the resolution capability of the system. Also, if the theoretical relationship between image contrast and line width is known, an assessment of how the zone plate optic performs can be made in terms of a comparison between the expected and measured contrast versus line width.

The fabrication of appropriate test gratings for the x-ray microscope does not present a significant problem, as the technology which is used to produce zone plates can also produce suitable test objects. The image of one quadrant of a zone plate test object is all that is needed to test the line width/contrast and stigmatic performance of the optic. Here the contrast or visibility (V) is defined as

$$V = \frac{I_{max} - I_{min}}{I_{max} + I_{min}} \quad , \tag{3}$$

where I_{max} and I_{min} are the maximum and minimum intensities in the image of the grating. A number of factors affect the visibility of the zone plate test object. These are: the x-ray absorption of the zones of the test object, the profile of these zones and their mark to space ratio. Of these the dominant factor is that of absorption (see fig.3). The visibility function is also highly sensitive to the degree of coherence with which the zone plate optic is illuminated. If the temporal coherence satisfies the condition of equation 1, small changes in the spectral purity have little effect on visibility. However the visibility response shows considerable variation to small changes in the spatial coherence of the illuminating radiation. Fig. 4 shows the calculated visibility versus line width under a range of spatial coherence conditions.

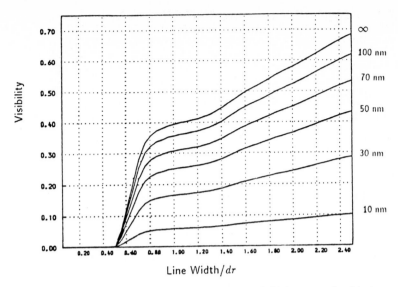

Fig. 3 Visibility of grating objects for a range of thicknesses of gold at a wavelength of 3.5 nm. The abscissa scale is normalised to the finest zone of the zone plate optic.

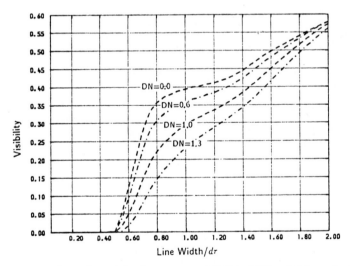

Fig. 4 Effect of spatial coherence on visibility (DN = 1.0 is equivalent to $S\Theta_{zp} = 1.0 \lambda$).

4 Conclusion

The knife edge experiment provides a quick and simple measurement of probe size. However the accuracy of the method is reduced when the probe size approaches a few tens of nanometres. This is because the effect of the edge profile becomes more pronounced. Also the depth of focus of the probe becomes smaller than the thickness of knife edge material required to completely absorb the incident x rays. The interpretation of the results obtained with the grating method is not so sensitive to object edge profile as the knife edge method. Visual inspection of the image of the grating gives an immediate indication of the resolution capability of the zone plate microscope. Under the conditions specified by equations 1 and 2, a line width of 0.65 dr should be visible for a practical test object when the average recorded flux is 200 photons per pixel or more, if the zone plate optic is well formed. Further information on the use of both the knife edge and zone plate test objects may be obtained from reference [4] where the subject has been discussed at length.

5 Acknowledgements

The author acknowledges the following persons: M.T. Browne, J.M. Kenney A.G. Michette and J. Kirz for discussions relevant to the subject of this paper.

6 References

1 Rarback H. et al., A Scanning x-ray microscope with 75 nm resolution, Rev. Sci. Inst. January 1988.

2 Morrison G.R. et al., Early Experience With the Kings College-Daresbury X-Ray Microscope, these proceedings.

3 Oho E. et al., Measurement of Electron Beam Diameter by Digital Image Processing. Journal of Electron Microscopy Technique, 2, 463-469 (1985).

4 Buckley C.J. The Fabrication of Gold Zone Plates and their use in Scanning X-ray Microscopy. Doctor of Philosophy Thesis, Kings College, London University, chapters 7 and 8, (1987).

5 Dainty C. and Shaw R., Image Science, Academic Press, 232-244, 1974.

Using Radiachromic Films for Soft X-Ray Dosimetry

G.D. Guttmann, B.L. Henke, and J.A. Kerner

Center for X-Ray Optics, Lawrence Berkeley Laboratory, Berkeley, CA 94720, USA

1 Introduction

The soft x-ray contact microscopy station at LBL images wet, living biological specimens with a resolution better than 1000 Å. High doses of soft x-rays are required to record a latent image onto a polymer resist which is then developed and examined by electron microscopy. In order to interpret the resulting image, it is necessary to determine whether the structure of the cell has changed during the exposure. We would therefore like to know which components of the cell change, how much radiation is necessary to produce the change, and whether the structural changes affect the soft x-ray micrograph. Before these questions may be answered, we must establish a dosimetric standard for soft x-ray microscopy. The soft x-ray contact microscopy station was calibrated by using radiachromic nylon film which in turn was standardized with a known source at our laboratory.

2 Materials and Methods

The radiachromic film was obtained from Far West Technology, Inc., Goleta, California, USA (Model No. FWT-60-00). This film is impregnated with a dye, hexahydroxy-ethyl aminotriphenylnitrile, which turns blue upon exposure to x-rays or ultraviolet light [1]. The film, used primarily within the food industry for calibrating food irradiation machines, has a sensitivity which is well matched to the dosage levels (10^3 to 10^7 rads) typically used in our imaging experiments [2].

For calibration, the radiachromic film was exposed to unfiltered characteristic x-ray line radiation from a demountable anode source and the dose was determined using an absolutely calibrated flow-proportional counter [3]. A series of films placed at different distances from the source were exposed for a fixed time period while the proportional counter measured the integrated flux. Corrections for counter efficiency, transmission of thin windows, and continuum were estimated by pulse-height analysis.

Two sources of radiation were used. The first was a graphitized copper anode which produced carbon K (277 eV/44.6 Å) x-rays as well as a contribution from continuum amounting to 20% of the total dose. The second source was an aluminum anode which produced both oxygen K (525 eV/ 23.6 Å) and aluminum K (1487 eV/8.34 Å) lines in an energy flux ratio of 0.2/1, with a continuum contribution amounting to 7% of the total dose. The anode voltage was maintained at the approximate operating value used for the soft x-ray contact microscopy station.

To calibrate the x-ray contact microscopy station, the radiachromic film was placed in a trial radiobiology chamber and exposed for different periods of time to soft x-rays from a carbon K source operating at 5.0 kV and 40 mA.

The optical density of the radiachromic films was read on a densitometer (Model No. FWT-91R, Far West Technology, Inc., Goleta, CA, USA) before and after irradiation. The film sensitivity required that the optical densities be measured at 510 (high scale) and/or 600 nm (low scale).

3 Results

Dose-response curves for the radiachromic film were obtained by plotting the change in optical density at 510 nm and 600 nm against the absolute dosage as seen in Fig. 1. The points obtained with the carbon and aluminum anodes fall on virtually the same line, illustrating the lack of an energy dependence in the response of the film.

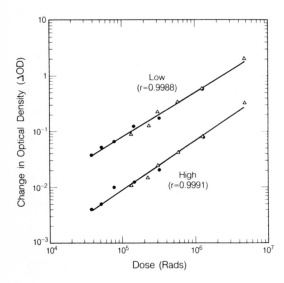

Figure 1. Dosimetry for a radiachromic nylon film The sources were (●) carbon K (277 eV) and (△) aluminum K (1487 eV). Low (600 nm) and high (510 nm) refer to the wavelength where the film's optical density was measured. "r" is the correlation coefficient.

Figure 2 illustrates the change in optical density of the film at 510 and 600 nm for varying periods of exposure in the contact microscopy instrument. Using Fig. 1, we can determine what length of exposure is required to attain a prescribed dose.

A calculation was done to determine if the film absorbed all or part of the dose. The thickness of the film is 2 mils (51 μm); its density is 1.14 g/cm^3, and the elemental composition (mass %) is 65.6% carbon, 12.8% oxygen, 11.1% nitrogen, and 10.5% hydrogen [4]. The film's thickness is 18 radiation lengths for carbon K x-rays, 59 for oxygen K x-rays, and 4.33 for aluminum K x-rays [5].

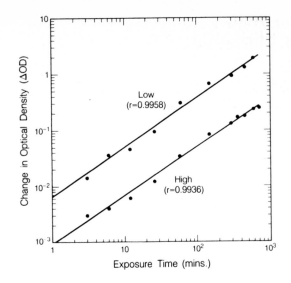

Figure 2. The change in optical density of radiachromic nylon film versus exposure time from the soft x-ray contact microscopy station. The source was carbon K_α (277 eV). The apparatus was operated at 5 kV and 40 mA. Low (600 nm) and high (510 nm) refer to the wavelength where the film's optical density was measured. "r" is the correlation coefficient.

4 Summary

Radiachromic nylon films are very useful for high dose measurements, and are versatile in that they detect soft x-rays as well as gamma rays or high energy x-rays. We conclude that the dose-response characteristic of this radiachromic nylon film is independent of energy for photons whose energies are in the 250-1500 eV range. This film will be useful for soft x-ray dosimetry and for calibrations in soft x-ray molecular radiobiology.

5 Acknowledgements

We wish to thank Phil Batson for his technical assistance with the trial radiobiology chamber. This project was supported by the U.S. Department of Energy under Contract No. DE-AC03-76-SF00098.

6 References

1. K.C. Humphreys, W.O. Wilde, A.D. Kantz: Radiat. Phys. Chem. 22, 291 (1983)
2. K.C. Humphreys, A.D. Kantz: Radiat. Phys. Chem. 9, 737 (1977)
3. B.L. Henke, M.A. Tester: Advances in X-Ray Analysis 18, 76 (1975)
4. Private communication with Far West Technology, Inc., Goleta CA USA
5. B.L. Henke, P. Lee, T.J. Tanaka, R.L. Shimabukuro, B.K. Fujikawa: At. Data Nucl. Data Tables 27, 1 (1982)

A Detector System for High Photon Rates for a Scanning X-Ray Microscope

P. Guttmann and B. Niemann

Universität Göttingen, Forschungsgruppe Röntgenmikroskopie,
Geiststraße 11, D-3400 Göttingen, Fed. Rep. of Germany

Abstract

To use the advantages of a scanning x-ray microscope it is necessary to have a detector system with a high detective quantum efficiency (DQE). The system must be able to detect photon rates greater than 10^6 photons/s, which occur at the detector of the Göttingen scanning x-ray microscope at the electron storage ring BESSY in Berlin.

Examinations and developments of a photon counting detector system with a proportional counter are described. The limitations of such a system are given. In the case of very high photon rates in the scanning spot, a detector system with a GaAsP-Schottky photodiode will be used.

1 Introduction

A detector system for the Göttingen scanning x-ray microscope should

a) be sensitive in the wavelength range 2.3 nm $\leq \lambda \leq$ 4.5 nm,
b) have a high detective quantum efficiency (DQE), to minimize the radiation dose transferred to the specimen,
c) be able to measure photon rates higher than 10^6 photons/s,
d) and be capable to be read out each millisecond, which is the aspired integration time for one image point.
e) Energy resolution is not required.

2 Limitations of a photon counting detector system

The highest count rate measurable with a photon counting detector system is limited by its dead time. If the photon rate incident to a photon counting detector system is increased more and more, the measurable count rate reaches a maximum. The reciprocal of this maximum count rate is defined as the dead time. For a photon counting detector system used at a synchrotron radiation facility, not only the dead time limits the maximum measurable count rate, but also the time distance t_B of the bunches [1], but this is not important if the dead time is much longer than t_B. For the Göttingen scanning x-ray microscope working at BESSY with the METRO electron optic with t_B = 2 ns the dead time of the photon counting detector system must be shorter than 10 ns for a highest measurable photon rate of 10^7 photons/s, if count rate losses of up to 10 % are allowed.

Proportional counters can have a DQE of up to 100%, if very thin, low absorbing windows are used and if the counter gas absorbs all photons. According to the gas amplification low energy x-ray photons deliver a measurable signal.

The classical application of a proportional counter is to analyse this signal, i. e. to measure the height of the output pulses, which is directly proportional to the energy of the investigated photons. In this classical application the upper count rate limit must be defined as the highest count rate providing a stable pulse height spectrum. This limit is usually lower than that given by the above-mentioned definition, which considers only the number of pulses.

In a cylindrical proportional counter the gas gain M can be expressed with the Diethorn approximation [2]:

$$ M = \exp\left[\frac{\ln 2}{\Delta U \cdot \ln(b/a)} \cdot U_0 \cdot \ln\left[\frac{U_0}{k \cdot p \cdot a \cdot \ln(b/a)} \right] \right] \qquad (1) $$

with ΔU = 40.3 V and k = 74.21 $V \cdot cm^{-1} \cdot Torr^{-1}$ for methane, gas pressure p, cathode radius b and anode radius a. U_0 is the potential between the anode wire and the cathode.

The collection times T_i and T_e for positive ions and electrons in a cylindrical proportional counter are given by [3]:

$$ T_{i(e)} = \frac{p \cdot \ln (b/a)}{\mu_{+(e)} \cdot U_0 \cdot 2} \cdot (b^2 - a^2) \qquad (2) $$

with positive ion mobility μ_+ and electron mobility μ_e.

In proportional counters, where the photons are incident perpendicular to the anode wire, the occurring space charge density is given by [4]:

$$ \rho = \frac{N_p \cdot M \cdot p \cdot \ln(b/a) \cdot e}{2 \cdot \pi \cdot \mu_+ \cdot U_0} \cdot \frac{n}{L_a} \qquad (3) $$

where n is the absorbed photon rate, e is the charge of one electron, N_p is the number of generated electrons per photon, N_p = photon energy/W, W is the electron-ion pair generation energy (for methane W = 29.2 eV) and L_a is the active length, where the charge is collected. The anode wire has the length $L \geq L_a$. If photons enter the counter parallel to the anode wire, $\frac{n}{L_a}$ has to be replaced by $\frac{dn(l)}{dl}$, which is the number of absorbed photons per length dl, l is the position on the anode wire. $\frac{dn(l)}{dl}$ is exponentially depending on l and can easily be calculated. At high photon rates the space charge reduces the electric field and therefore reduces the gas gain, which is reduced in the same manner as if the voltage at the anode wire would be reduced by [4]

$$\delta U = \frac{\rho \cdot (b^2 - a^2)}{4 \cdot \pi \cdot \epsilon_0} \tag{4}$$

with $\epsilon_0 \approx$ 8.85 pF/m being the dielectric field constant.

To decrease the collection times T_i and T_e it is necessary to:
a) use a gas with highest electron and ion mobility,
b) reduce the gas pressure,
c) reduce the cathode radius b, but it should remain b $\gg$ a.

In a cylindrical proportional counter the voltage signal on the anode wire induced by the electrons is very much smaller than that induced by the positive ions [4], because the electrons pass a smaller potential difference than the positive ions. If the shunt capacity is very small and the load resistance R_L, which supplies the bias voltage to the anode wire, is very large, the pulse shape occurring at the detector output, induced by the movement of the positive ions, is given by [3]

$$U(t) = \frac{Q}{2 \cdot \pi \cdot \epsilon_0 \cdot L} \cdot \ln\{1 + ((b/a)^2 - 1) \ t/T_i\} \tag{5}$$

with $Q = M \cdot N_p \cdot e$ and $t \leq T_i$. At a time $t_{1/2} = (a/b) \cdot T_i$, which is usually much shorter than T_i, the signal reaches approximately one half of its final amplitude $U(T_i)$.

Connecting the proportional counter to a preamplifier with a fast rise time and long decay time, only the fast rise of the voltage signal $U(t)$ will accurately be reproduced. The risetime of a charge sensitive preamplifier can be shortened by increasing its feedback capacitor C_F, but its output signal $U_{out} \sim 1/C_F$ will decrease and the signal to noise ratio of U_{out} will increase.

The maximum count rate n_{max}, which a preamplifier with ac-coupling can process, is very much higher than with dc-coupling, and if pile-up losses of 1% are tolerated, is given by [3]

$$n_{max} = \frac{2}{\tau_D} \cdot \left[\frac{U_m - U_{out}}{2.6 \cdot U_{out}} \right]^2 , \tag{6}$$

156

where U_m is the maximum output voltage of the preamplifier and $\tau_D = R_F \cdot C_F$ is the decay time of the output pulse U_{out}. Note that n_{max} depends linear on $1/\tau_D$, but quadratic on $1/U_{out}$. $U_{out} \ll U_m$. n_{max} can be increased by reducing the gas gain M of the proportional counter, because $U_{out} \sim M$. n_{max} can also be increased by decreasing the preamplifiers feedback resistor R_F. This changes mainly the pulse shape and not the pulse height, which is acceptable.

3 Design of a detector system with a proportional counter

A fast gas flow proportional counter was constructed and tested [5]. To get a pulse height as high as possible even at short shaping time constants, the geometrical design was chosen in order to get short collection times T_e and T_i. Methane, which is a very fast gas, with an electron mobility of $\mu_e = 10^7$ cm$^2 \cdot$Torr$\cdot$V$^{-1} \cdot$s^{-1} and a positive ion mobility of $\mu_+ = 1.72 \cdot 10^3$ cm$^2 \cdot$Torr$\cdot$V$^{-1} \cdot$s^{-1} was used. The cathode radius is b = 10 mm, the anode radius is a = 20 μm.

The photons enter the counter parallel to the anode wire, therefore a large active length $L_a \approx L$ is possible, which leads to a low space charge density. The length of the anode wire was chosen to L = 100 mm to get a high quantum efficiency of theoretically 73.9% at λ = 4.5 nm and at 95 mbar gas pressure. This low gas pressure is the differential pressure on the entrance window of the proportional counter, when it is connected to the evacuated scanning x-ray microscope. This low pressure is advantageous, because a thin, low absorbing entrance window can be used then.

The usable detector bias range is between 1400 V and 1600 V. With a detector bias of 1450 V we calculated: T_e = 15 ns, T_i = 90 μs, $U(T_e) = 0.33 \cdot U(T_i)$ and M = $2.3 \cdot 10^5$. With these values we calculated for 10^7 detected photons/s a space charge density of $\rho \leq 1.8 \cdot 10^{-11}$ As$\cdot$cm^{-3} and $\delta U \leq 16$ μV. This voltage drop δU is very small and leads to no significant change of the gas gain, thus the space charge does not affect the detector operation.

The maximum voltage drop across the load resistance R_L, which supplies the bias voltage to the anode wire, must be small, otherwise the detector bias voltage and consequently the gas gain M varies strongly with varying photon rate. With R_L = 10 MΩ the gas gain will decrease about 30% at a detected photon rate of 10^7 photons/s at λ = 4.5 nm. This alteration of the pulse height spectrum can be tolerated because energy resolution is not required.

The parameters of the proportional counter system were measured as follows: The proportional counter was connected to a charge sensitive preamplifier, whose output was connected to a pulse shaping amplifier, followed by a 300 MHz discriminator and a ratemeter. The output pulses of the shaping amplifier are nearly Gaussian and have a FWHM of about two times the shaping time constant. The shaping time constant should be matched to the electron collection time T_e and was selected to be 20 ns. At $\lambda = 4.5$ nm it resulted in:

a) dead time of the system: 49 ns, which limits the maximum measurable photon rate to $2 \cdot 10^6$ photons/s, if a count rate loss of up to 10% is allowed.
b) quantum efficiency: 44% including 50% transmission of detector window.
c) rise time of the preamplifier output voltage step: 10 ns.
d) capacity of the proportional counter: 13 pF.
e) capacity of the counter plus input capacity of the preamplifier: 15 pF.

4 Detector system with a photodiode

Another possibility is to use a semiconductor photodiode with an integrating electronical amplifier [5]. For example a GaAsP-Schottky photodiode, Hamamatsu type G 1126-02, is a highly efficient detector. The responsive quantum efficiency was measured to $(60 \pm 7)\%$ at $\lambda = 4.5$ nm and $(69.1 \pm 0.3)\%$ at $\lambda = 2.4$ nm. This photodiode shows a good linearity over a wide range of photon rates. Photon rates up to $5 \cdot 10^9$ photons/s are measured without reaching saturation in diode current. Referring to the saturation current, as specified by the manufacturer, it should be possible to work with photon rates up to 10^{13} photons/s at $\lambda = 4.5$ nm. The minimum photon rate which can be detected is limited by the noise. The noise equivalent power of the diode is NEP = 10^{-14} W/$\sqrt{\text{Hz}}$ at room temperature. This value may be decreased, if the diode will be cooled. To get each millisecond a measurement point, a bandwidth of 1 kHz should be used and therefore a minimum photon rate of about 10^4 photons/s at $\lambda = 4.5$ nm is necessary. To record an object with 10% (1%) contrast in the image with a signal to noise ratio of 3:1, about $2 \cdot 10^6$ ($2 \cdot 10^8$) photons/s are necessary by using an integration time of 1 ms per image point.

5 Conclusions

The dead time of the proportional counter system must be reduced to increase the upper count rate limit. This can be achieved by reducing the gas pressure in the proportional counter, which decreases the electron collection time. Shaping times below 20 nsec will be used. To achieve the same quantum efficiency the length of the chamber has to be increased. Photon rates up to about $4 \cdot 10^7$ photons/s can be measured, if the dead time is 2 ns, the time distance of the bunches is 2 ns and 10% count rate losses are tolerated.

The upper count rate limit of a photon counting system decreases the DQE the closer this limit is approached. On the other hand the NEP of charge integrating systems decreases the DQE at low photon rates, but they can handle much higher photon rates with high DQE than counting detectors. Reaching photon rates of more than 10^6 photons/s in a scanning x-ray microscope semiconductor photodiodes with an integrating electronical amplifier will be a good alternative.

Acknowledgements

Appreciations are given to our colleagues of the Göttingen x-ray microscopy group, especially to G. Schmahl who makes this work possible. This work has been funded by the German Federal Minister for Research and Technology (BMFT) under the contract number 05 320 DA B 7.

References

1. U.W. Arndt: J. Phys. E: Sci. Instrum. __11__, 671 (1978)

2. W. Diethorn: A Methane Proportional Counter System For Natural Radiocarbon Measurements, Thesis (Pittsburgh, Pennsylvania, 1956)

3. W. Franzen, L.W. Cochran: In Nuclear Instruments and Their Uses, ed. by A.H. Snell, Vol. I (London, 1962)

4. J. Hendrix: In Uses of Synchrotron Radiation in Biology, ed. by H.B. Stuhrmann (London, 1982)

5. P. Guttmann: In Soft X-Ray Optics and Technology, SPIE Proceedings Vol. 733, 496 (1987)

Multislice Calculations
of Dynamic Soft X-Ray Scattering

A.R. Hare, G.R. Morrison, and R.E. Burge

Department of Physics, King's College London, The Strand,
London WC2R 2LS, UK

1. Introduction

X-rays incident upon a plane of atoms undergo partial transmission, absorption, and diffraction. The incident wave may also be scattered by subsequent atomic layers in a specimen, and so the final transmitted beam includes contributions from multiply-scattered components. Such effects are included in full dynamical scattering calculations, but are neglected in the kinematical approximation when, for soft x-rays, simple exponential absorption is generally assumed.

Rigorous solutions for diffraction of electromagnetic radiation have been obtained for a few simple objects such as spheres [1], ellipsoids, and long circular cylinders (see, *e.g.* [2]). However the analytic solutions are very complicated even for these special cases.

The multislice method of COWLEY and MOODIE [3] is a general numerical technique for calculating the dynamic scattering of high-energy electrons. This method has now been adapted to model the diffraction of soft x-rays. The initial results presented are for diffraction by spheres, but this method is applicable to an arbitrary scattering object whose optical constants at the required wavelengths are known.

2. The Multislice Method

A 3-dimensional specimen is decomposed into a series of 2-dimensional slices of thickness Δz. The phase and amplitude change due to propagation through one slice is represented as taking place at a single plane, with the wave then propagating a distance Δz in vacuum to the next slice by Fresnel diffraction. In the limit of an infinite number of slices with $\Delta z \to 0$, this is consistent with a quantum mechanical description of the interaction [4].

The complex refractive index $n_r(x,y)$ of the specimen is calculated from the semi-empirical atomic scattering factors of HENKE et al. [5]. The wavelength of "water-window" soft x-rays is large compared with atomic and molecular structure, so a uniform specimen may be treated as a homogeneous dielectric medium. The complex transmission of a single slice is

$$q(x,y) = \exp\left\{-i\,k\,n_r\,\Delta z\right\},\tag{1}$$

where $k = 2\pi/\lambda$ is the angular wavenumber at wavelength λ.

Propagation in the z-direction by Fresnel diffraction is given by convolution with the function

$$p(x,y) \;=\; \exp\left\{-ik\,\tfrac{x^2+y^2}{2\,\Delta z}\right\} \tag{2}$$

assuming the paraboloidal approximation for small-angle scattering [4]. Hence the wavefunction ψ_n at the nth slice is calculated recursively from the previous slice by

$$\psi_n \;=\; \left[\psi_{n-1}\cdot q_{n-1}\right] \otimes p_{n-1,n} \quad, \tag{3}$$

where $\otimes$ denotes the convolution integral. For computational efficiency, the convolution is evaluated in Fourier space as

$$\psi_n \;=\; \mathcal{F}^{-1}\left\{\mathcal{F}\{\psi_{n-1}\cdot q_{n-1}\}\cdot\mathcal{F}\{p_{n-1,n}\}\right\} , \tag{4}$$

where $\mathcal{F}$ denotes Fourier transformation. The discrete Fourier transform of the propagation function is obtained by sampling the analytic result

$$P(u,v) \;=\; \exp\left\{i\pi\lambda\,\Delta z\,(u^2+v^2)\right\}, \tag{5}$$

where $P(u,v) = \mathcal{F}\{p(x,y)\}$. The other transforms are performed using FFT algorithms [6].

The slice thickness and sampling are related by the constraint that spread of the wave due to Fresnel diffraction within each slice should not significantly exceed the (x,y) resolution required. Hence from (2)

$$k\,\frac{(\Delta x)^2+(\Delta y)^2}{2\,\Delta z} \;\ll\; \pi \tag{6}$$

$$i.\,e. \quad (\Delta x)^2 + (\Delta y)^2 \;\ll\; \lambda\,\Delta z \;. \tag{7}$$

The discrete convolution of an N-sample and an M-sample array requires an $(N+M)$ element array for the result if no data is to be lost. Strictly applied to multislice calculations, this would lead to an unacceptably large array after many slices. In practice, it has been found sufficient to begin with an array considerably larger than the slice thickness and specimen dimensions require for the first slice, and then to maintain this size throughout the calculation. Blurring of the transmission function (1) by convolution with a small window reduces the FFT bandwidth and minimizes "ringing" at edges in the specimen, also helping to reduce the size of the arrays required.

Scattering under broadband spatially-coherent illumination has been simulated by making repeated monochromatic calculations at small wavelength intervals ($\sim 0.5\,\text{Å}$) and storing the cumulative intensity. The exit wavefunction from the specimen may be propagated to the image plane of the model system by Fresnel diffraction through vacuum, photoresist, or other homogeneous media.

3. Results

Intensity distributions have been calculated for spheres of vinyl acrylic copolymer latex ("haloflex", manufactured by Imperial Chemical Industries Plc, U. K.) under spatially-coherent soft x-ray illumination, with subsequent propagation in vacuum and PMMA photoresist. The optical constants were calculated assuming a formula $(C_2H_3Cl)_n$ and density 1.4 $g\,cm^{-3}$ for haloflex, and $(C_5H_8O_2)_n$ with density 1.2 $g\,cm^{-3}$ for PMMA. The array sizes required restrict the calculations to 1-dimensional slices through 2-dimensional specimens, so the "sphere" is represented by slices through a circular cylinder.

Figure 1 shows intensity profiles calculated by multislice iteration and exponential absorption for a haloflex sphere, with subsequent propagation into vacuum. The illumination was assumed spatially-coherent and uniformly intense over the range 20–44 Å. This was modelled by making 33 monochromatic calculations spanning this range, with appropriate optical constants for each wavelength [5], and averaging the intensity distributions. The 0.23 μm diameter "sphere" was embedded in a 12 000 element 1-dimensional array representing a 0.8 μm field to prevent aliasing, and 20 slices were taken for each wavelength. The complete multislice calculation required $\sim$8 hours cpu-time on a Digital Equipment Corporation VAX 11/785 computer.

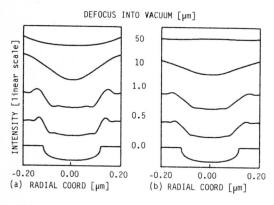

Figure 1. 0.23 μm diameter haloflex "sphere" illuminated by 20–44 Å plane-wave x-rays and defocused into vacuum:
(a) multislice calculation
(b) exponential absorption

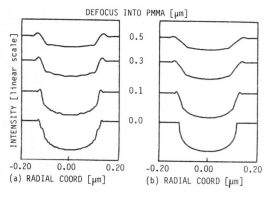

Figure 2. 0.23 μm diameter haloflex "sphere" illuminated by 20–44 Å plane-wave x-rays and defocused into PMMA photoresist:
(a) multislice calculation
(b) exponential absorption

The multislice calculation predicts a much more prominent bright fringe at the edge of the shadow of the sphere than that obtained by defocusing the output from the exponential absorption calculation. A similar result is obtained after propagation through $\leq 0.5\,\mu$m of PMMA (Fig. 2), and is strongly reminiscent of the bright fringe observed experimentally by MYRING and CLARKE [7] in contact microradiographs of similar polystyrene spheres. The multislice calculation is in much better agreement with the observed Fresnel diffraction around the sphere. In both Fig. 1 and Fig. 2 it is clear that simple exponential absorption does not adequately describe the transmission of soft x-rays through even these simple specimens.

4. Conclusions

The multislice approach provides a general numerical method for calculating the soft x-ray transmission of a specimen, including diffraction effects within the specimen which are neglected in models assuming only exponential absorption.

The calculations for vinyl spheres and the reported results [7] demonstrate that significant soft x-ray diffraction effects can occur in specimens of sub-micron thickness, so that it is invalid to model such specimens by a simple multiplicative transmission function. This has important consequences for the interpretation of high resolution patterns generated in contact microradiography and microlithography, since even when there is intimate contact with the recording medium there may be no simple relation between the recorded intensity distribution and the object structure.

The results shown in Fig. 1 suggest that if the wavefunction at the exit surface propagates in vacuum even a distance of only a few μm before being recorded, then the diffraction effects will make the difficulties of image interpretation even greater.

5. Acknowledgements

Thanks are due to W. J. Myring for valuable discussion of these calculations and comparison with experimental results. The computer simulations were performed at the King's College London Computer Centre and Physics Research Computer Centre. ARH is supported by a SERC(CASE) studentship in conjunction with Vacuum Generators (Scientific) Ltd, U. K. This paper was typeset using TEX.

6. References

1. G. Mie: Ann. die Physik 30, 377 (1908)
2. M. Born, E. Wolf: Principles of Optics, 6th. ed., (Pergamon, Oxford 1980) p. 664
3. J. M. Cowley, A. F. Moodie: Acta Cryst. 10, 609 (1957)
4. J. M. Cowley: Diffraction Physics, 2nd. ed., (North-Holland, Amsterdam 1981) p. 225
5. B. L. Henke, P. Lee, T. J. Tanaka, R. L. Shimabukuro, B. K. Fujikawa: Atomic Data and Nuclear Data Tables 27, 1 (1982)
6. E. O. Brigham: The Fast Fourier Transform, (Prentice-Hall, Englewood Cliffs, N. J. 1974)
7. W. J. Myring, D. T. Clarke: private communication

Fabrication and Focal Test of a Free-Standing Zone Plate in the VUV Region

H. Kihara[1], Y. Shimanuki[2], K. Kawasaki[2], Y. Watanabe[3], S. Ogura[3], H. Tsuruta[4], and Y. Nagai[5]

[1] Jichi Medical School, School of Nursing, Minamikawachi, Tochigi 329–04, Japan
[2] Department of Oral Anatomy, School of Dental Medicine, Tsurumi University, Tsurumi 2–1–3, Yokohama 230, Japan
[3] Canon Research Center, CANON INC., Morinosato-Wakamiya 5–1, Atsugi, Kanagawa 243–01, Japan
[4] Department of Materials Science, Faculty of Science, Hiroshima University, Hiroshima 730, Japan
[5] Laboratory of Molecular Biology, School of Veterinary Medicine, Azabu University, Fuchinobe, Sagamihara, Kanagawa 229, Japan

1. Introduction

Much effort has been put into developing X-ray microscopy in the wavelength region between 2.37 and 4.47 nm (absorption edges of oxygen and carbon, respectively), because of the high contrast of biological materials against water. In the longer wavelength region, however, relatively few efforts have been made. We have studied the feasibility of sorting out the importance of the development of X-ray microscopy in longer wavelength region, and point out possiblities of the utilization of VUV light, such as the use of the phosphor absorption edge [1,2].

To develop the X-ray microscope in the VUV region, the use of a free-standing zone plate is inevitable. Thus, we have fabricated a free-standing zone plate, and tested its focal and magnifying features at the Institute for Molecular Sciences (Okazaki, Japan)[3]. The test system was used for the observation of mesh and hard tissues (bones and teeth).

2. Fabrication procedure of zone plate

The original fabrication procedure has been reported elsewhere [1]. It was improved in several points. The present characteristics are: $n=312$, $f=150$ mm at 8 nm light, $dr_n=0.98$ μm, Au thickness of 2 μm with a central mask of 0.2 mmϕ.

3. Test system for zone plate

Focal and magnifying features of the fabricated free-standing zone plate have been tested at the beam line 6A2 of UVSOR at the Institute for Molecular Sciences (Okazaki). At UVSOR, a dedicated source of synchrotron radiation, soft X-rays of 100 mA with the critical wavelength of 5.69 nm (750 MeV) are supplied [3]. The zone plate was placed downstream of a plane-grating monochromator [4] through which monochromatized light from 8 nm to the visible region is utilized [4]. The test system is illustrated in Fig. 1. The beam

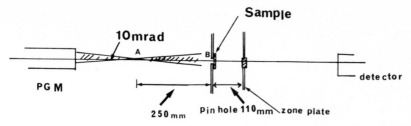

Fig. 1. A diagram of the test system for a free-standing zone plate.

from the monochromator was focused at point A with 1 mm square, and diverged with 10 mmø rad. Photons passing through a pinhole (point B) were used for the experiments. Samples (meshes or hard tissues) were fixed directly on a pinhole with silverpaste. The magnifying ratio was changed by altering distances between the zone plate and the detectors. To detect the focused images, multichannel plate (Hamamatsu Photonics, MCP-F2222) [5] or photographic films were used. For the first experiment, an Agfa 10E56 film for holographic imaging was used, as it had the finest grain size. However, the sensitivity of the film is too low to allow detection within a few hours. Then we carried out preliminary tests of several films under the same conditions using X-rays from CMR apparatus (Softex, type ESM). Results are summarized in Table 1, which shows a film for electron microscope, MEM, is the most sensitive. In fact, the contrast and grain size of MEM is the best, comparing Fuji EM film and Kodak EM film. In our case (developer Kodak D-19, 4 min; fixer, Kodak fixer, 10 min), grain size of a few microns was attained.

Table 1. Comparison of exposure time

Kodak 649-0	120 min
Agfa 10E56	60 min
Fuji Minicopy HR-II	2 min
Mitsubishi	6 s

4. Results

In Fig. 2, a photograph of an image of Cu 25 μm mesh (TAAB Lab. Equip.) is shown. The mesh was mounted on a 0.4 mmø pinhole. MEM films were used to detect images focused by the zone plate. The observed wavelength was 10.5 nm. Magnifying power was set at 10. The dark parts in the upper left corner are due to zero-order diffraction, and the mesh image at the center is due to first-order diffraction. Judging from the sharpness of the mesh edges, a resolution of a few microns is attained, which is comparable to the grain size of the film. The mesh images were defocused on increasing the wavelength of the incident beam. A typical defocused picture is shown in Fig. 3a in comparison with a focused picture (Fig. 3b). Hard tissues (bones and teeth) were observed. An example of a human thigh bone with many small holes (bone cavities) is shown in Fig. 4. The area other than cavities gave scarce information, probably because of its thickness (ca. 40 μm). A technique to prepare thinner specimens is necessary.

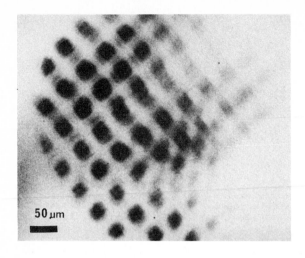

Fig. 2. Cu 25 μm
mesh pattern. The
picture was taken
with 10-fold magni-
fication. 94.8 nm.
2 hours with MEM
film.

50 μm

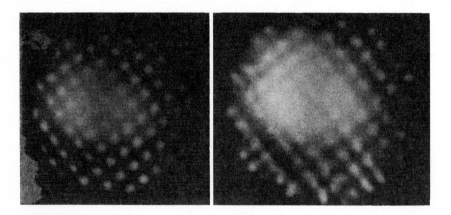

Fig. 3. Cu 25 μm mesh pattern (negative) detected by a multi-channel
plate. The fluorescent image on the screen was taken by a camera set
outside the vacuum chamber with Kodak 400 film. (a) 9.57 nm (in
focus). (b) 10.7 nm (out of focus).

5. Acknowledgement

The authors are grateful for help and encouragements from Prof. M.
Watanabe , K. Fukui, O. Matsudo and other members of staff of the
Institute for Molecular Sciences. The authors also appreciate
continuous encouragement from Profs. N. Kato and F. Oosawa. This work
was supported by a grant in aid from the Ministry of Education,
Research and Science in Japan.

166

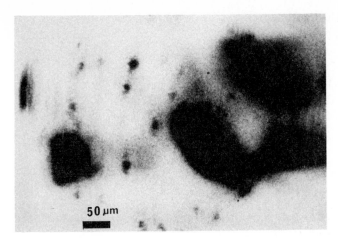

Fig. 4. Compact bone of human thigh with Haversian canals (big black holes) and bone cavities (small holes). The picture was taken with MEM film for 210 min. 9.48 nm.

6. References

1. Y. Nagai, Y. Nakajima, Y. Watanabe, S. Ogura, K. Uyeda, Y. Shimanuki, and H. Kihara: in "X-ray Microscopy. Instrumentation and Biological Applications" (ed. by P. C. Cheng and G. J. Jan), in press, Springer-Verlag (1987)
2. P.C. Cheng, H.B. Peng, K.H. Tan, J.Wm. McGowan, R. Feder, D.M. Shinozaki: in "X-Ray Microscopy" (ed. by G. Schmahl and D. Rudolph), pp.285-293, Springer-Verlag (1984)
3. M. Watanabe: Nucl. Inst. Methods, <u>A246,</u> 15-17 (1986)
4. K. Seki, H. Nakagawa, K. Fukui, E. Ishiguro, R. Kato, T. Mori, K. Sakai, and M. Watanabe: Nucl. Inst. Methods, <u>A246,</u> 264-266 (1986)
5. S. Matsuura, S. Umebayashi, C. Okuyama and K. Oba: IEEE Trans. Nucl. Sci., <u>NS-32,</u> 350-354 (1985)

Scattering Measurements
of Soft X-Ray Mirrors

K. Nakajima[1] and S. Aoki[2]

[1]Seiko Instruments Inc., Matsudo Chiba 271, Japan
[2]Institute of Applied Physics, University of Tsukuba,
 Tsukuba, Ibaraki 305, Japan

1. Introduction

Soft x-ray mirrors have been used as focussing elements in x-ray microscopes and other x-ray optics. Mirror surfaces must be figured very accurately and must be as smooth as possible to suppress unwanted x-ray scattering by surface irregularities. Since the production of supersmooth films for the soft x-ray multilayer mirror is our main concern, the most straight forward way to examine this quality is to measure the x-ray scattering in the same wavelength band which the mirror has been designed for.

In this paper, surface irregularities of the flat multilayer mirror are measured by using the scanning tunneling microscope (STM), the stylus profilometer(TALYSTEP), the reflectivity and the angle-resolved scattering with wavelength 8.34Å(Al-Kα).

2. Theory

The angular distribution of the scattered x-ray from the mirror surface is calculated by vector theory[1,2]. At first, we assume a sinusoidal surface wave whose spatial wavelength is ℓ. Then the diffraction condition is

$$n\lambda = \ell(\sin\theta_s - \sin\theta_i), \qquad\qquad (1)$$

where θ_i and θ_s is the angle of incident and diffracted beams, respectively and λ is the wavelength of x-ray and n is the order of diffraction. If the scattered beam is distributed over a wide range, the surface irregularity is composed of surface waves with various spatial wavelengths.

The squared spatial Fourier components of surface roughness $Z(x)$ which is called the power spectral density function $W(p)$ can be represented by

$$W(p) = \frac{\ell}{2L} \mid \frac{\ell}{2\pi} \int_{-L}^{+L} \exp(ipx)Z(x)dx \mid^2, \qquad\qquad (2)$$

where $p=(2\pi/\ell)$ is the wave number and -L and +L are the edges of the semiinfinite square scatterer. We measure directly the surface roughness and derive $W(p)$. There exists a one-to-one

168

Table 1   Surface roughness values for three mirrors

Mirror	rms roughness($\overset{\circ}{\text{A}}$) TALYSTEP	STM	Peak to valley($\overset{\circ}{\text{A}}$) TALYSTEP	STM
Si	8	—	56	—
1000$\overset{\circ}{\text{A}}$-Mo single layer	16	1.6	67	26.4
Mo/Si multilayer	20	5.0	87	47.5

relationship between the angular distribution of scattered
x-ray and the power spectral density function. The intensity
of radiation scattered into the solid-angle $d\omega_s$ for nonmagnetic
surface is given by

$$\frac{1}{Ir}(\frac{dI}{d\omega})_s \, d\omega_s = (\frac{L}{2\pi})^2 \cdot 4k^4 \cdot W(p) \cdot \cos\theta_i \, \frac{\cos\theta_s}{(\cos\theta_s + \sqrt{\varepsilon - \sin^2\theta_s})} \, d\omega_s \,, (3)$$

where Ir is specular reflected intensity, $d\omega_s = \sin\theta_s d\theta_s$, $k = 2\pi/\lambda$
and ε is the effective permittivity of the surface material.

3. Experimental

Multilayer films are made of molybdenum and silicon, and the
substrate is single crystalline silicon. Each mirror has a
cross section 35x40mm^2 and a thickness of 10mm. These films are
prepared by electron beam evaporation in a conventional vacuum
system with pressures around 10^{-7} Torr. Thickness is controlled by an oscillating quartz monitor.

Figure 1 shows schematically the geometrical arrangement for
the angle-resolved x-ray scattering and the reflectivity
measurement. The scattering chamber is evacuated to less than
10^{-2} Torr. The x-ray source having an Al target is operated at
13KV and 7μA. The incident x-ray beam is defined by a slit with
100μm wide. The mirror is mounted on a remotely controlled
holder to permit the adjustment of the position of the mirror.
The incident and reflected beam are detected by a proportional
counter with a slit of 90μm width. The angular distribution of
scattered x-rays is recorded by measuring the scattering
intensity as a function of its angular position. The scattered
x-ray intensity is measured in a range of ±60' around specular

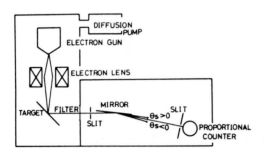

Fig.1 Schematic of set-up
used for x-ray scattering
and the reflectivity
measurement.

reflection at every 2' interval of the angular position. The reflectivity of the mirror is measured as a function of the grazing angle of the x-ray beam to the mirror surface.

4. Results

Mo/Si multilayer is synthesized on the 1000Å-Mo coated film and has 90Å of d-spacing. This multilayer coated film also contains 5 layer pairs of thicknesses t_{Mo}=45Å and t_{Si}=45Å. Figure 2 shows topographs of the 1000Å-Mo single layer film and the top of the molybdenum surface of a Mo/Si multilayer film with the STM. There exist micro-irregularities on the order of angstroms on the surface of a Mo single-layer film, while crystal grains of 200-300Å size appear on a Mo/Si multilayer coating. Table 1 shows the rms roughness values for three mirrors. The spatial wavelengths (=ℓ) measured by the TALYSTEP and the STM are more than 6.3μm and 70Å, respectively[from the equation(2)]. There appears to be little correlation between the rms roughness values measured by two methods.

Figure 3 shows the reflectivities for Si, Mo single layer film and Mo/Si multilayer film. The reduction of the reflectivity due to the surface irregularities is calculated by

$$R=R_0 \exp[-(4\pi z \cos\theta_i/\lambda)^2], \tag{4}$$

where R_0 is the reflectivity of a smooth surface; z is the rms surface roughness determined by the TALYSTEP[3]. The reflectivity curves deviate from the calculated ones with the increase of the grazing angle. This may be due to the reason that the reflectivity becomes insensitive to the surface roughness at the very small angle. The angular distributions of x-ray scattering from multilayer mirror are shown in fig.4. Since the scattering becomes sensitive to the irregularity in the shorter spatial wavelength with the increase of the grazing angle, the scattered intensity at the large scattering angle in Fig.4(b) is large compared with that of fig.4(a).

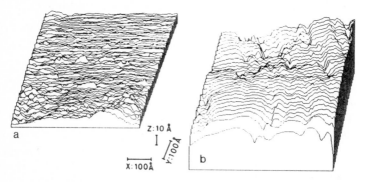

Fig.2 Topographs of Mo single layer film (a) and the top of Mo surface of Mo/Si multilayer film (b) with the STM.

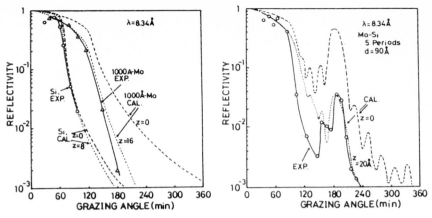

Fig.3 Reflectivities of AlKα for Si and Mo single layer film and Mo/Si multilayer film.

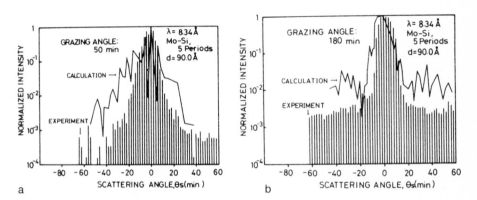

Fig.4 Scattered beam profiles for a Mo/Si multilayer film at the grazing angle of 50' and 180'.

5. Summary

We have synthesized Mo/Si multilayer mirror by the electron beam evaporation and evaluated the characteristics. Mirror surfaces are composed of the spatial surface wavelength of 70 Å or more. The spatial wavelengths become longer by evaporating. The surface roughness also becomes worse by the growth of crystal grains. Surface irregularities in the shorter spatial wavelengths have effect on the reflectivity and the angular distribution of scattered intensity with increase of the grazing angle.

171

References

1. P.A.J de Korte and R.Laine: <u>Appl. Opt.</u> 18, 236 (1979)
2. E.L.Church, H.A.JENKISON and J.M.Zavada: <u>Opt. Eng.</u> 16, 360 (1977)
3. E.Spiller, A.Segmüller and R.P.Haelbich: <u>Annals New york Academy of Sciences</u> 188 (The New York Academy of Sciences, 1980)

Scattering, Absorption, and a Detailed Look at the Field Near an Absorbing Particle

D. Sayre

IBM Research Center, Yorktown Heights, NY 10598, USA

1. INTRODUCTION

A radiating system of charges which is radiating into field-free space is necessarily an exporter of energy, and must therefore have access to energy to keep radiating. However, if it is radiating into a region of space already occupied by a strong field of exactly its own frequency, then it may be either an importer or exporter of energy, depending on the phase relationship of its field to the external field. The effect can be large enough that in the case of a passive system of charges not having access to an energy source, enough energy can be imported from the external field to power the radiation by the system, and even to supply further energy if the system is dissipative as well. If we consider these two importations of energy separately, we see the system as a radiation scatterer in the first, and as a radiation absorber in the second, with the second role being dependent on the first in this way of viewing the matter. The efficiency of the energy extraction process is such in the soft x-ray region that the absorbed energy can considerably exceed the scattered energy.

In this paper we go over these ideas in more detail for the case where the external field is a plane-wave field and the radiated or scattered field is spherical. In addition we will examine the field, and the energy flow in the field, in detail in the region of a system closely resembling a carbon atom as the frequency of the external field passes through the carbon absorption edge.

2. BASIC RELATIONS

We consider a plane-wave field $\exp(-i2\pi z/\lambda)$. We imagine a sphere of arbitrary radius $r = R$ centered at an origin at $z = 0$, and note that in the presence of the plane-wave field alone, the integrated energy flow entering the sphere is zero. We now place at the origin a vanishingly small system radiating a spherical-wave field $(a/r)\exp(-i2\pi r/\lambda)$, $\mathbf{a} = a\exp(-i\alpha)$, α the phase-delay of the spherical field relative to the plane-wave field, both fields measured at the origin. The addition of the spherical field to the plane-wave field produces a field for which the integrated energy flow into a sphere surrounding the radiating system may not be zero. In fact we find:

Inward energy flow = $2\lambda a \sin \alpha - 4\pi a^2$.

The second term is the energy exported by the spherical wave. The first term is the imported energy arising from the interaction between the plane and spherical waves, and is seen to be largest when the spherical wave lags the plane wave by 90°. We may summarize the situation thus:

Received from plane wave $= 2\lambda a \sin\alpha = \sigma_e$
Radiated out in spherical wave $= 4\pi a^2 = \sigma_s$ (1)
Balance dissipated $= 2\lambda a \sin\alpha - 4\pi a^2 = \sigma_a$.

At the right we have identified the expressions with the appropriate cross-sections (extinction, scattering, and absorption) for the system. That the expression are cross-sections follows from the fact that they have the dimensions of area and that the plane wave has unit amplitude. Since $a \sin\alpha$ is $\mathrm{Im}(\mathbf{a}^*)$, the expression for σ_e can be recognized as a special case of the optical theorem.

We note that the ratio of energy received from the plane wave to energy expended in the spherical wave is of the order of λ/a, provided that α is not too far from 90°. For atoms, as we shall see below, this ratio is of the order of λ/r_0, or 10^6 in the soft x-ray region. (Here r_0 is the classical radius of the electron.) This justifies the statement made earlier that large energy effects can result from radiating into a strong external field. In this connection we may note that when $\alpha \approx -90°$, large amounts of energy are <u>added</u> to the plane-wave field (case of stimulated emission).

We note also that a relation involving σ_a, σ_e , and σ_s, but not a, is easily derived from (1):

$$\sigma_a < \sigma_e \leq \lambda\sqrt{\sigma_s/\pi} \quad .$$

The relation is illustrated for the carbon atom in Fig. 1.

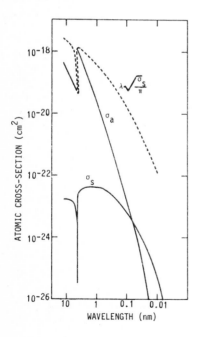

Fig. 1. σ_s, $\lambda\sqrt{\sigma_s/\pi}$, and σ_a for carbon

3. THE FIELD NEAR A CARBON ATOM

In the remainder of the paper we shall examine the carbon atom in more detail. The behavior of **a** for a radiating system consisting of a single harmonically bound electron is shown in Fig. 2(a). The magnitude is zero at zero frequency and r_0 at infinite frequency; the phase lag increases with increasing frequency from zero to 180°, passing through 90° at resonance. The behavior of a carbon atom, with its 6 quantum-mechanically bound electrons, is more complex, but has been tabulated in the x-ray region by HENKE [1] and is shown in Fig. 2(b). Essentially 2 of its electrons are making the transition of Fig. 2(a) in this frequency region, its 4 more loosely bound electrons having already done so at lower frequencies. The radiation from the 2 electrons initially opposes, then reinforces, the radiation from the 4. Because of this, atomic scattering cross-sections have a local minimum, rather than maximum, near their soft x-ray absorption edges (Fig. 1), the depth of the minimum varying with the atom species and absorption edge.

Given the information in Fig. 2(b), it is straightforward to compute the field distribution in the neighborhood of a carbon atom at different frequencies in the soft x-ray region (Fig. 3). The field effects consist of alternating shadow and brightness in confocal paraboloids, with shadow generally occupying the downstream paraboloid. In the immediate vicinity of the absorption edge, however, there is perceptible hollowing of the downstream shadow, and in the case of carbon there is even a brief wavelength interval in which inversion of shadow and brightness occurs. There is a minimum in the size of the field effects just below the absorption edge, in the case of carbon producing almost complete disappearance of these effects at $\lambda = 4.46$ nm. Field effects thus vary considerably with wavelength. Since contact micrographs are based on these fields, interpretation of contact images should be done with these facts in mind.

The maps in Fig. 4 are similar, but show the inward-directed component of the energy flow in the vicinity of a carbon atom (energy flow = $-2\pi\lambda E^2 \text{grad}\varphi$, E = field magnitude, φ = field phase). Throughout most of the wavelength region the energy for the absorbing particle is supplied mainly from the downstream region of the field, supporting the intuitive view that absorbers more frequently operate by preventing downstream energy escape than by enhancing upstream energy intake.

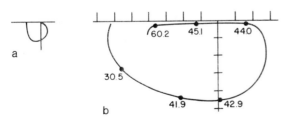

Fig. 2. Behavior of **a**: (a) for a harmonically bound electron, (b) for a carbon atom. Both figures are drawn to the same scale (axis marks in units of r_0). Numbers in (b) give wavelengths in Å.

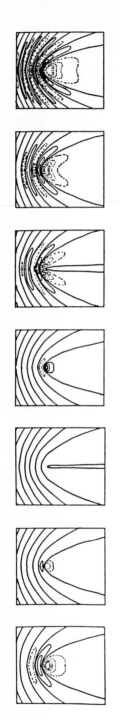

Fig. 3. Field intensity in the neighborhood of a carbon atom. The incident beam is directed vertically downward. The maps cover 18nm vertically and 12nm horizontally; the atom is located one-third of the distance down the maps. The wavelengths, from left to right, are 6.02, 4.51, 4.46, 4.40, 4.29, 4.19, and 3.05nm. The maps show $(I_C - I_0)/I_0$, where I_C and I_0 are the field intensity with and without the carbon atom present. Contours are at intervals of 5×10^{-6}, with negative contours dotted. The paraboloidal solid contours are the zero contour lines.

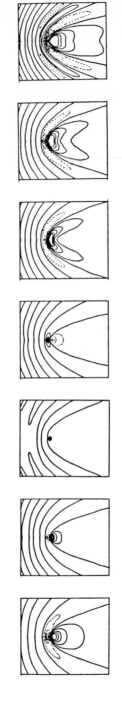

Fig. 4. Like Fig. 3, but showing the change in the inward-directed component of energy flow when a carbon atom is present. Contours are at intervals of 3×10^{-6}.

Finally, although we have spoken of Figs. 3 and 4 as referring to carbon atoms, it is more accurate to speak of them as referring to systems closely resembling carbon atoms. This is because atoms are treated approximately in Henke's tables, as spherically symmetric point structures.

REFERENCES

1. B.L. Henke, P. Lee, T.J. Tanaka, R.K. Shimabukuro, B.K. Fujikawa: In
   Low Energy X-Ray Diagnostics, ed. by D.T. Attwood and B.L. Henke, AIP Conf.
   Proc., Vol.75 (American Institute of Physics, New York 1981) p.340

X-Ray Zone Plate with Tantalum Film for an X-Ray Microscope

M. Sekimoto[1], A. Ozawa[1], T. Ohkubo[1], H. Yoshihara[1], M. Kakuchi[1], and T. Tamamura[2]

[1]NTT LSI Laboratories, Atsugi-shi,
  Kanagawa 243–01, Japan
[2]NTT Opto-electronics Laboratories, Atsugi-shi,
  Kanagawa 243–01, Japan

1. Introduction

X-ray microscope is an useful tool in biology because of the advantages of an inherent contrast mechanism. The performance of an x-ray microscope depends on the quality of its zone plates. The requirements for zone plate material are 1) x-ray absorptivity, 2) mechanical strength, and 3) ease of fine pattern fabrication. Hitherto, zone plates have usually been fabricated by electroplating of Au [1-3]. However, large diameter free-standing zone plates with high resolution have not been fabricated due to their low mechanical strength and complicated fabrication process. Recently we have developed an x-ray mask with Ta absorber for fine pattern replication [4]. Tantalum can meet the requirements for the free-standing zone plate material.

This paper describes a fabrication technique for a x-ray zone plate using a Ta absorber, and optical experimental results at Photon Factory.

2. Tantalum Zone Plate

Tantalum was chosen as the pattern material of free-standing zone plate. The properties of Ta are compared with that of Au in Table 1. Its advantages are high x-ray absorptivity and its strength which is four times that of Au. In addition, reactive ion etching (RIE) can be applied to fabricate fine patterns.

Table 1   The properties of Ta and Au

	Ta	Au
X-ray absorptivity(dB/µm)		
λ = 2 nm	179	167
λ = 3 nm	465	419
Tensile strength(kg/mm²)	53.2	13.3
Capability of RIE	possible	impossible

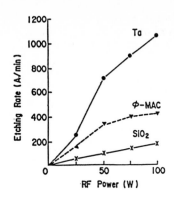

Fig.1 Reactive ion etching characteristics using $CBrF_3$. Gas flow rate is 15 sccm, and etching pressure is at 0.03 torr.

Figure 1 shows the reactive ion etching characteristics of Ta, resist, and SiO_2 using $CBrF_3$ gas. Tantalum can be etched at a high speed, and high selectivity can be attained using SiO_2 as the mask.

The parameters of the largest zone plate which we have tried are as follows; the outer diameter, the innermost zone diameter, the outermost zone width, and zone number are 2 mm, 62.9 µm, 0.25 µm, and 2000, respectively. Its focal length is 424.5 mm at the wavelength of 2.33 nm.

3. Fabrication

Figure 2 shows the fabrication process for the free standing zone plate using RIE of Ta. The process begins with the deposition of a 0.5 µm thick SiN film on a Si wafer by LP-CVD. This film acts as a temporary support substrate. Then a 0.5-1.0 µm thick Ta film is deposited on the SiN film by RF sputtering, and a 0.3 µm thick SiO_2 is deposited on the Ta film by ECR-CVD. After resist pattern formation by electron beam lithography, the resist patterns are transferred to SiO_2 film by RIE using C_2F_6 gas.

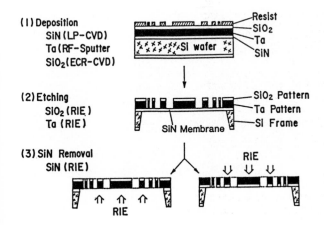

Fig.2 Fabrication process of free-standing zone plate

Then, Ta patterns are formed by RIE using SiO_2 as the etching mask. Next, the Si wafer is removed from the back with KOH solutions. Finally, the SiN temporary substrate is removed either from the front or the back by RIE using CF_4.

Resist patterns are written with a focused electron beam exposure system operating at 30 kV. The system has 18 bit D/A converters for precise positioning, and a circle generator which consisted of a deflection system to produce almost complete circles of 2 mm diameter, and software to facilitate the programming of many circle patterns. Resist patterns of 0.25 μm width were easily obtained in 0.5 μm thick positive resist ϕ-MAC. Figure 3 is a photograph of Ta zone patterns. The 0.25 μm width Ta patterns are quite well defined. To achieve better yield of zone plates, it was concluded from experiments that SiN substrates had to be etched from the back of the Ta patterns. The lower yield for front removal is probably due to stress building up in the grooves of the substrate as they are being etched. For the back removal, no grooves are etched in the substrate. Figure 4 shows a 2 mm diameter, free-standing zone plate with a Ta absorber. The outermost zone width is 0.25 μm. The circular zones are supported by 24 radial struts. In our new process, such a large free-

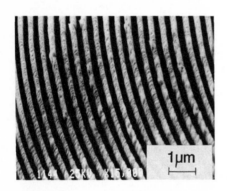

Fig.3 Tantalum patterns by RIE using $CBrF_3$. Pattern width is 0.25 μm.

Fig.4 Free-standing zone plate with Ta absorber. Outer diameter is 2 mm, and outermost zone width is 0.25 μm

standing zone plate with fine outermost zones can be easily fabricated at high yield.

4. Performance

Zone plates with Ta absorber were applied to construction of the imaging x-ray microscope by Kagoshima et al. [5]. Figure 5 shows a magnified image of test samples we took in the resolution experiments. 0.3 μm lines & spaces patterns (0.6 μm pitch grating) can be resolved. This result agrees well with the theoretical resolution limit of the zone plate with 0.25 μm width outermost zone $(1.22 \Delta Rn \doteqdot 0.3)$. Detailed experimental results are described in another paper of this volume [5].

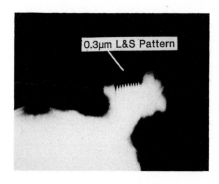

0.3μm L&S Pattern

Fig.5

Magnified image of 0.3 μm lines & spaces pattern focused at off-axis region to avoid overlapping with zeroth radiation.

5.Aknowledgements

The authors are grateful for the help and encouragements of Prof. M. Ando, and Dr. H. Maezawa of the Photon Factory of the High Energy Physics Laboratory, and Prof. S. Aoki, and Y. Kagoshima of the University of Tsukuba.

6.References

1. G. Schmahl, D. Rudolph, P. Guttmann, and O. Christ: In X-Ray Microscopy, ed. by G. Schmahl and D. Rudolph, Springer-Verlag, Berlin, p.63, Vol.43 (Series in Optical Sciences 1983)
2. D.P. Kern, P.J. Houzero, P.J. Coane, and T.H.P. Chang: J. Vac. Sci. Technol., Vol.B1, p.1096, 1983
3. H. Aritome, H. Aoki, and S. Namba: Jpn. J. Appl. Phys., p. L406, 1984
4. M. Sekimoto, A. Ozawa, T. Ohkubo and H. Yoshihara: Extend Abstracts of 16th Conference on Solid Devices and Materials, Kobe, p.23 1984
5. Y. Kagoshima, S. Aoki, M. Kakuchi, M. Sekimoto, H. Maezawa, K. Hyodo, and M. Ando: A zone plate soft x-ray microscope using undulator radiation at the photon factory (this volume)

Experimental Demonstration of Producing High Resolution Zone Plates by Spatial-Frequency Multiplication

W.B. Yun and M.R. Howells

Argonne National Laboratory, Argonne, IL 60439, USA,
Center for X-Ray Optics, Lawrence Berkeley Laboratory,
Berkeley, CA 94720, USA

1. INTRODUCTION

In an earlier publication [1], the possibility of producing high resolution zone plates for X-ray applications by spatial-frequency multiplication was analyzed theoretically. The theory predicted that for a daughter zone plate generated from the interference of mth and nth diffraction orders of a parent zone plate, its primary focal spot size and focal length are one $(m+n)th$ of their counterparts of the parent zone plate, respectively. It was also shown that a zone plate with the outermost zone width of as small as $13.8nm$ might be produced by this technique. In this paper, we report an experiment which we carried out with laser light $(\lambda = 4166\text{Å})$ for demonstrating this technique. In addition, an outlook for producing high resolution zone plates for X-ray application is briefly discussed.

2. THE EXPERIMENTAL DEMONSTRATION

The demonstration can be logically divided into two parts. In the first part, an interference pattern of two diffraction orders of a parent zone plate was recorded. In the second part, the primary focus of the recording (daughter zone plate) was recorded and its focal length was measured. Satisfactory demonstration was thought to have been made when the measured focal spot sizes and focal lengths of several daughter zone plates (each generated from different combination of m and n) were in good agreement with those predicted by theory.

The arrangement of optics used in recording the daughter zone plates is shown in Fig. 1. The parent zone plate used was one demagnified from an original zone plate drawn by means of a computer plotter. Some relevant parameters of the parent zone plate are listed in Table 1. The inner stop was purposely made to make the two-beam interference possible

Table 1   Parameters of the parent zone plate

focal length	f(4166A)=59cm
radius of inner stop	3.56mm
radius of the outer stop	4.86mm
number of open zones	24
maximum zone index	96
zonewidth of the outer-most zone	25.3 μm

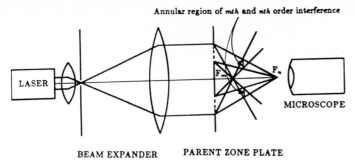

Annular region of mth and nth order interference

LASER

MICROSCOPE

BEAM EXPANDER   PARENT ZONE PLATE

Fig. 1 Schematic illustration of the arrangement used in recording the daughter zone plates. The microscope aids in the inspection and accurate location of the daughter zone plate patterns.

[1]. A laser beam magnified by the beam expander to a size larger than the size of the parent zone plate was used for illumination. The interference pattern of the mth and nth order diffractions, whose intensity distribution follows the zone plate geometry, was recorded on high resolution holographic film plate (Ilford type TE), which had a resolving power of 3000 lines/mm.

With the help of the microscope (Fig. 1), the intensity distribution of the optical field behind the parent zone plate (e.g., the foci of the parent zone plate and the interference patterns) could be visually studied. Sharp focusing was observed for the foci F_1, F_2, F_3, F_4 and F_5. For foci with diffraction order higher than 6, aberrations increased quickly with the diffraction order. For foci with diffraction order higher than 8, aberrations became so strong that a clear focal spot could not be observed. This is probably related to the degree of imperfection of the parent zone plate. The observed results may be explained with the knowledge that the requirements on the degree of perfection of a zone plate (e.g., the accuracy of placement of zone positions and the circularity of zones of the zone plate) for producing good focus increases with diffraction order.

The scale on the optical bench, on which the observing microscope could slide back and forth, was used for measuring distances of interest. The focal length of the parent zone plate was measured to be $59\pm.5cm$, which was in good agreement with that calculated (Table 1). A daughter zone plate corresponding to the mth and nth order interference pattern was recorded by placing the emulsion side of the recording film plate at the theoretical distance, $2f/(n+m)$ away from the parent zone plate [1]. Exposure time was properly set to obtain optimal fringe visibility in the daughter zone plate.

For measuring the focal length of a daughter zone plate and recording its primary focus on a film plate, the same arrangement as that shown in Fig. 1 was used except where the parent zone plate was replaced by the daughter zone plate. In Table 2, the measured focal lengths of several daughter zone plates together with those predicted by theory are listed. The notation $n\&m$ in the column under daughter zone plate is used to denote that the corresponding daughter zone plate was generated from the interference of mth and nth diffraction orders of the parent zone plate. The error in the measurement of the focal lengths was mainly due to the rough scale on the optical bench. The recorded primary foci of the daughter zone plates are shown in Fig. 2b. For convenience of comparison, the microphotographs shown in Fig. 2 were arranged so that the foci in the same column should have the same focal spot size according to the theory. All the microphotographs were magnified 320 times from the original recordings for convenience of observation.

Table 2 Focal lengths of the daughter zone plates

| Daughter | Focal Length | |
Zone PLate	Measured (mm)	Theory (mm)
0 & 2	29.5± .5	29.5
0 & 3	19.5± .5	19.6
0 & 4	14.5± .5	14.7
0 & 5	12.0± .5	11.8
1 & 2	19.5± .5	19.6
1 & 3	14.5± .5	14.7
1 & 4	12.0± .5	11.8
-1 & 4	19.5± .5	19.6
2 & 3	12.0± .5	11.8
3 & 7	6.0± .5	5.9
4 & 6	6.0± .5	5.9

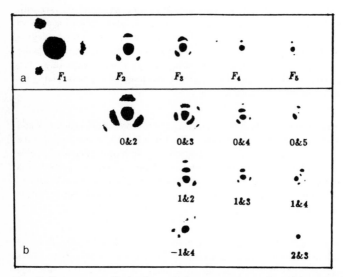

Fig. 2 Comparison of the foci of the parent zone plate (a) with the primary focus of its daughter zone plates (b). The F_i in (a) is used to denote the ith focus of the parent zone plate. The notation n&m in (b) indicates that the corresponding daughter zone plate was obtained through nth and mth order interference.

We see from Fig. 2 that quality of primary foci of the daughter zone plates resulting from 0 & 2, 0 & 3, 0 & 4, 1 & 2, 1 & 3 and 1 & 4 are fairly good. Note that all foci in a column have similar size, as expected from the theoretical analysis. Due to limitation in our setup, the foci of the daughter zone plates resulting from 3rd & 7th order interference and 4th & 6th order interference were not recorded. Sharp focusing from those two daughter zone plates, however, was observed with the microscope. Recalling that the 10th order focus of the parent zone plate suffers very severe degradation, we think that a defect cancelation effect may have played some role in the generation of the daughter zone plates. In other words, even if two foci involved in generating a daughter zone plate have some defects, the daughter zone plate generated may still have good focusing property. In addition, we also notice that the foci involved in generating the two zone plates have less aberration than the 10th order focus.

It is interesting to note the similarity between the process of our demonstration and the process of hologram recording and reconstruction. However, except for the daughter zone plates generated from zeroth order and mth order interference, the illuminating beam in our demonstration, which can be viewed as the reconstruction wave in hologram reconstruction, is different from that used at recording, where one of the two interfering beams has served as the reference beam. This leads to the results that the primary focal spot of a daughter zone plate recorded from zeroth order and mth order interference is the same as the mth order focus of the parent zone plate, while the primary focus of the other daughter zone plates is smaller than any one of the foci involved. The demagnification in the latter case is a well known result in holography.

3. SUMMARY

The principle of the techique is experimentally demonstrated. Improvement in zone plate spatial resolution up to a factor of 10 was achieved. In order to apply the technique to produce zone plates with spatial resolution down to the level of $10nm$, it is necessary to use a X-ray source of adequate coherent flux for illumination. With some positive findings from the case study[2] of a soft X-ray version of the experiment described in this context, we believe that the technique may ultimately play an important role in producing high resolution zone plates for X-ray application.

REFERENCES

1. W.B. Yun and M.R. Howells: J. Opt. Soc. Am. A, 4(1), 34(1987).

2. W.B. Yun: Ph. D. thesis (State University of New York at Stony Brook, 1987).

Overview of Activity in X-Ray Microscopy in Japan

H. Kihara

Jichi Medical School, School of Nursing,
Minamikawachi 3311–159, Tochigi 329–04, Japan

ABSTRACT

The activity in X-ray microscopy research in Japan with regard
to sources, optical systems, detectors and their biological
applications is reviewed.

1. INTRODUCTION

Activity in X-ray microscopy in Japan is divided into two
phases, before and after the utilization of synchrotron
radiation, as is the case in all other countries. Research in
this field has accelerated since the Photon Factory facility at
the National Institute for High Energy Physics was opened to
users in 1982. Since then, several research projects for the
development of X-ray microscopy have started. However, I should
say that activity in this field in Japan is not in full
operation yet, with some exceptions, considering our potential
capability. Therefore, this review also includes such potential
capabilities as well. This review is composed of three parts:
sources, optics and detectors.

2. SOURCES

2.1. Synchrotron radiation

In Table 1, synchrotron radiation (SR) facilities in Japan are
summarized [1-9]. The largest SR facility in Japan is the Photon
Factory of the National Laboratory for High Energy Physics,
located at Tsukuba, which was constructed in 1982, and has been
operated with 2.5 GeV, more than 200 mA and 41 beam lines. The
critical wavelength of synchrotron radiation with bending
magnets at Photon Factory is 2.98Å [1]. An undulator is
installed at beam line 2, whose spectral distribution and main
parameters are reported elsewhere in this volume [13]. Studies
of X-ray microscopy at the Photon Factory take place using this
beam line.
UVSOR of the Institute for Molecular Science, located at
Okazaki, is the newest SR source with 15 beam lines [2]. The
critical wavelength of sychrotron radiation at UVSOR is 56.9 Å.
Beam line 6A2 of UVSOR is used for test experiments on zone
plates.

TABLE 1

Synchrotron Radiation Facilities in Japan

Name	Energy (GeV)	Current (mA)	Critical Wavelength (nm)	Beam Line	XM exp.	Construction Year	Ref.
Photon Factory (Tsukuba)	2.5	>200	0.298	41	Yes	1982	[1]
UVSOR (Okazaki)	0.6 (0.75max)	100 (500max)	5.69	15	Yes	1983	[2]
SOR-RING (Tanashi)	0.38	100	11.2	5		1975	[3]
TERAS	0.7 (0.8max)	250	3.3	5		1981	[4]
NIJI-I (ETL/Tsukuba)	0.23 (0.6max)	340	32.2	(1)		1986	[4]
SUPER-SOR (Tsukuba)	1.0			16		planned	[5]
HISOR (Hiroshima)	1.5					planned	[6]
6GeV.SR (Harima ?)	6.0					planned	[7]
Tohoku-ring (Sendai)	1.5					planned	[8]
Baby-SOR (Tanashi)	0.65	300	1.0			under construction	[9]

The other three SR rings, the SOR-RING of the Institute for Solid State Physics, and TERAS and NIJI-I of the Electrotechnical Laboratory are also in operation [3,4].

Success in various fields of SR research has stimulated the construction of new SR sources. A 6GeV ring [7], Super-SOR [5], HISOR [6] and Tohoku-ring [8] are now planned. The Tohoku-ring is planned to incorporate a special Wiggler for producing intense far-infrared and millimeter waves, in addition to conventional insertion devices.

Another interesting source is a compact SR-ring [9]. It is of special interest for X-ray microscopy, as it is capable of operation at many laboratories for in-house use. The most characteristic feature of the machine is its bending radius of 0.50 m, which nonetheless supplies 650 MeV with current of 300 mA. Its critical wavelength is 1.0 nm. This SR-ring is under construction.

2.2 Other sources

A laser-induced plasma source at the Institute of Laser Engineering, Osaka University, has a strong potential for use in X-ray microscopy because of its high brightness (10^{12} W/cm^2 < I < 10^{17} W/cm^2) of X-ray flux. Basic characteristics of this bright X-ray source have been studied and are reported elsewhere [10]. Other sources, such as a Z-pinched X-ray source, may also be under development although the author is not aware of any.

3. Optical system

Aoki and his collaborators have been engaged in this field, and have made progress in improving a Wolter-type camera [11], X-ray holography [12], and test experiments of zone plates [13]. The first and the last topics are reported in other reports in this volume [11,13]. They have constructed a Wolter-type focussing mirror with resolution of 0.5 μm [11]. Concerning holography experiments, they have succeeded in taking diffraction patterns using the undulator at the Photon Factory, and in reconstructing a three-dimensional image by He-Ne laser [12]. They have also succeeded in taking pictures of meshes and gratings using two zone plates and the undulator. The resolution is less than 0.6 μm. Details of the fabrication of zone plates are reported elsewhere in this volume by Sekimoto et al. [14], and details of test experiments are reported by Kagoshima et al. [13].

Aritome et al. have been engaged in fabricating a zone plate with very high resolution [15]. Usually, fabrication of zone plates by electron beam lithography meets a difficulty at a size of about 500 Å because of the proximity effect due to electron scattering in a resist. To avoid this difficulty, a partially transparent zone plate has been constructed, which has wide opaque zones, reducing the proximity effect. In this configuration, the spatial resolution of the zone plate is still determined by the width of the outermost zone. A zone plate with a transparent zone width of 25 nm and an opaque zone of 150 nm has been fabricated. Details of the procedure are reported elsewhere in this volume [15].

We have been fabricating and testing another type of zone plate. Specifications and fabrication procedure of this type of zone plate are reported elsewhere, as well as that fabricated by ion beam lithography [16].

Multilayers for soft X-rays are also being extensively studied by several groups; Namioka et al. of Tohoku University, Yamashita et al. of Osaka University, and others. Their activities and progress made are reviewed elsewhere [17].

Shinohara has been working on the method of contact X-ray microscopy by replica [18]. Surface relief on a PMMA resist was replicated by plasma-polymerization film, and was observed by transmission electron microscopy. As the resolution of the replica method is less than 2 μm, the possibility of obtaining an optimal resolution of 5 nm is suggested. His results on chromosomes are reported elsewhere in this volume [18].

Projection X-ray microscopy has been tried by Yada of Tohoku University [19]. He modified a scanning electron microscope for the X-ray microscope. Using the converted instrument, an intense electron beam is exactly focused on a thin film target and resolution of 0.2-0.3 μm is easily obtained with 10-30 kV accelerating voltage. By extensive study of image contrast, he showed that Ti is very suitable as target material for most biological objects like microscopic insects or plant cells. An advantage of this system is that it is possible to get a stereo view of the specimen by tilting the specimen stage. Recent results and details are reported elsewhere in this volume [19].

A projection X-ray microscopy of Cosslett-Nixon type has been constructcd by Ohkawa et al. so as to obtain X-ray magnified Laue diffraction images and also projection shadow images having resolution of the order of 0.5 μm [20].

Nakazawa of the National Institute for Research in Inorganic Materials has tested the reflectivity of glass tube [21], and has found it good. Intensity through glass tube is 30- to 40-fold higher than that without glass tube in his experimental setup. This capability will hopefully be used to get more intense light just in front of the specimen for the scanning X-ray microscope.

Kohra has proposed the possibility of using asymmetric diffraction for X-ray microscopy [22]. The idea has been applied in dental research by Kuriyama of NBS [23]. With use of asymmetric diffraction in two directions, two-dimensional magnification will be also available [22].

Suzuki of Hitachi Co. and his colleagues use SR for the development of monochromatized X-ray computer tomography [24]. In their system, monochromatized X-ray photons pass through a specimen and are detected by a one-dimensional photodiode array. Using asymmetric Bragg diffraction of x5.0 magnification, an image with spatial resolution of less than 10 μm was obtained. A subtracted image below and above the absorption edge of molybdenum was obtained.

4. Detectors

Two potential activities in the development of detectors are briefly reviewed: imaging plate and zooming tube.

4.1. Imaging plate [25]

This type of detector, originally developed for diagnostic radiography, has been used for SR experiments. In the system, an X-ray image is temporarily stored in the phosphor screen as a distribution of quasi-stable color centers without any substantial fading for several days. The stored image is read out by a scanning He-Ne laser beam which releases photostimulated luminescence (PSL; ca. 390 nm) from color centers. The luminescence is converted into a timeseries of digital signals via a photomultiplier tube, amplifiers and an A/D converter. The image is stored and reconstructed by a computer.

The detector system has the following specifications. (i) sensitivity: 10-50 times higher than conventional high-sensitivity X-ray film because of high absorption efficiency of the phosphor screen and very low background level, (ii) no counting rate limitation via integrating detection (iii) a

dynamic range which reaches 5 orders of magnitude, (iv) a
spatial resolution of about 150 x 150 μm^2 in FWHM (which will be
improved to a few tens of micrometers), (v) an effective area
size larger than 251 x 200 mm^2, (vi) a uniform response of less
than 1.6 % in relative deviation all over the effective area,
and (vii) no appreciable image distortion (less than 1.0 %) in
two orthogonal directions.
The imaging plate was successfully applied for a diffraction
study of muscle contraction [25].

4. 2. Zooming tube [26]

Another type of detector, the X-ray zooming tube, is also
promising as a detector of X-ray microscopy. The zooming tube is
composed of two parts: a photosensitive plate (CsI or Au) and
electron multiplier. The photoefficiency of the zooming tube is
sometimes more than 100 % in VUV regions. As the emission is a
multiphotoelectric event, such a high photoefficiency is not so
surprising. The system is essentially the same as that employed
by Polack for photoelectron X-ray microscopy [27]. It is also
useful   combined with a grazing incident mirror system or
imaging zone plate microscopy.

5. Conclusion

There are many attempts at improvements X-ray microscopy going
on in Japan. Many of them have just started. I hope that all
these efforts will contribute in pushing this field up to the
international level, and more scientific results (not only the
improvement of the apparatus) will be reported in next meeting
on X-ray microscopy.

REFERENCES

1. Photon Factory Activity Report #4 (1986)
2. M. Watanabe: Nucl. Instr. Meth. Phys. Res. A246, 15 (1986)
3. T. Miyahara et al.: Particle Accelerators 7, 163-175 (1976)
4. TELL-TERAS Activity Report (1980-1986), Electrotechnical
   Lab., Tsukuba, Japan (1987)
5. T. Ishii and G. Isoyama: Technical Report of ISSP, Ser.A (No.
   1607), 1-13 (1985)
6. I. Endo, T. Ohta and M. Tobiyama: Proceedings of the Sixth
   Symposium on Accelerator Sci. Technol., in press
7. Report of 6GeV.SR Plan (1986)
8. Plan Report of Tohoku University Ring in Common Use for
   Linac, Pulse, Stretcher and Synchrotron Radiation (1987)
9. N. Takahashi: Nucl. Instr. Methods, 24/25 425-428 (1987)
10a) R. Kodama, K. Okada, N. Ikeda, M. Mineo, K.A. Tanaka, T.
    Mochizuki, and C. Yamanaka: J. Appl. Phys. 59 3050-3052
    (1986)
  b) T. Mochizuki, T. Yabe, K. Okada, M. Hamada, N. Ikeda, S.
    Kiyokawa and C. Yamanaka: Phys. Rev. A, 33(1), 525-539
    (1986)
  c) K.A. Tanaka, A. Yamaguchi, R. Kodama, T. Mochizuki and C.
    Yamanaka: J. Appl. Phys., to be published.

11. S. Aoki: this volume
12. S. Aoki and S. Kikuta: in "Short Wavelength Coherent Radiation: Generation and Application", ed. by D.T. Attwood and J. Boker (AIP, Proceedings No. 147, New York 1986), p.49
13. S. Kagoshima, S. Aoki, M. Kakuchi, M. Sekimoto, H. Maezawa, K. Hyodo and M. Ando: this volume
14. M. Sekimoto, A. Ozawa, T. Ohkubo, H. Yoshihara, M. Kakuchi and T. Tamamura: this volume
15a) H. Aritome, Y. Kasumi and S. Namba: this volume
   b) H. Aritome, K. Nagata and S. Namba: Microelectronic Engineering $\underline{3}$, 459-466 (1985)
16a) Y. Nagai, Y. Nakajima, Y. Watanabe, S. Ogura, K. Ujeda, Y. Shimanuki and H. Kihara: in "X-ray Microscopy. Instrumentation and Biological Applications" (ed. by P.C. Cheng and G.J. Jan), in press, Springer-Verlag (1987)
   b) H. Kihara, Y. Shimanuki, K. Kawasaki, Y. Watanabe, S. Ogura, H. Tsuruta and Y. Nagai: this volume.
17. T. Namioka: Rev. de Phys. Appl., in press.
18a) K. Shinohara, H. Nakano, M. Watanabe, Y. Kinjo, S. Kikuchi, Y. Kagoshima, K. Kobayashi and H. Maezawa:this volume
   b) K. Shinohara, S. Aoki, M. Yanagihara, A. Yagishita, Y. Iguchi and A. Tanaka: Photochem. Photobiol., $\underline{44}$, 401-403 (1986)
19. K. Yada and S. Takahashi: this volume
20a) T. Ohkawa, H. Hashimoto, C. Kaito and S. Urai: Jpn. J. Appl. Phys. $\underline{19}$, 2347-2353 (1980)
   b) S. Kozaki, T. Ohkawa and H. Hashimoto: J. Appl. Phys. $\underline{39}$, 3967-3976 (1968).
21a) H. Nakazawa: J. Appl Cryst. $\underline{16}$, 239-241 (1983)
   b) H. Nozaki and H. Nakazawa: J. Appl. Cryst., $\underline{19}$, 453-455 (1986)
22. K. Kohra: in "Collected Papers of Tsukuba International Meeting of X-ray Microscopy", 158-165 (1986)
23a) W.J. Boettinger, H.E. Burdette and M. Kuriyama: Rev. Sci. Instr., $\underline{50}$, 26 (1978)
   b) S. Takagi, L.C. Chow, W.E. Brown, R.C. Doblyn and M. Kuriyama: J. Dent. Res. $\underline{64}$, 866 (1985)
24a) Y. Suzuki, K. Sakamoto, H. Kozaka, T. Hirano, H. Shiono and H. Kohno: Jpn. J. Appl. Phys., submitted.
   b) K. Sakamoto, Y. Suzuki, T. Hirano and K. Usami: Jpn. J. Appl. Phys., submitted.
25a) Y. Amemiya, K. Wakabayashi, H. Tanaka, Y. Ueno and J. Miyahara: Science $\underline{237}$, 164-168 (1987)
   b) Y. Amemiya, T. Matsushita, A. Nakagawa, Y. Satow, J. Miyahara and J. Chikawa: Nuclear Instr. Methods, in press.
26. K. Kinoshita, Y. Suzuki and M. Kaneko: Proceedings of TV Society of Japan, 95-96 (1981)
27a) F. Polack, S. Lowenthal, D. Phalippou and P. Fournet: this volume
   b) F. Polack and S. Lowenthal: Scanning Microscopy Suppl. $\underline{1}$, 41-46 (1987)

Part III

X-Ray Microscopes and
Imaging Systems

The Stony Brook/NSLS Scanning Microscope

H. Rarback[1], D. Shu[1], Su Cheng Feng[1], H. Ade[2], C. Jacobsen[2], J. Kirz[2],
I. McNulty[2], Y. Vladimirsky[3], D. Kern[4], and P. Chang[4]

[1]National Synchrotron Light Source,
  Brookhaven National Laboratory, Upton, NY 11973, USA
[2]Physics Department, State University of New York,
  Stony Brook, NY 11794, USA
[3]Center for X-Ray Optics, Lawrence Berkeley Laboratory,
  Berkeley, CA 94720, USA
[4] IBM Research Center, Yorktown Heights,
  NY 10598, USA

1. INTRODUCTION

The Stony Brook / NSLS scanning microscope started operation in 1983. It is based on
the use of high resolution zone plates to create an x-ray microprobe; the specimen is then
scanned through the spot by a piezoelectrically driven stage. Synchrotron radiation from
the NSLS is used as the x-ray source, and the portion of the beam that is transmitted by
the specimen and detected by a flow proportional counter has been used to form the image.
The apparatus and results obtained with it during the first three years of operation have
been published elsewhere [1,2,3,4]. Ref.4 also includes a historical review of other work
in the field. Recent advances by other groups using scanning microscopes are presented
elsewhere in this volume [5,6].

During the past two years we have undertaken a program to rebuild the instrument
with substantially increased speed and resolution. The basis of higher speed is an x-ray
source of much greater brightness, the Soft X-ray Undulator [7]. A mini - version of this
device was used during the period between November '86 and February '87 [8,9], and
this brief period gave us a glimpse of the performance to be expected when the full scale
undulator starts operation in 1988.

The basis of higher resolution is a new generation of zone plates fabricated at IBM as
part of a joint LBL - IBM program in the development of high resolution x-ray optical
elements [10,11]. During the run with the mini - undulator we used zone plates with
outermost zone widths of 70 nm and 50 nm.

To accommodate the higher imaging speed and resolution, we completely rebuilt the
scanning stage and the control system. In this paper we describe the new instrument.
Some of the results obtained with it are presented elsewhere [12,13,14].

2. THE SCANNING MICROSCOPE

The instrument was mounted on a massive granite table with pneumatic supports to minimize vibrations. In addition, a lucite box was built around the microscope to reduce acoustic noise.

A toroidal grating monochromator provided the temporal coherence required for the zone plates. The beam emerging from the exit slit of the monochromator had enough spatial coherence to be used directly, without additional collimation. The X-ray wavelength chosen for the experiments was 3.2 nm with 0.2% bandwidth.

As in our previous work, the zoneplate, the stage and the specimen were in an atmospheric environment. To bring the beam out into the atmosphere we used two windows. A 150 nm aluminum barrier, located near the monochromator exit slit, separated the UHV and the high vacuum regions of the beamline. At the downstream end of the beamline was a 120 nm thick silicon nitride window, 200 μm x 200 μm in size, able to withstand a pressure difference of one atmosphere, while transmitting 70% of the incident x-ray flux. The distance from the exit slit to the zoneplate was 3 m. In the region between the two windows a photodiode was available to monitor the flux.

The flange that held the exit window was coupled to the beamline via welded bellows, and was supported on an xyz positioner. This arrangement allowed us to center the window in the beam, and to move it close to the zone plate which was mounted on another manipulator. The upstream face of the flange was coated with phosphor around the window and could be observed through a viewport to facilitate the alignment.

The zone plate and the order sorting aperture (OSA) downstream of it were mounted together on an xyz positioner. Their relative position was adjusted using a pair of fine screws, while the distance between them was preset with shims. With zone plate focal lengths of approximately 1 mm, the spacing was set to 3/4 mm to assure some clearance between the aperture and the specimen. The axial position of the zone plate/aperture system was driven by a stepping motor for focusing the instrument, while the position of the specimen stage was held fixed in that direction. This arrangement was chosen to simplify the design of the scanning device.

The scanning stage incorporated stepping motor driven coarse motion, piezoelectric flexure for fine motion and a laser interferometer to monitor the stage position in two dimensions. The specimens were mounted onto small stainless steel holders which were held to the stage by a small permanent magnet. The stepping motors were used primarily for moving the desired specimens into the range of the fine scan. They could be actuated from the computer terminal or, more conveniently, using an electronic trackball controller. The rough alignment of the specimen was carried out using this trackball while looking through a 40X optical microscope, with cross hairs centered on the zone plate focus.

The piezoelectric flexure elements had a range of 100 μm for an input voltage of - 1000V. A high voltage operational amplifier converted the low voltage signals generated from a CAMAC based programmable waveform generator to produce the raster scan. The non-linear and hysteretic nature of piezoelectric devices made it necessary to continuously

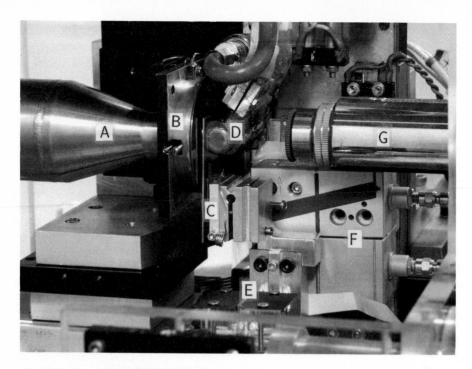

Fig.1 The Stony Brook/NSLS Scanning X-ray Microscope: (A) Vacuum snout with Si_3N_4 window at the end, (B) zone plate/OSA assembly, (C) specimen holder, (D) proportional counter, (E) laser interferometer, (F) piezoelectric translator, (G) alignment telescope.

monitor the position of the stage. A two-dimensional laser interferometer was custom designed and built for this purpose [15,16]. The interferometer was able to follow the stage over large scale motion (several mm), and provided a least count resolution of 1/10 fringe, or 31.6 nm.

Soft x rays transmitted by the specimen were detected by a flow proportional counter. The counter had cylindrical geometry with a 1 mm square side window (120 nm silicon nitride coated with 40 nm of aluminum). The anode wire was 12.5 μm diameter gold coated tungsten, and was operated at 1200 V. The gas filling consisted of 20% helium and 80% P10 (90% argon, 10% methane) at atmospheric pressure, and was flowed through at a rate of about 8 liters/hour. Due to the low capacitance of the counter, we were able to operate with a standard commercial counting chain at a rate approaching 1 MHz. At routine rates of 200 kHz the pulse height peak due to 400 eV x rays was clearly separated from noise, and it was possible to set a pulse height window on the single channel analyzer to discriminate against the 1200 eV third harmonic that was passed by the monochromator.

The proportional counter was mounted on a manual xyz positioner with which it could be positioned in the beam just behind the specimen, or removed to provide visual access to the specimen via the 40X optical microscope.

To reduce absorption losses in the air space between the vacuum window and the proportional counter, all the components were brought to close proximity. The total air path was about 2 mm, or about half an absorption length. For simplicity, we did not fill the lucite box with helium, to further reduce the losses.

3. ELECTRONICS AND COMPUTER CONTROL

The microscope was under the control of a DEC PDP 11/73 microcomputer through a CAMAC interface. The interaction with the operator was through a menu - driven software package. This allowed the operator to position the specimen, to set up the scan parameters (step size, number of rows and columns, scan speed, etc.), and to monitor the progress of the scan.

In contrast to our previous aquisition system, data was taken "on the fly". We used multichannel scalers and memory in the CAMAC crate itself. While the scan along one line proceeded at roughly constant speed, proportional counter counts were accumulated and the interferometer was monitored. After a preset number of interferometer counts the channel into which the x-ray counts were accumulated was incremented to start the next picture element. The contents of the multichannel scaler, corresponding to one line of the scan was transferred to computer memory only at the end of the line. Simultaneously with the data, a clock was read into a second multichannel scaler to monitor the "dwell time" per pixel, or the speed of the scan. We also used the photoelectron signal from the 150 nm aluminum window to monitor the flux incident on the zone plate.

Image data were stored in computer memory. The information was typically saved on disk after completion of the scan, transferred to a MicroVAX II computer, and displayed on a color graphics monitor.

4. PERFORMANCE

Installation of a new zone plate required that we align the order sorting aperture on the optic axis defined by the zone plate. Initial alignment was performed off-line under an optical microscope, while final alignment made use of the x-ray microscope itself. A 5 μm diameter pinhole was mounted on the specimen stage to explore the radiation pattern emerging from the zone plate/aperture combination. When properly aligned, the pattern is a hollow cone of illumination that is defined by the zones of the zone plate with the first order focus forming the apex. A scan of this pattern downstream of the focal plane should produce an annular shaped image. Misalignment manifests itself by deviations from circular symmetry.

Focusing of the microscope was accomplished by locating a sharp feature in the specimen, and making a series of scans across this feature with different axial locations of the zone plate. Line scans (such as Fig.2) were used to find the sharpest edge. Similar scans across artificially sharp structures, i.e. a mesh, were used to determine the resolution of the particular zone plate used. Based on the study of such scans we determined that

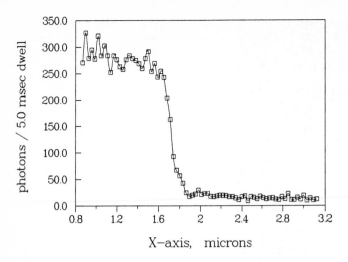

X–axis, microns

Fig. 2: Line scan of a knife edge test across a sharp edge of a nickel mesh, showing a resolution of 75 nm. Similar scans were used for focusing.

the resolution of the ZP70 is better than 95 nm and the resolution of the ZP50 is better than 75 nm. In view of the uncertainty of the edge quality and undamped high frequency vibrations, these values are consistent with the zone plates performing at the expected resolution.

Under typical operating conditions we had a count rate of about 200 kHz. An average of 400 counts/pixel corresponds to 2 ms "dwell time", or a net counting time of a little over two minutes for a 256 x 256 pixel picture. With time necessary for accelerating and decelerating the stage, for "fly back" to the start position on each line, and for computer overhead, the actual time per picture was in the neighborhood of 4 - 5 minutes.

To assure temporal stability of the incident flux we depended on the feedback system that was built to stabilize the electron beam in the region of the undulator [16,17]. Due to the extremely tight collimation of the X-ray beam derived from the undulator, even small fluctuations in electron orbit are reflected in large flux variations on the specimen. In fact, prior to the completion of the feedback system we were not able to usefully operate the microscope. The feedback was highly effective in eliminating low frequency variations and long term drift. Unfortunately, the bandwidth of the system was limited to about 10 Hz, while there was significant oscillation in the electron orbit at 12 Hz. This oscillation, with a period corresponding to several pixels, generated an artificial tweed - like pattern of artifacts on many of our images. The random nature of the amplitude of these oscillations made it difficult to remove the effect by off–line image processing. Fortunately, the amplitude was rarely larger than 5% of the average count rate. Nevertheless, the elimination of this problem is of high priority for future use of the instrument.

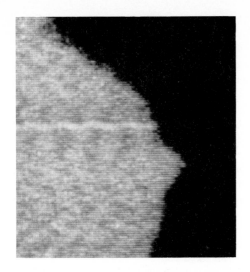

Fig.3: Part of a nickel mesh used for resolution tests, with the open area to the left. It shows a tweed-like pattern, created by beam instabilities

Although some of the components of the microscope require significant improvement before artifact-free images can be made with it in a rapid and convenient way, the running period with the mini-undulator already demonstrated its power in high resolution imaging. The necessary improvements are under way and should be completed in 1988.

5.ACKNOWLEDGMENTS

Many people contributed to the work reported here. The mini-undulator was proposed by J.Hastings, and implemented with particular dedication by the NSLS staff. We acknowledge J.Hamos, S. Rothmann and S. F. Fan for providing us with specimens. The NSLS is supported by the Department of Energy division of Material Science and Chemical Sciences. Work at Stony Brook is supported by National Science Foundation Grants BBS 8618066 and DBM 8410587. The Center for X-Ray Optics is supported by the Department of Energy.

REFERENCES

1. H. Rarback, J. M. Kenney, J. Kirz, M. Howells, T. H. P. Chang, P. J. Coane, R. Feder, P. J. Houzego, D. Kern, and D. Sayre: in _X-Ray Microscopy_, edited by G. Schmahl and D. Rudolph, (Springer, Berlin, 1984) p.203
2. J. M. Kenney, C. Jacobsen, J. Kirz, H. Rarback, F. Cinotti, W. Thomlinson, R. Rosser, and G. Schidlovsky: J. Micros.$\underline{138}$, 321(1985)
3. C. Jacobsen, J. M. Kenney, J. Kirz, R. J. Rosser, F. Cinotti, H. Rarback, and J. Pine: Phys. Med. Biol.$\underline{32}$, 431 (1987)
4. J. Kirz and H. Rarback: In Rev. Sci. Instr. $\underline{56}$, 1, (1985)
5. B. Niemann, P. Guttmann, R. Hilkenbach, J. Thieme and W. Meyer- Ilse: This volume

6. G. R. Morrison, M. T. Brown, C. J. Buckley, R. E. Burge, R. C. Cave, P. Charalambus, P. J. Duke, A. R. Hare, C. P. B. Hills, J. M. Kenney, A. G. Michette, K. Ogawa, A. M. Rogoysky, T. Taguchin: This volume

7. C. Jacobsen and H. Rarback: In International Conference on Insertion Devices for Synchrotron Sources, edited by R. Tatchyn and I. Lindau, SPIE Proc. $\underline{582}$, 201 (1985)

8. H. Rarback, C. Jacobsen, J. Kirz and I. McNulty: This volume

9. H. Rarback, C. Jacobsen, J. Kirz and I. McNulty: In the Proc. of the 5th National Conference on Synchrotron Radiation Instrumentation, Madison, Wisconsin (1987) (to be published in Nucl. Instr. and Meth.)

10. Y. Vladimirsky, D. Kern, T. H. P. Chang, D. T. Attwood, N. Iskander, S. Rothman, K. McQuaide, J. Kirz, H. Ade, I. McNulty, H. Rarback, D. Shu: In the Proc. of the 5th National Conference on Synchrotron Radiation Instrumentation, Madison, Wisconsin (1987) (to be published in Nucl. Instr. and Meth.)

11. S. A. Rishton, H. Schmid, D. P. Kern et al.: In the Proceedings of the 31st International Symposium on Electron, Ion and Photon Beams (to be published in J. of Vacuum Science and Technology B, Jan/Feb 1988)

12. S. S. Rothman, N. Iskander, K. McQuaid, H. Ade, D. T. Attwood, T.H.P. Chang, J. H. Grendell, D. P. Kern, J. Kirz, I. McNulty, H.Rarback, D.Shu and Y.Vladimirsky: This volume

13. S. F. Fan, H. Rarback, H. Ade and J. Kirz: This volume

14. H.Rarback, D. Shu, S. C. Feng, H. Ade, J. Kirz, I. McNulty, D.P.Kern, T.H.P.Chang, Y. Vladimirsky, N. Iskander, D. Attwood, K. Quaid and S. S. Rothman: Rev. Sci. Instr. (to be published about Jan. 1988)

15. D. Shu et al.: Proc. of the 5th National Conference on Synchrotron Radiation Instrumentation, Madison, Wisconsin (1987) (to be published in Nucl. Instr. and Meth.)

16. H. Rarback, D. Shu, H. Ade, C. Jacobsen, J. Kirz, I. McNulty, and R. Rosser: In X-ray Imaging II, SPIE Proc. $\underline{691}$, 107 (1986)

17. R. J. Nawrocky, J. Bittner, H. Rarback, L. Ma, P. Siddons and L.-H. Yu: Proc. of the 5th National Conference on Synchrotron Radiation Instrumentation, Madison, Wisconsin (1987) (to be published in Nucl. Instr. and Meth.)

18. S. Krinsky: This volume

Early Experience with the King's College – Daresbury X-Ray Microscope

G.R. Morrison[1], M.T. Browne[1], C.J. Buckley[1], R.E. Burge[1], R.C. Cave[1],
P. Charalambous[1], P.J. Duke[2], A.R. Hare[1], C.P.B. Hills[1], J.M. Kenney[1],
A.G. Michette[1], K. Ogawa[1], A.M. Rogoyski[1], and T. Taguchi[1]

[1]Department of Physics, King's College, The Strand,
London WC2R 2LS, UK
[2]SERC Daresbury Laboratory, Warrington, WA4 4AD, UK

1. Introduction

Within the last 10 years a number of research groups have developed working x-ray microscopes which use zone plates as high resolution diffractive optical elements and synchrotron radiation to provide the intense monochromatic soft x-ray flux needed to achieve tolerable imaging times. The instruments developed before 1984 were described in a review by KIRZ and RARBACK [1], while more recent progress at the NSLS (Brookhaven, USA), at BESSY (Berlin, FRG) and at the Photon Factory (Japan) is described elsewhere in these proceedings [2]. In this article we shall outline some of the early results obtained with the King's College–Daresbury scanning transmission x-ray microscope (STXM) during its first experimental run at the Daresbury synchrotron radiation source (SRS) in August-September 1986, immediately prior to the SRS shutdown for the installation of a new high brightness lattice (HBL) source for the electron storage ring.

2. The X-ray Microscope

Many of the important design criteria for the x-ray microscope have been discussed previously by KENNEY et al [3], so only a brief description of the main features of the completed instrument will be given here. The SRS was operating at 2 GeV with a beam current ≈ 200 mA and the STXM was installed on undulator beamline 5U. The 10-period undulator enhances the broadband soft x-ray output from the synchrotron [4] and quasi-monochromatic radiation ($\lambda/\Delta\lambda \approx 1500$) is produced by an SX700 type monochromator [5] consisting of a plane mirror, a plane grating, and a spherical mirror. Initially operation was confined to wavelengths within the "water window" (2.4–4.5 nm). The beam leaves the monochromator at an angle of 4° to the horizontal and a pair of orthogonal exit slits are used to define the effective x-ray source for the STXM. Alignment of the x-ray beam along the axis of the microscope is achieved with the aid of phosphor screens, and a secondary emission photodiode is allowed to intercept the outer edges of the beam so that changes in the x-ray flux during image acquisition can be monitored and allowed for when attempting any quantitative interpretation of the image contrast.

Focusing of the x-ray flux to form a small spot was achieved with zone plate replicas formed in 100 nm thick gold films on 100 nm Si_3N_4 substrates using the techniques of contact x-ray lithography [6]. The master zone plate pattern had been written using contamination lithography in a scanning transmission electron microscope [7] and had a diameter of 47 μm with an outer zone width of 75 nm. The overall focusing efficiency of the zone plate replicas into the positive first order is $\approx 3\%$, somewhat below the theoretical maximum for a perfect Fresnel zone plate but still a significant improvement on that possible when using a carbon master directly as the imaging element. The combination of a central stop formed from a 17 μm diameter gold disk on the zone plate itself and a coaxial collimating aperture 700 μm downstream is used to reduce the amount of undiffracted radiation which is incident on the specimen and which can cause an undesirable background signal. The alignment of the collimating aperture is quite critical to the success of this procedure and the three-axis mechanism used to achieve it can be seen in Fig. 1. Monolithic flexure joint hinges and a mechanical reduction of 10:1 are used to provide fine motion with minimal backlash.

When acquiring images a high resolution mechanical stage is used to move the specimen in a rectangular raster perpendicular to the x-ray beam. This stage is also shown in Fig. 1 and consists of an orthogonal pair of piezoelectric transducers providing fine motion with a minimum step size of 11nm over a field size of 45 μm. These are mounted on a stepper-motor driven coarse stage which allows the full area of a standard 3 mm diameter electron microscope grid to be explored and which also provides 5 mm of axial motion so that the specimen may easily be brought to the correct focus setting. The inherent nonlinearity of the piezoelectric transducers used in the fine stage is counteracted by an analogue feedback loop which uses

Figure 1. Zone plate collimator and specimen stage. The various parts are labelled as follows: (A) collimating aperture mount; (B) specimen mounting blade; (C) fine stage piezoelectric transducers; (D) coarse stage; (E) optical microscope for alignment

linear variable differential transformers (LVDT's) to sense the position of the stage, while the coarse stage drives have a ministepping option which allows a minimum position increment of only 0.08 μm, equivalent to $1/16^{th}$ of a full step.

The x-ray flux transmitted by the specimen is detected by a gas flow proportional counter mounted transverse to the optic axis and operated at atmospheric pressure with a 6:1 mixture of the gases P10 and helium. This detector has successfully operated at count rates up to $2 \times 10^5 \, s^{-1}$, although in practice the rates attainable during image acquisition were quite significantly less than this.

The main features of the stage control and data acquisition system are summarised schematically in Fig. 2. Overall control is provided by a BBC model B microcomputer which communicates with the hardware over the IEEE-488 bus, an industry standard bus chosen to allow any future upgrade of the computer system to be achieved without requiring any other significant modifications to the system. A suite of BASIC programs was written to provide a relatively user-friendly menu-driven environment for the operator, whereby all the necessary scan conditions can easily be established or modified. The time critical processes associated with data acquisition were all written in assembly language so that dwell times of 4 ms per pixel could be achieved before the software overhead became a limiting factor. All the important scan parameters are stored with the image, which is displayed as a count rate scaled to lie in the range 0–255 and can be viewed using either a linear grey scale or user-definable colour map.

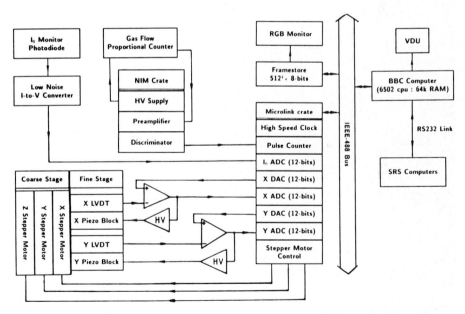

Figure 2. A schematic diagram of the stage control and data acquisition system

3. Results

The first specimens to be examined using the STXM were chosen for their suitability as test objects which would allow an assessment of the microscope performance in terms of the contrast, signal to noise ratio, and resolution that could be achieved. In this context it was important that the objects be relatively simple, well characterised, and in some sense typical of the form of specimen which might later prove to be of more than academic interest. Images from four such specimens are shown in Fig. 3.

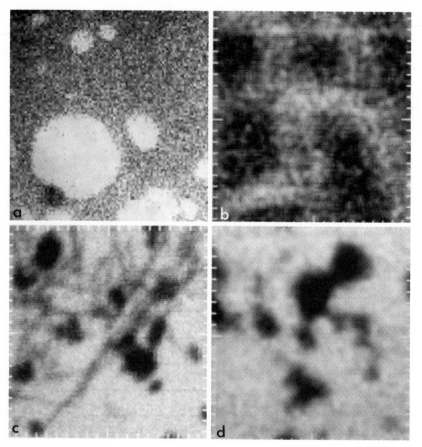

Figure 3. Test images obtained with the x-ray microscope operating at a wavelength of 3.1 nm. In each case the image size is 100 × 100 pixels and the field size is 4.9 μm
a) 100 nm aluminium on a holey carbon TEM specimen
b) Heavy-metal stained thin section of striated frog muscle, a TEM specimen supplied by Dr.M. Stewart, MRC Laboratory, Cambridge
c) Hydrated cement (Ca_3SiO_5) particles, supplied by Dr.R.J. Oldman, ICI plc
d) Vinyl chloride spheres, supplied by Dr.R.J. Oldman, ICI plc

204

The aluminium coated holey carbon specimen shown in Fig. 3a is perhaps the simplest specimen that was examined. Holey carbon films are used as standard test specimens in the transmission electron microscope (TEM) and are usually only a few tens of nanometres thick, but by evaporating a thin metal coating (usually 100 nm of aluminium) at normal incidence the absorption of soft x-rays is increased and image contrast is improved. Provided the metal coating is thin it is also likely that the hole edges will be significantly better defined than those associated with, say, the bar of a specimen support grid, and resolution measurements can be made by taking radial line scans across a hole edge. Figure 3b shows the heavy-metal stained section of a striated frog muscle which was examined to allow a direct comparison to be made with the capabilities of the TEM. This specimen is really too thin ($< 100\,\mathrm{nm}$) to exploit properly the potential of the x-ray microscope for imaging biological material, for although there is sufficient contrast to distinguish the broad A-bands the narrow Z-bands are much less clear. The other two specimens represented in Fig. 3 were both wet when placed in the x-ray microscope, although in the absence of a proper wet-cell it is likely that they both dehydrated during examination. They are examples of the type of non-biological specimens which the x-ray microscope should be particularly well suited to studying, since it should be possible to observe dynamically over a period of hours the changes which occur in the physical or chemical structure of the hydrated state, at a resolution which is beyond the capability of the optical microscope. Both these specimens should give good contrast within the "water window", because of the high calcium concentration in the cement and the carbon present in the vinyl chloride polymer. In addition the size of the vinyl chloride spheres ($\approx 200\,\mathrm{nm}$) means that they represent a good early test of the resolution of the STXM.

3.1 Assessment of Contrast and Signal to Noise Ratio

In each case the image contrast C is calculated using the formula $C = (I_m - I_s)/I_m$, where I_m and I_s are the mean signal levels measured over apparently uniform sub-regions of background medium and specimen respectively. In practice, signal fluctuations within any of these sub-regions can be accounted for simply in terms of the Poisson statistics associated with the detected x-ray photons, so the signal to noise ratio S/N is determined by the relation $S/N = (I_m - I_s)/(I_m + I_s)^{1/2}$. The contrast and signal to noise ratio were measured for three of the four images shown in Fig. 3, and these data are summarised in Table 1. The difference between the measured image contrast and that predicted using the soft x-ray scattering factors tabulated by HENKE [8] allows an estimate to be made of the zero order radiation incident on the specimen, and suggests that the total flux in the first order focused spot is approximately twice the total unfocused radiation passing through the collimating aperture, in agreement with direct measurements by KENNEY $et\ al$ [9].

Table 1. Measured contrast and signal to noise ratios for STXM images

Specimen	Contrast	S/N
Aluminium coated holey carbon	0.2	4
Hydrated cement	0.6	11
Vinyl chloride spheres	0.26	5

It is clear from both the visual appearance of the images in Fig. 3 and from the measurements presented in Table 1 that the signal to noise ratio was disappointingly low. This was due largely to insufficient coherent soft x-ray flux from the monochromator when attempting to achieve near diffraction-limited imaging conditions. The consequence of this was that the detected x-ray count rates were generally only a few kilohertz, necessitating dwell times of over 100 ms per pixel even to achieve the S/N ratios shown in Table 1.

3.2 Assessment of Resolution

The resolution was assessed by two methods: first, by imposing a circular aperture on the two-dimensional Fourier transform of the image and progressively reducing the range of spatial frequencies present in the image until a perceptible loss of image sharpness was observed; secondly, by direct measurement of the edge sharpness in images of metallised holey carbon films, such as that shown in Fig. 3a. The Fourier filtering method suggested a value for the resolution somewhere between 100 nm and 200 nm, but because of the low S/N ratios referred to earlier the edge profiles measured directly from single line traces were too noisy to confirm

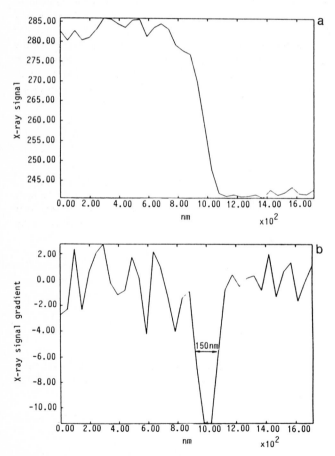

Figure 4. a) Average radial section across the hole shown in Fig. 3a   b) Average signal gradient obtained by differentiating the trace shown in Fig. 4a

this. However, by fitting circular arcs to the edges of possibly quite irregularly shaped holes and then taking radial averages over these arcs it is possible to reduce considerably the effects of noise and obtain average radial sections such as that shown in Fig. 4a. The distance over which the signal changes from 75% to 25% of the maximum intensity difference across the edge yields a direct estimate of 100 nm for the width of the line spread function, while Fig. 4b shows the average signal gradient across the hole edge, a function which is proportional to the line spread function itself. The full width half maximum of this curve is 150 nm, in satisfactory agreement with the value expected after convolving the coherent point spread function for the zone plate with the Gaussian image of the effective x-ray source.

4. Conclusions

During the first few weeks of operation the King's College–Daresbury x-ray microscope successfully produced its first images from a range of different specimens, with a resolution which was close to theoretical expectations for the imaging conditions used. The main operational problem was the low signal to noise ratio of the images obtained, but the installation of the HBL source at the SRS and the re-coating and cleaning of the elements of the monochromator on line 5U should considerably alleviate this difficulty. Work is also in progress to improve the specimen stage and facilitate a helium containment system for operation below the nitrogen K-edge, to upgrade the computer control system by replacing the BBC microcomputer with an IBM-PC/AT capable of much faster and more sophisticated on-line processing of the image data, and to design a wet cell with which it should be possible to study dynamic processes in aqueous media.

5. Acknowledgements

We are grateful to the Science and Engineering Research Council, UK, for their financial support, and would particularly like to thank Dr.M. Stewart of the MRC Laboratory, Cambridge, and Dr.R.J. Oldman of ICI plc for their help with the provision of suitable test specimens. ARH and AMR were supported by SERC postgraduate (CASE) studentships. This paper was typeset using TEX.

6. References

1. J. Kirz, H.M. Rarback: Rev. Sci. Instrum. 56 1-13 (1984)
2. See for example the papers in this volume by: H.M. Rarback *et al*, B. Niemann, G. Schmahl and D. Rudolph, Y. Kagoshima *et al*
3. J.M. Kenney, M.T. Browne, C.J. Buckley, R.E. Burge, R.C. Cave, P. Charalambous, P.J. Duke, A.G. Michette, G.R. Morrison, K. Ogawa, A.M. Rogoyski: to be published in X-ray Microscopy 86, (Springer, New York, 1987)
4. M.W. Poole, I.H. Munro, D.G. Taylor, R.P. Walker, G.V. Marr: Nucl. Instrum. Meth. in Phys. Res. 208 143-148 (1983)
5. H. Petersen: Optics Communications 40 402-406 (1982)
6. C.J. Buckley, M.T. Browne, P. Charalambous: in E-beam, X-ray and Ion-Beam Techniques for Submicrometer Lithography IV, Proc SPIE 537 213-217 (1985)

7. C.J. Buckley, R. Feder, M.T. Browne: in Short Wavelength Coherent Radiation: Generation and Applications, Am. Inst. Phys. Conf. Proc. 147 368-381 (1986)

8. B.L. Henke, P. Lee, T.J. Tanaka, R.L. Shimabukuro, B.J. Fujikawa: Atomic and Nuclear Data Tables 27 (1982)

9. J.M. Kenney, M.T. Browne, C.J. Buckley, R.E. Burge, R.C. Cave, P. Charalambous, P.J. Duke, A.R. Hare, C.P.B. Hills, A.G. Michette, G.R. Morrison, K. Ogawa: in preparation

The Göttingen Scanning X-Ray Microscope

B. Niemann, P. Guttmann, R. Hilkenbach, J. Thieme, and W. Meyer-Ilse

Universität Göttingen, Forschungsgruppe Röntgenmikroskopie,
Geiststraße 11, D-3400 Göttingen, Fed. Rep. of Germany

Abstract

A scanning X-ray microscope (SXM) is in operation at the BESSY electron storage ring. Images with 0.1 µm resolution have been obtained. The scanning time is less than 10 ms per image point at a measured photon rate of up to 10^6 photons/s. Images with some minutes total scan time are shown. The SXM is compared to the imaging x-ray microscope (XM).

1 Introduction

The scanning x-ray microscope uses the synchrotron radiation generated by the electrons in a bending magnet and can operate at higher count rates too, e.g. if it were connected to an undulator. It is designed to work also at higher spatial resolution, when zone plates with higher resolution are used.

The concept of the Göttingen scanning X-ray microscope has already been published earlier [1,2,3]. In the following text some additional details and results are described. The microscope consists of two parts: a zone plate linear monochromator and a computer controlled scanning stage.

2 The monochromator

The zone plate linear monochromator simply consists of three elements: an entrance pinhole near the synchrotron source, which is comparable to the entrance slit of a grating monochromator, a condenser zone plate, which is the dispersing element, and a monochromator pinhole, which corresponds to the exit slit of a grating monochromator. The source delivers polychromatic synchrotron radiation through the entrance pinhole, which is 100 µm in diameter and at 43 cm distance, and illuminates the condenser zone plate at 12 m distance.

The entrance pinhole is remote controlled. The pinhole adjustment is reproducible to about 50 µm. In the horizontal direction it can be moved about 25 mm, thus it can be removed from the source, and the first tests of the scanning X-ray microscope were done without the entrance pinhole being in the line of sight [3]. In the vertical direction the pinhole can be moved about 2 mm. The source position is always within this range. The optimum position of the pinhole in front of the source is adjusted as follows.

The synchrotron source emits visible light into a large angle, therefore the pinhole always transmits some light. This light is focussed with a quartz lens onto an ordinary photodiode and its current is measured. A maximum photodiode current can easily be found by adjusting the pinhole. Next the adjustment is repeated with x-rays, which are measured with the proportional counter [4] of the scanning stage. As the x-rays are much more collimated, this adjustment is more sensitive and accurate.

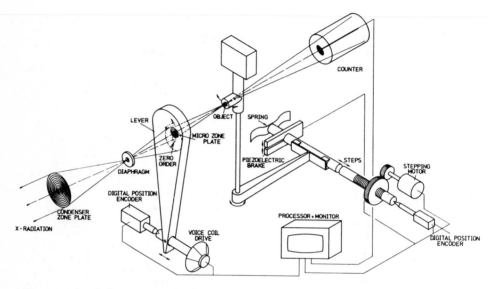

Figure 1. Schematic of the scanning X-ray microscope.

The condenser zone plate KZP4 (KZP5) [5] has a diameter of D = 2.8 (2.5) mm and images the entrance pinhole to a spot of d = 5 μm diameter. In the position of the image spot a monochromator pinhole of the image spot diameter d is placed. The radiation passing the pinhole has a monochromaticity M of approximately M ≈ D / 2·d, or, as was shown in detailed calculations [6], M = 200, corresponding to a coherence length L of L = M · λ . This is sufficient, because the connected scanning stage contains micro zone plate MZP3 with 250 zones [7], requiring a minimum L of roughly L ≈ 250 · λ/2.

The FWHM of the BESSY source is about 0.16 mm x 0.3 mm. Using the entrance pinhole, which is smaller than the source, the total flux delivered to the microscope is reduced to about 1/5. The heat load applied to the condenser zone plate

210

decreases by this factor, as does the spatial incoherent flux through the monochromator pinhole. However, the spatial coherent flux emerging from the monochromator pinhole in 1% bandwidth is not changing at all, but it is diverging in a larger solid angle from a smaller spot size, because the space phase volume remains constant.

This coherent flux has to be collected in order to produce a diffraction-limited spot size with the micro zone plate in the scanning stage behind the monochromator.

3 The scanning stage

The scanning stage is computer controlled. The computer concept already dates back to 1979 and still uses DEC LSI 11 microprocessors and a conventional CAMAC module system. The CAMAC controller contains a comparatively slow LSI 11/2 microprocessor. The CAMAC program is written in assembly language and the run time to go through the program loop is about 1 ms for each image point. This can be shortened by simply employing a faster model of the microprocessor, if necessary.

For each image point the processor calculates the intensity to be displayed on the monitor according to the calculated photon rate. Then a corresponding byte is transmitted to a host computer, which stores the bytes of one line. One byte indicating the location of a reference point is transmitted among the intensity bytes of each line.

The moment when a line has been scanned completely is indicated to the host. At once the host shifts all bytes of the whole line until the location of the reference point is centered. Next the host transmits the centered bytes to the image memory of the DE ANZA image display processor and the line can be seen on a monitor. Hereby the host works in an interrupt mode. Interrupts occur whenever the CAMAC controller has transmitted another byte to the host. Thus an image of the scanned object can be seen nearly in real time.

4 Calculated and measured photon flux

The next calculations are based on the following realistic parameters: 500 mA beam current, 0.5% bandwidth, 4.5 nm wavelength, 5% condenser zone plate efficiency and 0.16 mm x 0.3 mm FWHM source size.

The number of photons delivered from BESSY is about 2.3×10^{13} photons/s at 1 mrad horizontal angle and 1.2 mrad FWHM vertical angle. Of this flux 8×10^9 spatially coherent photons/s can be delivered to the monochromator according to the criterion given in [8]. The zone plate linear monochromator then delivers 4×10^8 spatially coherent photons/s.

In the microscope further flux losses occur at the micro zone plate MZP3, which has only 3% efficiency, and at the foils of the entrance window, the object chamber and the proportional counter entrance window, which are altogether 13% transmitting. Thus up to 1.6×10^6 photons/s can be expected as the measurable count rate with a scan spot of 60 nm and with an object on a thin foil in an evacuated scanning stage.

The flux losses can be decreased by using better transmitting foils and by improving the efficiency of the zone plates. Especially the efficiency of the micro zone plate has to be increased, for instance by processing it to a phase zone plate, this has already been done for the condenser zone plate [9].

5 Estimated Scan Time

In the following the SXM will be compared to the imaging x-ray microscope (XM) at BESSY [10]. The XM contains nearly the same optical components with about the same efficiencies and transmittances as the SXM. The XM contains a thin window at the front, a condenser zone plate of higher numerical aperture, a monochromator pinhole, an object chamber and a micro zone plate, and a detector, which is a photographic emulsion and has about the same efficiency (DQE $\leqslant$ 60 % [11]) as a proportional counter [4] in the SXM. Using the condenser zone plate KZP3 in the XM, which is at 15 m distance from the BESSY source, KZP3 collects about $U = 1/4$ of the photons delivered from the corresponding source point. The image spot of the source point produced by the condenser is about 6 μm in diameter and within this area A the radiation has a monochromaticity of about 700, which is $T = 3.5$ times better than the monochromaticity used in the SXM. The photons in the area A are used for imaging an object in the XM. Now we can easily calculate the ratio F of the scan time to the exposure time for an object of size A and for the same signal-to-noise ratio S/N. As the XM uses all photons focussed in the area A, F is simply $F = R /(U \cdot T)$, where R is the number of incoherent photons divided by the number of spatially coherent photons. For the parameters of BESSY given above we get $R = 3000$ and thus $F = 200$, if the SXM works as designed and really uses all spatially coherent photons. If an object of size A is imaged with 60 nm resolution we get 100 x 100 image points. A typical exposure time in an XM for an image with a good S/N ratio is of the order of 1 to 10 seconds, and therefore we need a total scan time of 3 to 30 minutes in the SXM, if it works optimally.

Nevertheless, the radiation dose applied to a specimen in a SXM is about 6 to 20 times less than in an XM, depending on the efficiency of the micro zone plates and the detectors.

6 Results

The maximum measured photon rate is 5×10^4 photons/s with micro zone plate MZP3 working at diffraction-limited

Figure 2: Spores *Dawsonia superba* imaged at 4.5 nm wavelength. Image size: 7 x 5 μm².

Figure 3 (left): Picture showing a part of a grating, scanned at a rate of 3.5 x 10⁴ photons/s and using the entrance pinhole. Grating constant: about 2.7 μm. All parameters are as described in sect. 4.

Figure 4 (right): The same object as Fig. 3, but scanned at a rate of 0.3 x 10⁶ photons/s using no entrance pinhole. Scan time: about 3 minutes.

resolution and with a dried specimen in the focal plane. This is about 30 times less than is calculated above. Further checks of the adjustment will be done to understand this discrepancy.

The image of Fig. 4 was obtained earlier [3] under the following conditions at 4.5 nm wavelength at 400 mA electron beam current: No entrance pinhole was used, therefore the monochromaticity M depends on the actual source size. It was about 0.3 x 0.6 mm² in size, resulting in M ≈ 70. Condenser zone plate KZP4 accepted about 0.27 mrad of the radiation and illuminated a monochromator pinhole 18 μm in diameter. Micro zone plate MZP3 produced a scan spot of 0.1 μm. The scanning system was evacuated. The object was surrounded by a small pinhole, which eliminated most of the minus first order radiation of the micro zone plate.

Each image is processed in the following manner: As there is background radiation superimposed on the image, a constant offset is subtracted from the data. Thus the minimum displayed intensity level is reduced to zero. Next, all data values are multiplied by a factor to use the full range of gray levels on the monitor. The image can be filtered before being photographed.

7 Future development

Up to now the microscope can handle only dried material. A special chamber will be constructed with which specimen can be examined at normal air pressure. The wavelength used will be changed from 4.5 nm to 2.4 nm, where water is more transparent.

8 Acknowledgement

We thank our colleagues very much for numerous discussions, especially D. Rudolph and G. Schmahl , who initialized this project, and V.Sarafis, who supplied the spores investigated in this work. This work was supported by the Stiftung Volkswagenwerk and has been funded by the German Federal Ministry for Research and Technology (BMFT) under contract number 05 320 DA B 7.

9 References

1. B. Niemann, G. Schmahl and D. Rudolph: "Status of the Scanning X-Ray Microscope", Proc. SPIE, Vol. 316, p. 106-108 (1981)

2. B. Niemann: "The Göttingen Scanning X-ray Microscope" in: X-Ray Microscopy, eds. G. Schmahl and D. Rudolph, Springer Series in Optical Sciences, Vol. 43, (Springer Berlin, Heidelberg 1984), p. 217-225

3. B. Niemann: "The Göttingen Scanning X-Ray Microscope", Proc. SPIE, Vol. 733, p. 422-427 (1986)

4. P. Guttmann and B. Niemann: "A Detector System for High Photon Rates for a Scanning X-Ray Microscope", this volume

5. J. Thieme: "Construction of Condenser Zone Plates for a Scanning X-Ray Microscope" in: X-Ray Microscopy, eds. G. Schmahl and D. Rudolph, Springer Series in Optical Sciences Vol. 43, (Springer Berlin, Heidelberg 1984), p. 91-96

6. J. Thieme: "Theoretische Untersuchungen über Kondensorzonenplatten als abbildende Systeme für weiche Röntgenstrahlung", Diplomarbeit, Göttingen, 1984

7. G. Schmahl, D. Rudolph, P. Guttmann and O. Christ: "Zone Plates for X-Ray Microscopy" in: X-Ray Microscopy, eds. G. Schmahl and D. Rudolph, Springer Series in Optical Sciences, Vol. 43, (Springer Berlin, Heidelberg 1984), p. 63-74

8. M. R. Howells: "Possibilities for X-Ray Holography Using Synchrotron Radiation" in: X-Ray Microscopy, eds. G. Schmahl and D. Rudolph, Springer Series in Optical Sciences, Vol. 43, (Springer Berlin, Heidelberg 1984), p. 318-335

9. R. Hilkenbach and J. Thieme: "Phase Zone Plates as Condensers for the Göttingen Scanning X-Ray Microscope", Proc. SPIE, Vol. 733, p. 422-427 (1986)

10. D. Rudolph, B. Niemann, G. Schmahl and O. Christ: "The Göttingen X-Ray Microscope and X-Ray Microscopy Experiments at the BESSY Storage Ring" in: X-Ray Microscopy, eds. G. Schmahl and D. Rudolph, Springer Series in Optical Sciences, Vol. 43, (Springer Berlin, Heidelberg 1984), p. 192-202

11. L. Jochum: "Detective Quantum Efficiency of Film Emulsions" Proc. SPIE, Vol. 733, p. 492-495 (1986)

Status of a Laboratory X-Ray Microscope

D. Rudolph, B. Niemann, G. Schmahl, and J. Thieme

Universität Göttingen, Forschungsgruppe Röntgenmikroskopie,
Geiststraße 11, D-3400 Göttingen, Fed. Rep. of Germany

X-ray microscopy experiments using x-ray optical systems are
performed with synchrotron radiation of electron storage
rings. The highest resolution obtained up to now in such
experiments is about 50 nm. The exposure time to take pictures
with this resolution is a few seconds. X-ray optical elements
are under development which allow to improve the resolution to
about 10 nm. For such high resolution experiments synchrotron
radiation sources will not be replaced by other devices in the
foreseeable future.

Nevertheless, for special applications it is attractive to use
pulsed laboratory x-ray sources for x-ray microscopy. Candidates of such sources are the plasma focus [1,2], laser produced plasma sources and possibly x-ray lasers. Ideally, such
sources allow to make x-ray images with one pulse which would
be of advantage, especially to obtain clear images of living
moving cells. In addition, it is important that an x-ray
microscope with such a source is small enough to fit in a
normal biological or physical laboratory.

A laboratory x-ray microscope including a plasma focus as
pulsed x-ray source is under development as a joint effort of
the Fraunhofer Institut für Lasertechnik, Aachen, the University of Göttingen and the Carl Zeiss company, Oberkochen and
Göttingen.

Figure 1 shows the schematic arrangement of the laboratory
x-ray microscope. The vertical setup of the x-ray microscope
is shown in the right part. The x-ray condenser images the
x-ray source into the object placed on a monochromator pinhole
covered with a foil. The combination of the condenser zone
plate and the monochromator pinhole acts as a linear monochromator [3]. The x-ray condenser has a central stop to
prevent zero-order radiation from the condenser reaching the
object and first order radiation of the condenser arriving on
the direct way, via zero order of the micro zone plate, at the
center of the image field [4]. The x-ray objective (micro zone
plate) generates an enlarged image in the image plane. This
image can be recorded either on a photographic emulsion or by
a CCD-camera [5].

The object and the micro zone plate are located in air or
helium under normal pressure whereas the other parts of the
microscope are under vacuum. Object and micro zone plate can

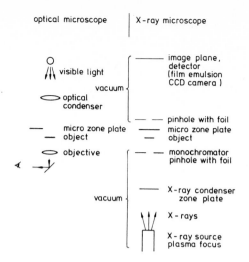

optical microscope | X-ray microscope

Fig.1 Schematic of a laboratory x-ray microscope

be turned from the x-ray microscope into an inverse optical microscope. This arrangement allows alignment of the object and prefocussing.

The plasma focus will be operated with nitrogen and is described in [2]. It is planned to perform the first experiments with a wavelength range around the Lyman α-line of the hydrogen like N VII at λ = 2,478 nm.

To evaluate the influence of the spectrum of the plasma focus on the quality of the x-ray image, modulation transfer functions (MTF) have been calculated under the following assumptions [6]: It is assumed that the x-ray spectrum consists of two components: A continuous spectrum, with constant intensity over the whole range of consideration, is superimposed by an emission line with a peak intensity of ten times the continuum intensity. The spectral width of the emission line at λ = 2,478 nm is assumed to be $\Delta\lambda$ = $2,478 \cdot 10^{-3}$ nm. The calculations were made for a micro zone plate (MZP5) which is under construction with the following parameters: radius of the innermost zone r_1 = 0,624 μm, width of the outermost zone dr_n = 18 nm, number of zones n = 300. The focal length for λ=2,45 nm radiation is f = 159 μm in the first diffractive order. Figure 2 shows modulation transfer functions for different values of the reciprocal value of the relative bandwidth of the spectrum of consideration. The different relative bandwidths can be obtained by appropriate arrangement of the linear monochromator.

The data show that a reciprocal relative bandwidth of $\frac{\lambda}{\Delta\lambda}$= 150= $\frac{n}{2}$ (n = number of zones of the micro zone plate) is sufficient for a good image quality, even half of this value can be tolerated.

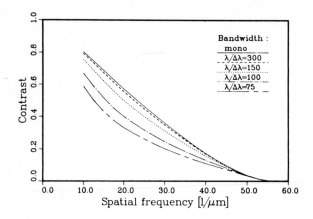

Fig.2 Modulation transfer function (MTF) for imaging with MZP5
and x-radiation of the plasma focus

In practice, the discussed relative bandwidths can be obtained
with a linear monochromator comprising a condenser zone plate
with a focal length of about 200 mm and a demagnification of
the x-ray source of 2 to 4. This leads to a tolerable size of
the laboratory x-ray microscope.

It is envisaged to get images with a lateral resolution
of $\leq$ 50 nm and to develop the spectral brilliance of the
laboratory source, the diffractive efficiency of the zone
plates and the detective quantum efficiency (DQE) of the
detector such that it will be possible to obtain images with
one pulse each.

Acknowledgements

This work is supported by the German Federal Minister for
Research and Technology (BMFT) under the contract number
13 N 5328.

References

1. G. Herziger: "X-Ray Emission from a 1kJ Plasma Focus"
   in: X-Ray Microscopy, eds. G. Schmahl and D. Rudolph,
   Springer Series in Optical Sciences, 43, (Springer,
   Berlin, Heidelberg 1984), p. 19-24

2. W. Neff, J. Eberle, R. Holz, F. Richter and R. Lebert:
   "A Plasma Focus as Radiation Source for a Laboratory
   X-Ray Microscope", this volume

3. B. Niemann, D. Rudolph and G. Schmahl: "Soft X-Ray Imaging Zone Plates with Large Zone Numbers for Microscopic and Spectroscopic Applications", Optics Communications, <u>12</u>, Nr.2, p. 160-163, (1974)

4. D. Rudolph, B. Niemann, G. Schmahl and O. Christ: "The Göttingen X-Ray Microscope and X-Ray Microscopy Experiments at the BESSY Storage Ring", see ref. 1, p. 192-202

5. W. Meyer-Ilse: "Application of Charge Coupled Detectors in X-Ray Microscopy", this volume

6. J. Thieme: "Theoretical Investigations of Imaging Properties of Zone Plates using Diffraction Theory", this volume

First Images with the Soft X-Ray Image Converting Microscope at LURE

F. Polack[1], *S. Lowenthal*[2], *D. Phalippou*[2], *and P. Fournet*[2]

[1]LURE, Université de Paris-Sud,
 F-91405 Orsay Cedex, France
[2]C.N.R.S., BP 43, F-91406 Orsay Cedex, France

1. INTRODUCTION

In X-ray contact microscopy, the image detector is usually a photoresist which is observed, after development, with an electron microscope. With this simple technique very high resolutions have been demonstrated, but quantitative measurements are very difficult to obtain [1,2]. Quantitative and real-time images can be produced on a high resolution image converter. In the fifties MOLLENSTEDT and HUANG realized such an apparatus which worked with a tungsten target X-ray tube [3]. A few years ago, we began the construction of a similar instrument to be used with soft X-ray radiation of the ACO storage ring [4].

This microscope was first set up at ACO in 1985. The resolution of the very first images was rather poor for various reasons, the main one being the alignment of the electron optics. During last year, the optical system has been subtantially modified. This improved instrument is described here, and its first images are shown. Finally we present some means of perfecting the microscope and sketches of future developments in connexion with the impending transfer of all the ACO experiments to LURE's new storage ring : Super ACO.

2. PRINCIPLE OF THE MICROSCOPE

Principles of X-ray microscopy by photoelectron conversion have been previously described in detail [4]. In the following we summarize the general ideas and main theoretical aspects.

2.1 Image Conversion

The object which is exposed to X-rays is immediately followed by a photoemissive layer. The X-ray illumination modulated by the object transmittance is proportionally converted into an electron image which is enlarged by means of electron optics. Image conversion must occur as near to the object as possible, but the photocathode layer cannot be directly deposited onto the object because the roughness of the surface would perturb the electron collection. Therefore the cathode layer is evaporated onto a thin membrane stretched on a holder ring and the object is fixed on the other side.

Resolution of the image is first affected by the Fresnel diffraction between object and detection points. It is ultimately limited by the thickness

of the support foil, the radius of the point spread function being $r_g \approx \sqrt{\lambda d}$.
Resolution is secondly limited by electron diffusion in the cathode layer.
The mean diffusion length can be evaluated from measurements of the escape
depth e of photoelectrons: r_s = e . Direct measurement of the escape depth of
typical cathodes has given e = 40 A for gold, e = 250 A for CsI [5].

Quantum efficiency η of the conversion process is in close approximation
the probability of photon absorption in the active layer

$$\eta = e\ \mu(\lambda),\qquad\qquad (1)$$

where μ is the linear absorption coefficient. By appropriate choice of photocathodes (high μ) the efficiency can be maintained between 3 % and 10 % over
the soft X-ray range (5-100 A).

2.2 Electron Image Collection and Electron Optics

The photoelectrons, emitted in 2π solid angle, are first collimated by a
uniform electric field $\mathscr{E}$ ($\simeq$ 50 kV/cm), which accelerates the electrons
axially but does not change their transverse velocities. The inclination of
an electron trajectory is thus reduced by

$$\alpha'/\alpha_0 = \sqrt{\varphi/V},\qquad\qquad (2)$$

where φ [eV] is the emission energy and V [volts] the acceleration potential
(V ~ 20 kV) . The slower the electrons, the more efficient the aperture reduction. It results that only low energy secondary electrons will be imaged
through the low aperture of electron optics (α' < 10^{-2}) ; fortunately secondary electron emission is usually over 90 % of the photocurrent.

The acceleration field gives a virtual image, with unit magnification, of
the emission distribution at the cathode plane. However, because electron
trajectories initially have very large aperture angles, and because the reduction of this aperture depends on the emission energy, the image is affected by spherical and chromatic aberration. This aberration cannot be optically corrected but its influence can be reduced :
- First, by choosing materials with a small energy distribution width w : resolution being proportional to $w/\mathscr{E}$, insulators like alkali halides
(w < 2 eV) should be preferred to metals or semiconductors [5].
- Second, limiting the image aperture α' in the electron optics cross-over
reduces the radius of the aberration spread function r_a but also the image
collection efficiency. As an order of magnitude the resolution r_a = $w/\mathscr{E}$
(r_a = 40 nm with w = 2 eV and $\mathscr{E}$= 50 kV/cm) is reached with still 10 % collection efficiency.
- Third, for high resolution, better collection efficiency could be obtained
by electron energy filtering in the electron optics [4].

Reduction of the collection efficiency affects the brightness of the final image, but as the absorption of one photon gives many secondaries, it
has little effect on the above defined quantum efficiency.

3. DESCRIPTION OF THE MICROSCOPE

3.1 Illumination

The microscope is set at the end of the 12 m "A8" beam line of the ACO storage ring. Its first element is a toroïdal mirror which focusses a demagnified image of the source onto the specimen. This nickel coated mirror of 0.9 m focal length is used at 4° of grazing incidence and eliminates the hard X-ray component of the ACO spectrum, approximately above 700 eV. Visible light, UV and radiations over 40 A are absorbed by a 0.8 µm thick aluminum foil at the entrance of the object chamber. The 2×10^{14} photons/s that ACO delivers on the condenser aperture in the 20-40 A bandwidth are focussed in a 300 µm spot. This large dimension comes from the coma of the very asymmetrical lay-out dictated by space restrictions. Illumination of the object itself is estimated to be 10^8 photons/µm^2/s. (Figs. 1,2)

3.2 Conversion Stage

Object and cathode layer are presently supported by a 0.5 µm polyimide foil stretched on a glass ring and mounted in a conductive specimen holder. This is inserted in the high tension stage provided with vertical and horizontal movements. In the specimen holder an annular fluorescent screen can be inserted for illumination adjustments. Field searching is done through an optical viewfinder. Electron collection and focussing are separate functions in this emission microscope. The lens formed by the object stage and the grounded anode only collects the electrons. It gives a virtual image and cross-over, respectively 2 and 10 mm behind the cathode plane. Because of the diverging power of the anode hole, image magnification at this stage is 0.6.

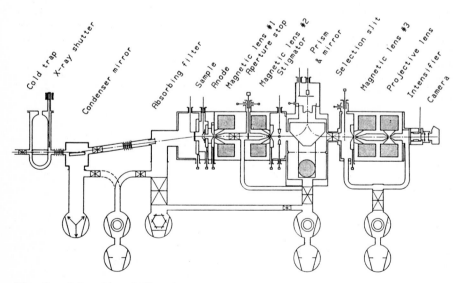

Fig. 1. Schematic of the microscope.

Fig. 2. The X-ray microscope at ACO.

3.3 Electron Optics

The electron optics is composed of two pairs of magnetic lenses horizontally laid on an optical bench. Between them a vacuum chamber provides space for the magnetic prism of an electron energy filter which is not yet in operation. The optical system has been accordingly designed. First a weak lens transfers the image to the object focal plane of lens 2. The cross-over is well outside the lens and, due to the low image magnification ($\sim$ 3), large aperture stops can be used. A stigmator is also set in this plane. The second lens acts as an objective giving a ×20 to ×100 magnification at the symmetry plane of the prism. Lens 4 is a projective lens and produces the final magnification up to ×2000. Lens 3 is a transfer lens associated with the electron filter. When this filter is not in use, either it is used in a telescopic mode with lens 2, or the prism chamber is dismounted and lens 3 not excited.

The horizontal lay-out that synchrotron radiation calls for is a source of trouble for alignment, vacuum, assembling and disassembling operations. These last points are well solved by the low play air bearings on which each optical element in mounted and which are activated for column interventions. For alignments however, first experiments have compelled us to provide the two first lenses with a movable pole piece.

3.4 Image Observation

In the microscope's first version, the magnified image was observed directly on a fluorescent screen. We used gold cathodes at that time and, above ×100 magnification, images were very faint. Since, an image intensifier of gain 30 has been added and we use CsI cathodes which make an overall gain of 250.

Images up to ×1000 magnification are easily observed. The photographic camera records the intensifier output through a tandem objective system of 2.8 F number. At ×500, typical exposures range from 1/4s to 30s according to the electron optics aperture and ACO current. The resolution of the observation system is estimated as 80 mm^2 on 20 mm field diameter.

3.5 Vacuum

Valves divide the microscope into four compartments with individual pumping. The condenser is ion pumped to $\sim 10^{-7}$mbar and protected from hydrocarbon contamination of the beam line by a cold trap. Pressure of the object chamber is held under 10^{-6}mbar by a cryopump to prevent high tension breakdown. To avoid the complication of an air lock, the object chamber is opened at specimen interchange. Pressure in the electron optics is $\sim 10^{-5}$mbar.

4. EXPERIMENTAL RESULTS

Only test objects have been imaged with the microscope. Diatoms have been used for the well-known reason that they present reproducible structures at different resolution scales, although their thickness may be a source of trouble in contact microscopy.

We began with 200 A gold cathodes evaporated on the 0.5 µm polyimide membranes. Objects prepared in this manner never gave good images. They were always blurred by strong astigmatism. After irradiation the membranes had a wrinkled aspect which is believed to result from thermal damage. Reliefs in the cathode surface locally distort the acceleration field and produce an effect similar to that of glass inhomogeneities in visible optics. It is therefore of crucial importance that the cathode surface remains smooth and flat. Later on, we changed to a cesium iodide cathode (of thickness between 500 and 1000 A) deposited on a 1000 A conductive layer of aluminum. Beside the expected gain in luminosity these objects were also found flatter. The best results were obtained when both faces of the polyimide membrane were covered with aluminum layers.

Figure 3 shows images of two different diatoms obtained with this cathode make-up. The sets of large transverse apertures (with 1.2 µm period in Fig. 3a, 0.8 µm in Fig. 3b) are well resolved. In these apertures the second periodic structure (0.3 µm period in both diatoms) is not resolved, though some substructures begin to appear in Fig. 3.b. The point resolution is thus estimated as about 0.3 µm. Presently, this resolution is chiefly limited by incompletely corrected astigmatism coming partly from lens imperfections and in particular from irregular curvatures of the cathode support membrane. The thickness of the objects used, which have a basket shape of about 5 µm thickness, may also play a role. The images of Fig. 3 also show that CsI cathodes are not structureless as they ideally should be. The background grain seems to be related to microcrystalline disorientation in the vacuum deposited CsI layer, but the mechanism of contrast generation is not clearly understood. It is interesting to notice that this grain increases under X-ray exposure as shown in Fig. 4.

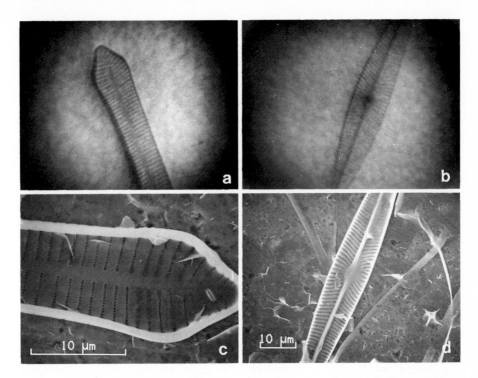

Fig. 3. Two diatoms (3a and 3b) X-ray imaged on CsI photocathodes. Exposure time 4s. Electron magnification ×300. SEM images of diatoms of the same types (3c and 3d) are given for structure comparison.

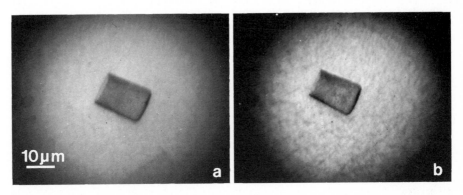

Fig. 4 Two images of the same diatom on CsI cathode taken at different times of X-ray exposure :
4a - freshly illuminated sample.
4b - same sample after 15 min of X-ray exposure, showing grain growth.

5. PLANS FOR THE FUTURE

Images which have been obtained are encouraging but far from the expected limits of resolution. Many improvements can be made in a more or less short time.

1- The weak point of the microscope is the support foil which must be very thin and still remain perfectly flat when the object is deposited. Silicon nitride is expected to have better characteristics than polyimide. It will be tried as soon as possible.

2- A gain of eight in luminosity has been observed between gold and CsI, however, at 20 A we are still on the low absorption side of Cs and I L edges. Better efficiency can be expected from RbBr or RbI which also have narrower emission energy distribution (w $\leqslant$ 0.8 eV instead of 1.6 eV for CsI) [5]. Means of reducing the cathode granularity must be investigated.

3- The illumination system also should be perfected. We require first the capability of large field illumination for alignment purposes ; second we look for monochromatic and tunable illumination which should be of great interest in X-ray microscopy.

The closure of ACO at the end of this year gives us the opportunity to make important changes. Compared to ACO, super ACO will have a 100 times greater brilliance, but moreover we expect to set the microscope on a specialized beam line. The design of such a line is presently viewed as follows. A large bandpass monochromator ($\lambda/\Delta\lambda \simeq 300$), probably grazing incidence (without an entrance slit since the source ($\sigma = 0,15$ mm) should be small enough) will cause a slight demagnification. A second optical element, also a toroïdal mirror, would augment this demagnification to about 1/4 giving about the same illumination as we presently have on ACO but monochromatic. The mirror could also be defocussed in order to increase the illuminated field during alignments.

With such an apparatus we think that it would become possible to use the absorption dependance of materials with wavelength. This "color effect" has been, and still is, a major aspect of visible microscopy of biological objects, either looking at natural specimens or at specimens specifically stained to point out particular structures. These two aspects should also be considered in soft X-rays, as we can, for instance, imagine immunolabelling with different atoms (colloïdal gold, silver, ...) on the same object or using edge structures for microanalysis [6,7].

REFERENCES

1. R. Feder, V. Mayne-Banton, D. Sayre, J. Costa, B.K. Kim, M.G. Baldini and P.C. Cheng : In X-ray Microscopy, ed. by G. Schmahl and D. Rudolph, Springer Ser. Opt. Sci., Vol. 43 (Springer, Berlin, Heidelberg 1984) p.279
2. P.C. Cheng, KH. Tan, J.W. Mc Gowan, R. Feder, H.B. Peng and B.M. Shinozaki : IN X-ray Microscopy, ed. by G. Schmahl and D. Rudolph, Springer Ser. Opt. Sci., Vol. 43 (Springer, Berlin, Heidelberg 1984) p. 285
3. L.Y. Huang : Z. Phys. 149, 225 (1957)
4. F. Polack and S. Lowenthal : In X-ray Microscopy, ed. by G. Schmahl and D. Rudolph, Springer Ser. Opt. Sci., Vol. 43 (Springer, Berlin Heidelberg 1984) p. 251

5. B.L. Henke, J. Liesegang, S.D. Smith : Phys. Rev. <u>B 19</u>, 3004 (1979

6. P.C. Cheng : <u>ESRF Workshop on X-ray microscopy</u>, EMBL, Heidelberg, Dec. 1986, unpublished.

7. J.M. Kenney, C. Jacobsen, J. Kirz, H. Rarback, F. Cinotti, W. Thomlinson, R. Rosser and G. Schidlowski : J. Microsc. <u>138</u>, 321 (1985)

Phase Contrast X-Ray Microscopy
– Experiments at the BESSY Storage Ring

G. Schmahl, D. Rudolph, and P. Guttmann

Universität Göttingen, Forschungsgruppe Röntgenmikroskopie,
Geiststraße 11, D-3400 Göttingen, Fed. Rep. of Germany

1. Introduction

Up to the end of 1986 it was always stated that to a good
approximation soft x-ray photons participate in only one
reaction with matter, i.e. absorption of the photon in an
atom, because the atomic cross sections for inelastic and
elastic scattering are 4 to 5 orders of magnitude smaller than
the atomic cross section for photoelectric absorption. But
this is only half the truth.

The atomic scattering factor, which describes the interaction
of soft x-rays with matter, consists of two parts. The
imaginary part, if_2, describes the photoelectric absorption
whereas the real part, f_1, describes the phase shift. In the
Henke tables /1/ one can see that for many elements f_1 is
considerably larger than f_2. This means that phase shift is
the dominating process for many elements and wavelengths. In
addition, the differences of f_1-values for two elements are in
general larger than the differences of f_2-values. This
observation was the motivation for two of us to propose phase
contrast x-ray microscopy /2/.

2. Experiments

Preliminary experiments for phase contrast x-ray microscopy
have been performed with the Göttingen x-ray microscope at the
BESSY storage ring /3/. These experiments were done with
partially coherent illumination of the object as shown in
Fig.1.

The x-ray phase plate in the back focal plane of the micro
zone plate consists of a metal layer (silver or gold) on a
thin polyimide layer providing a phase shift of about $\frac{\pi}{2}$ for 4.5
nm radiation (positive phase contrast for x-rays). The focal
ratio of the micro zone plate with a diameter of 55.6 micrometer is about three times that of the condenser zone plate.

The zero order radiation of the object is phase shifted by $\frac{\pi}{2}$
by the phase plate with a diameter of about 20 micrometers.

Because of this comparably large diameter of the phase plate
not only the direct radiation of the object is phase shifted

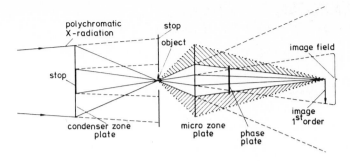

Fig.1 Phase contrast x-ray microscope with partially coherent illumination

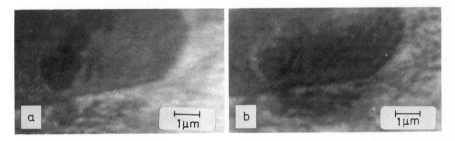

Fig.2 Part of a human fibroblast imaged with
a) amplitude contrast
b) phase contrast

but also part of the diffracted radiation of the object, namely diffracted radiation from object structures of low spatial frequency.

To illustrate x-ray phase contrast imaging, a part of a human fibroblast was imaged at first with the x-ray microscope without a phase plate (Fig.2a) and then with the same arrangement using a micro zone plate with a gold phase plate, 150 nanometers thick (Fig.2b). The cells were critical point dried.

The picture of Fig.2a was made with an x-ray magnification of 250 and an exposure time of 3 seconds with an electron beam current of the storage ring of 390 mA. The picture of Fig.2b was made with an x-ray magnification of 250 and an exposure time of 30 s with an electron beam current of 245 mA. The longer exposure time of Fig.2b is caused by the absorption of the gold phase plate.

Fig.3 shows part of a nucleus of a human fibroblast, critical point dried, imaged at 4.5 nm with the phase contrast arrangement shown in Fig.1 with a silver phase plate of 120 nm thickness. The phase plate did not have the optimum thickness

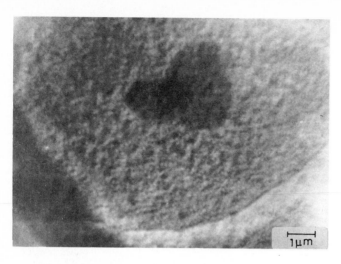

Fig.3   Part of a nucleus of a human fibroblast imaged with
        phase contrast

of 150 nm for this wavelength. The picture of Fig.3 was made
with an x-ray magnification of 250 and an exposure time of 30s
with an electron beam current of 240 mA. Further results,
especially phase contrast images of wet cells, are reported in
/4/.

3. Phase contrast x-ray microscopy with coherent illumination

To obtain the full phase contrast advantage according to
Zernike /5/ coherent illumination of the object is required.
One possible arrangement for x-rays is shown in Fig.4. The
collector and the pinhole act together as a linear
monochromator. The condenser illuminates the object with a
nearly parallel quasi-monochromatic beam. This arrangement
allows a better separation of the direct light of the object
and the diffracted radiation of the object in the back focal
plane of the micro zone plate. Thus, the diameter of the phase
plate can be much smaller and the diffracted radiation is not
affected by the phase plate. A disadvantage of this
arrangement is the necessity of three zone plates instead of
two as shown in Fig.1. But three zone plates will be tolerable
for future work when phase zone plates with high efficiency
can be used.

As already mentioned in /2/, phase contrast x-ray microscopy
offers the possibility of extending the usable wavelength
range to about 0.5 nanometer. For the wavelength range
0.62 nm $\leq \lambda \leq$ 5.74 nm calculations of the contrast of several
object structures, e.g. protein in water, have been performed
which show phase and amplitude contrast. For these
calculations, a formulation of the image contrast was used

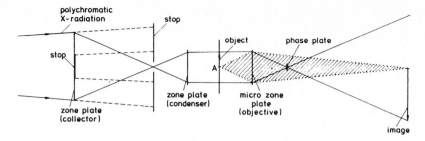

Fig.4 Phase contrast x-ray microscope with coherent illumination

which includes absorption of the object and phase plate as well as the phase shift of both /6/. In addition, dosages were calculated which are necessary to image structures with a given signal to noise ratio. Though not completed, the calculations show the following results for wet cell x-ray microscopy:

1. For wavelengths slightly larger than the carbon K-edge the amplitude contrast of organic material surrounded by water is very low. In contrast to that, the phase contrast for wavelengths slightly larger than CK_α is very high and can be used for imaging with dosages lower than those necessary for imaging with amplitude contrast in the wavelength range between the K-edges of carbon and oxygen (water window).

2. Even in the water window wavelengths are found for which imaging in phase contrast is better concerning contrast and dosage.

3. Imaging with phase contrast can be extended to wavelengths of about 0.5 nm with a dosage comparable to that necessary for imaging with amplitude contrast in the water window.

Acknowledgements

We thank P. C. Cheng, State University of New York at Buffalo, USA, for providing us with the human fibroblasts and the staff of the Berliner Elektronenspeicherring GmbH (BESSY) for providing good working conditions. This work has been funded by the German Federal Minister of Research and Technology (BMFT) under contract number 05 320 DAB.

References

1. B.L. Henke, P. Lee, T.J. Tanaka, R.L. Shimabukuro, B.K. Fujikawa: "The Atomic Scattering Factor, $f_1 + i\,f_2$, for 94 Elements and for the 100 to 2000 eV photon energy region", in: Low Energy X-Ray Diagnostics – 1981, ed. by D.T. Attwood and B.L. Henke, AIP Conf. Proc. No. 75, p. 340–388 (1981)

2. G. Schmahl, D. Rudolph: "Proposal for a Phase Contrast X-Ray Microscope", in: X-Ray Microscopy - Instrumentation and Biological Applications, ed. by P.C. Cheng and G.J. Jan (Springer, Berlin, Heidelberg), in press

3. D. Rudolph, B. Niemann, G. Schmahl and O. Christ: "The Göttingen X-Ray Microscope and X-Ray Microscopy Experiments at the BESSY Storage Ring", ed. by G. Schmahl and D. Rudolph, Springer Series in Optical Sciences, Vol. 43 (Springer, Berlin, Heidelberg, 1984) p. 192-202

4. N.G. Nyakatura, W. Meyer-Ilse, P. Guttmann, B. Niemann, D. Rudolph, G. Schmahl, V. Sarafis, N. Hertel, E. Uggerhøj, E. Skriver, J.O.R. Nørgaard and A.B. Maunsbach:"Investigations of Biological Specimens with the X-Ray Microscope at BESSY", this volume

5. F. Zernike: "Das Phasenkontrastverfahren bei der mikroskopischen Beobachtung", Zeitschr. f. techn. Physik, Nr.11, p. 454-457 (1935)

6. H. Beyer:"Theorie und Praxis des Phasenkontrastverfahrens" Akademische Verlagsgesellschaft, Frankfurt a.M. (1965)

X-Ray Fluorescence Imaging with Synchrotron Radiation

M. L. Rivers

Department of the Geophysical Sciences,
The University of Chicago, 5734 S. Ellis Avenue,
Chicago, IL 60637, USA

1. INTRODUCTION

The micro-distribution of trace elements is of great interest in fields such as geochemistry, biology and material science. The synchrotron x-ray fluorescence microprobe provides a technique to quantitatively measure trace element compositions at individual points and to construct semiquantitative two dimensional maps of trace element compositions [1,2,3].

Most of the papers in this volume deal with soft x-ray microscopes for imaging biological materials in the "water window". Soft x-ray microscopes face stiff competition from optical microscopy on the one hand, which offers great simplicity and resolution of 0.25 μm or better, and from electron microscopy on the other hand, which has a resolution as good as 1 nm, but requires that the specimen be placed in vacuum. The hard x-ray microprobe has much less competition: electron microprobes have spatial resolutions of 1 μm or so, but the detection limits are typically above 50 ppm. Ion microprobes have much lower detection limits, but quantitative results are difficult to achieve. Samples must also be placed in vacuum, and are destroyed during the analysis. Even the very simple scanning x-ray microprobe described in this paper has proven to be a very useful tool for earth scientists and biologists.

2. EXPERIMENTAL

Experiments were performed on beamline X-26C at the National Synchrotron Light Source, Brookhaven National Laboratory. The microprobe used was a "first generation" device with no focussing optics or monochromators installed. Rather, white radiation from the bending magnet was collimated with two pairs of stepper motor controlled slits located 0.5 m from the sample. The electron beam source dimensions are 800 μm (horizontal) $\times$ 200 μm (vertical) for $\pm 1\sigma$. The distance from the electron beam to the slits was 20 m, and thus the minimum spot size on the sample is 20 μm $\times$ 5 μm $\pm 1\sigma$. Actual beam sizes used were typically 30-50 μm.

The low energy photons in the incident beam were removed by the 500 μm Be windows in the beamline, and by the 20 cm air path between the end of the beamline and the sample. Additional filtering was often used, typically 100-400 μm of aluminum to reduce the flux of low energy photons which would contribute scattered background counts but not contribute to fluorescing interesting trace elements. The theoretical incident spectrum is shown in Fig. 1. The calculated flux for all photons with energies between 5 keV and 20 keV is 2×10^7 photons/sec/μm^2, or about 2×10^{10} photons/sec for the 30-50 μm beam diameters normally used. These calculated fluxes compare favorably with those measured by Underwood *et al.* [4] for their focussing microprobe which uses multilayer coated Kirkpatrick-Baez mirrors.

Fluorescent photons were detected with a 30 mm^2 Si(Li) detector positioned in the horizontal plane, 90° to the incident beam, to minimize the scattered background [5]. The detector has a resolution of about 160 eV at 5.9 keV and a maximum count rate of about 5000 counts/sec under normal operating conditions.

The sample is mounted on a Klinger Scientific translation/rotation stage with 100 mm of travel in the X and Y (scanning) directions and 25 mm travel in the Z (focussing) direction. The rotation stage is used to orient crystalline materials to eliminate Bragg diffraction peaks [6]. The stages have step sizes of 1 μm and backlash under 3 μm. The sample is observed with a Nikon SMZ-10 binocular zoom microscope equipped with a high resolution DAGE/MTI black and white TV camera. The field of view on the TV monitor is 2-10 mm, and the optical resolution on the monitor is about 10 μm.

The principle system electronics consist of a MicroVAX II computer, a CAMAC crate and a Nuclear Data ND9900 multichannel analyzer (MCA). The CAMAC crate houses 16 channels of intelligent stepper motor control, a frequency divider and scalers. The ND9900 is comprised of display controller and acquisition interface boards which plug directly into the MicroVAX Q-bus, and an external color monitor and keyboard. The acquisition interface can be connected to up to 4 NIM analog-to-digital converters (ADCs) and/or 8-input multi-channel scalers (MCSs).

The closely coupled nature of the MCA and the MicroVAX II minimizes the overhead involved in transferring spectra to the computer for processing and storage. The ND9900 display controller contains a significant amount of local intelligence, and thus permits

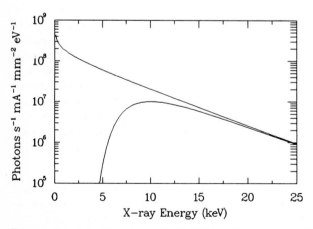

Fig. 1: Calculated flux from the X-26C bending magnet. Upper curve is the flux before any windows or absorbers. Lower curve is the flux downstream of a 100 μm Al filter

real-time spectral display and manipulation during scanning with minimal load on the MicroVAX host.

3. SCANNING

The x-ray microprobe can collect data in a variety of modes. The simplest is to move the specimen to a location to be analysed and to collect a fluorescent spectrum at that point. This "non-scanning" mode is in fact the way the microprobe was employed most of the time. It permits long collection times, typically 300 seconds per point, and hence low detection limits, typically 0.5-5 ppm by weight, depending upon the matrix and elements analysed [3,7].

Data can also be collected in one of 3 different "scanning" modes, sampling over a 1 or 2 dimensional array of pixels. The x-ray beam cannot be rastered electrostatically or magnetically as an electron or ion beam can, so the sample must be mechanically scanned in front of the stationary beam. In the first scanning mode a complete fluorescent spectrum is stored as a separate disk file at each pixel. The integrated ion chamber current at each pixel is stored with the spectrum for normalization purposes. This mode is often used simply to automate the collection of spectra at many points, without constructing an image of the sample. It maximizes the information recorded at each point but requires a large amount of disk storage for high resolution scans, and requires substantial post-processing to construct an image. The stage motion in this mode is discontinuous, with the computer moving the stages to a new pixel, stopping, and then starting data acquisition. Each spectrum must be read from the ND9900 and saved to disk before moving to the next point. The overhead involved in these operations is about 0.5-1 second, so the minimum practical total time per pixel is about 1-2 seconds.

In the second scanning mode the user defines regions-of-interest (ROIs) at the fluorescent lines of the elements to be measured. Background windows on either side of the peaks may also be defined. At each pixel the program determines the background corrected peak areas for each ROI and stores only these values. The shaded areas in Fig. 2 show the peak areas which would be stored for a spectrum similar to those from the sample in Fig. 3. The data are normalized by dividing the net peaks areas by the integrated ion chamber current at each pixel. Up to 32 ROIs be defined in the spectrum, and the data for all of the ROIs for all of the pixels are stored in a single disk file. The overhead per pixel is again about 0.5-1 second, due to the discontinuous stage motion, the communications overhead required to read the spectrum, and the processing time to extract and normalize the peak areas.

In the third scanning mode the output of the Si(Li) detector amplifier is fed into one or more single channel analyzers (SCAs), whose output is then fed into the multichannel scaler. The window on the SCA is set to the energy of a fluorescent line of interest. In this mode scanning is continuous, with the channel advance signal to the MCS being derived from the stepper motor pulses on the scanning stage. The advantage of this

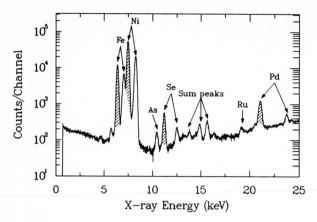

Fig. 2: Fluorescent spectrum of the mineral pentlandite ($Fe_{4.5}Ni_{4.5}S_8$) from an ore deposit. Shaded regions are the net peak areas which are shown in the images in Fig. 3

scan mode is speed, the maximum scan rate being limited only by the counting statistics and not the system overhead. There are several disadvantages of this scan mode as well. It is hardware intensive, requiring an SCA and MCS channel for each element to be analysed. Background correction and ion chamber normalization also require MCS channels and post-processing if they are to be performed. Finally, it is difficult to set the SCAs accurately when there are closely spaced peaks in the spectrum.

The time required to acquire an image with the current scanning microprobe is not really limited by the overhead described for the second scan mode, and MCS scanning does not appreciably reduce the time required to collect a trace element image. The scan rate is actually limited by the nature of the detector, which can only handle a total count rate of about 5000 counts/second. On most samples most of the counts are either due to scattered x-rays or to the fluorescence of major elements. The scattered x-ray signal is minimized by the polarized nature of the synchrotron radiation and the detector geometry, but is still typically thousands of counts per second for thick targets and a 30 μm beam diameter. The major element fluorescence can sometimes be minimized by placing an absorber such as Kapton or aluminum in front of the detector to preferentially absorb lower energy photons from the major elements and pass the higher energy photons from the trace elements. In many samples, however, when the detector is counting at its maximum rate there are only a few photons per second from the trace element of interest, and it is thus necessary to count for a few seconds to achieve reasonable statistics. The detection limits for thick samples with 300 second count times are typically about 1 ppm by weight. Since the detection limit scales with the square root of the count time, a trace element image with a collection time of 1 second will have detection limit of about 17 ppm at each pixel. This is a reasonable detection limit, but it implies a scan time of nearly 3 hours for a 100 × 100 pixel image. Most of the images we have made to date have been lower resolution, about 50 × 50 pixels.

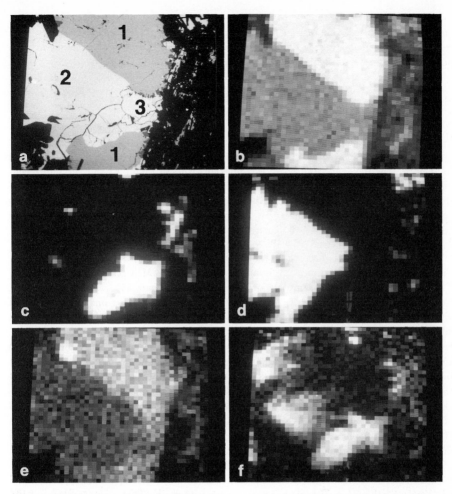

Fig. 3a: Backscattered electron image of a sulfide sample from an ore deposit.

Fig. 3b: Fe Kα fluorescence image

Fig. 3c: Ni Kα fluorescence image

Fig. 3d: Cu Kα fluorescence image

Fig 3e: Se Kα fluorescence image

Fig. 3f: Pd Kα fluorescence image

The data from scan modes 2 and 3 are saved as image files on disk in a very general file format common to several beamlines at the NSLS, including X-1, the soft x-ray microscope. Images can contain many different data values at each pixel, e.g. the peak area of different fluorescent lines, the incident and transmitted ion chamber currents, etc. Each data value can be stored in any of 8 numerical data types; signed or unsigned 8, 16, or 32 bit integers, and 32 or 64 bit floating point. The data can be compressed with linear predictive coding, which allows 2 or 4 byte integers to be compressed to fewer bytes, or run-length limited compression, which compresses runs of identical pixel values. The compression mechanisms are completely loss-free and the different data values at each pixel can be compressed with either algorithm.

4. RESULTS

A spectrum of the mineral pentlandite ($Fe_{4.5}Ni_{4.5}S_8$) from an ore deposit is shown in Fig. 2. This spectrum was collected for 300 seconds with a 30 μm beam diameter. The Si(Li) detector was placed 40 mm from the sample with a 6 mm diameter collimator in front of it. The incident x-ray beam was filtered with 410 μm of Al and the fluorescent photons were filtered with 150 μm of Al. These filters served to reduce the intensities of the Fe and Ni lines while attenuating the trace element lines of interest (As, Se, Ru, Pd) much less. This sample contains about 200 ppm of Se and Pd, and the detection limits are about 3 ppm for each.

Six scanning images of a rock section from a Finnish ore deposit are shown in Fig. 3. Fig. 3a is a backscattered electron image made with a CAMECA electron microprobe. The major minerals in the sample (numbered in Fig. 3a) are 1) pyrrhotite ($Fe_{1-x}S$), 2) chalcopyrite ($CuFeS_2$) and 3) pentlandite ($Fe_{4.5}Ni_{4.5}S_8$). This image is 512 × 512 pixels with a pixel size of 2.5 μm. Figures 3b-3f are scanning x-ray images, made with a 30 μm beam and a collection time of 2 seconds per pixel. The images are 40 × 40 pixels, for a total scanned area of 1.2 × 1.2 mm. The images of Fe, Ni and Cu (major elements) could have been made on an electron microprobe with much greater spatial resolution but poorer signal to noise. Note in Fig. 3c the large pentlandite grain in the lower center, the small grain in the top center and the single Ni rich pixel in the lower left. The large chalcopyrite grain on the left seems to contain a region with low copper near the center of the grain (Fig. 3d). The selenium is contained mainly in the Fe rich pyrrhotite phase (Fig. 3e) where it seems to be homogeneous, e.g. present in solid solution. There is an unidentified Se rich phase in the upper left of the image. The concentration of Pd (Fig. 3f) is much higher in the Ni rich pentlandite than in the chalcopyrite and pyrrhotite. Note the Pd rich area in the lower left of the image. This region corresponds to the single Ni rich pixel and the Cu poor region in the chalcopyrite grain. This is interpreted to be a grain of pentlandite under the surface of the chalcopyrite. The Pd Kα x-rays, 21.2 keV, are sampling much deeper ($\approx$100 μm) than the Ni or Cu x-rays ($\approx$10 μm), and thus the hidden grain is more evident in the Pd image. There are two unidentified grains with high Pd contents at the top left and top center of the image. The Pd content of these grains is too low to be detected with the electron microprobe.

5. FUTURE IMPROVEMENTS

Improvements planned for the x-ray fluorescence microprobe at X-26 in the near future will significantly improve its imaging capabilities. An 8:1 ellipsoidal grazing incidence mirror [8] will increase the flux at the sample by more than a factor of 1000. This will permit the use of a curved crystal, wavelength dispersive spectrometer (WDS), rather than the Si(Li) detector for many applications. The WDS detector has much better resolution and hence will have lower background and will reduce the problem of overlapping peaks. It does not suffer from the count rate limitations due to scattered background and major element fluorescence, so scans can be made much faster. Finally, the mirror will permit the use of a monochromator in the beam, which will lower the scattered background under the peaks and allow selective excitation of particular elements.

6. ACKNOWLEDGMENTS

Research supported in part by NSF Grant No. EAR-8618346. Research carried out at the National Synchrotron Light Source, Brookhaven National Laboratory, which is sponsored by the U.S. Department of Energy, Division of Material Sciences and Division of Chemical Sciences under Contract No. DE-AC02-76CH00016.

7. REFERENCES

1. K. W. Jones, W. M. Kwiatek, B. M. Gordon, A. L. Hanson, J. G. Pounds, M. L. Rivers, S. R. Sutton, A. C. Thompson, J. H. Underwood, R. D. Giauque, Y. Wu: In *Advances in X-ray Analysis*, **31** (1987)
2. S. R. Sutton, M. L. Rivers, J. V. Smith: *Nucl. Instr. Methods*, **B24/25**, 405 (1987)
3. A. L. Hanson, K. W. Jones, B. M. Gordon, J. G. Pounds, W. M. Kwiatek, M. L. Rivers, G. Schidlovsky, S. R. Sutton: *Nucl. Instr. Methods*, **B24/25**, 400 (1987)
4. J. H. Underwood: "X-ray Microscopy and Microprobing with Reflective Optics", this volume
5. C. J. Sparks: In *Synchrotron Radiation Research*, ed. by H. Winick and S. Doniach, (Plenum, New York, London 1980) p. 459
6. S. R. Sutton, M. L. Rivers, J. V. Smith: *Analytical Chemistry* **58**, 2167 (1986)
7. B. M. Gordon, K. W. Jones: *Nucl. Instr. Methods*, **B10/11**, 293 (1987)
8. K. W. Jones, P. E. Takacs, J. B. Hastings, J. M. Casstevens, C. D. Pionke: *Soc. Photo-Optical Engineers*, **749**, 37 (1987)

Synchrotron X-Ray Microtomography

K.L. D'Amico, H.W. Deckman, B.P. Flannery, and W.G. Roberge

Corporate Research Laboratories,
Exxon Research and Engineering Co., Annandale,
NJ 08801, USA

We present results obtained with a new form of 3-D microscopy. The technique is based on tomographic methods and we refer to it as 3-D Microtomography. When used with a tunable X-ray source, it is a powerful diagnostic and research tool for a wide variety of materials studies problems. It is capable of producing maps of the interior structure and chemical composition of samples approximately 0.5-1.0 mm in size, with spatial resolution in the map of the density variations approaching 1.0 micron.

1. INTRODUCTION

1.1 Computed Tomography

Tomography is a technique for determining the internal structure of an object.[1] The name derives from the Greek word for "slice", and connotes a technique which uses views of an object from different angles to try to recover the depth information lost when doing a conventional shadow radiograph of the object. The physical basis for the technique is the determination of the absorption of the penetrating radiation by the sample as a means of measuring the density profile. The intensity of the transmitted radiation is compared with the intensity of the incident radiation and a map of the attenuation coefficient versus position in the sample is determined.

While it is possible to do a tomographic measurement using only film as the detector with corotating sample and detector stages,[2] the most significant advances in the field were made by CORMACK and HOUNSFIELD.[3,4] With the further development of computed tomography (CT), tomographic scans are now routinely done with commercially available equipment. Using a conventional X-ray tube and an array of detectors, it was possible to develop an instrument which, when coupled with a suitable computer facility for mathematically inverting the data, was capable of producing maps of the density variations in a large object (e.g. a patient). The data so obtained consist of a cross-sectional slice of the density variation within the object for a plane which is normal to an axis of rotation. The technique (a CAT scan) has become a standard medical diagnostic tool. In addition, the technique has become widely used for non-destructive testing of large objects.[5]

The characteristics of a typical medical device are that it is capable of giving a cross-sectional image of an object the size of a human with millimeter scale spatial resolution in the density profile, it can produce this scan in a matter of minutes, and the map of the object can be made with about 1% accuracy in the determination of the density variations.[1]

240

1.2 Microtomography

An obvious extension of this technique would be to develop the capability
to observe features in smaller objects with higher spatial resolution.
This has been pursued on several different fronts. GRODZINS has outlined
the method of using a collimated X-ray beam which is defined by a pinhole-
type aperture to probe the sample.[6,7] The resolution of the device is
then determined by the size of this aperture. This technique requires that
one raster the beam over the sample at each of the view angles to cover the
entire data set necessary to determine the sample's density map. If more
than one single plane is to be viewed, the sample must further be
translated along the rotation axis to the plane desired. This technique
has been applied to samples using a Synchrotron X-ray source, where both
the transmitted beam[8] and fluorescence from Fe in the sample[9] have been
detected.

A second technique has used a different scheme for the data acquisition.
This has been developed at the Photon Factory, and involves the use of a
linear, diode array detector.[10] Thus the projection of the attenuation
of the X rays by the sample is measured by many detector elements
simultaneously across the full width of the sample. For this device the
detector consists of about 1000 diode elements, each of size about 25
microns, upon which the X rays impinge. Thus the resolution of the
reconstructed image is determined by the diode size of 25 microns. Again,
this allows for the detection of only a single plane through the sample at
a time. If one wants information on many consecutive planes of the sample,
the procedure must be repeated for each plane. This technique is somewhat
more efficient than the pinhole system, however, in that it allows for
parallel detection within a plane of the sample.

Still another scheme for data acquisition has been applied to
synchrotron measurements. This technique uses a Charge Coupled Device
(CCD) detector, which gives an area rather than a linear arrangement to the
diode elements.[11] This has the advantage that data from many stacked
planes can be collected in parallel. The phosphor employed with the device
allows for about 50-60 micron resolution to be achieved.

1.3 3-D Microtomography

We have developed a system similar to the latter in that a multi-element
area detector has been used.[12] Ours is also a CCD based unit, which
employs a novel high-spatial resolution phosphor plate to improve the
inherent resolution of the CCD.[13] The phosphor acts as the X-ray
detecting transducer, producing visible light which is focussed on to the
CCD. Without the high-resolution phosphor the device would permit
resolution comparable to that of both the diode array and the CCD systems
above. The phosphor used has a spatial resolution of better than 3
microns. When combined with a lens system for imaging the light produced
by the X-ray beam in the phosphor, it is possible to vary the resolution of
the system from 25 microns to near 1.0 micron.

The system is capable of giving 3-D information because of the
organization of the detector elements. Because the area detector allows
for the simultaneous acquisition of projection data from a multiplicity of
stacked planes, the system can efficiently obtain the data necessary to
produce a full 3-D map of the density. Of course the first two systems
above can do the same thing, but ours achieves at least a 300-fold gain in

speed over those because of the 300 rows of detector elements. Our system is different from that of KINNEY et al.[11] in that higher spatial resolution is achievable, owing to the high resolution phosphor.

2. EXPERIMENTAL TECHNIQUES

2.1 Detector and Sample Stages

As mentioned above, the detector is based on a CCD with 330x512 elements.[13] The device is maintained at low temperature to reduce thermal noise in the detector wells. X rays strike the specially fabricated phosphor and produce visible light; this light is imaged onto the CCD by a lens system.

The sample manipulator is made up of a precision turntable atop a precision translation stage (Klinger Instruments). Since we are attempting to achieve micron scale spatial resolution in the reconstructed images, both of these stages must have motions which are reproducible at the sub-micron level. The stages have been carefully chosen and checked to satisfy this criterion.

2.2 Mathematical Data Reduction

The projection data are converted to tomographic reconstructions by means of a Direct Fourier Inversion (DFI) algorithm. ROBERGE and FLANNERY have discussed the implementation of this algorithm,[14] which is superior to the more common Filtered Back Projection technique used for most tomography applications.[1] Because of the large quantity of data obtained with our 3-D capability, it is necessary to use the faster data reduction afforded by the DFI procedure. The details of the mathematics involved in the data reduction procedure have been previously described.[12,14]

2.3 X-ray Beamline

Synchrotron experiments have been carried out at the EXXON Beamline X10A. This beamline is located at the National Synchrotron Light Source at Brookhaven National Laboratory, Upton, New York, USA. The beamline is operated in the focussed mode, collecting about 2.5 milliradians from the horizontal fan of the radiation produced by a bending magnet. The beam is focussed in both the vertical and horizontal directions by a 0.6 meter long Pt-coated glass mirror and monochromatized by two Si(111) crystals. In such a configuration the beamline produces about $1x10^{11}$ monochromatic X rays per second into a Gaussian spot with a full width at half maximum of 0.75 mm in the vertical and 1.0 mm in the horizontal.

2.4 Data Acquisition Procedure

Data are acquired as a set of projection images (radiographs) of the object being scanned. For a typical 0.5 mm size sample, and a detector resolution of about 3 microns, a set of data consists of 450 individual 2-D projection images of the sample taken through 180 degrees. In addition, the sample must be repeatedly removed from the beam so that the incident flux can be determined. This is carried out with the translation stage and is done once for every 5-10 exposures. Individual radiographs consist of about

1.5x10⁵ data pixels; the radiographs are stored on magnetic tape along with
the calibration frames for later reduction by a separate computer system.

The criteria that determine the number of view angles, the X-ray energy
used for the measurement and the exposure times have been discussed
elsewhere.[12]   Typical exposure times are a few seconds per view angle and
calibration; dead time in the electronics adds a few more seconds per
exposure.   This results in a typical data acquisition time of 1-2 hours per
sample with a current of about 80 milliamperes in the 2.5 GeV storage ring
at the NSLS.

3. DATA

3.1 Tomographic Data

Figure 1 shows representative data which illustrate the capabilities of
the technique.[12]   The sample is two different concentrations of a water
solution of cupric sulfate contained in three glass tubes.   The two outer
ones are Debye-Scherrer capillary tubes and the inner is a hand-made
capillary.   The scale of the figure gives the dimensions of the tubes.   The
cupric sulfate solution in the outer and inner tubes is at a concentration
of 25 g/100 ml and the intermediate tube has a concentration of 2.5 g/100
ml.

The tomographic data were collected at two different X-ray energies.
The left panel shows the results for data collected at an energy of 8.88
keV and the right panel for an energy of 9.08 keV.   These energies are 100
eV below and above, respectively, the Cu K-absorption edge at approximately
8.98 keV.   For Fig. 1, the images are displayed with a grey scale such that
the darker the image, the greater the attenuation coefficient of the
material.   The two pictures have had their grey scales adjusted such that
the capillaries are the same, since the absorption coefficient of the glass
does not vary significantly between these two energies.

Several features of Fig. 1 are noteworthy.   The sample size is about
0.75 mm.   The walls of the Debye-Scherrer tubes, which are nominally 10
microns thick, provide a good measure of the resolution in the
reconstructed image.   The current resolution for a reconstructed image is
about 10 microns, which is larger than the 3 micron resolution of the

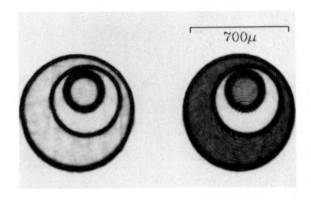

700μ

Figure 1 shows the
cross-sectional image
of a sample of a water
solution of cupric sul-
fate as discussed in
the text. The left and
right panels show the
results of data taken
at 8.88 keV and 9.08
keV, respectively. The
scale for sample is
shown above the right
panel

detector pixels. We have achieved better than 3 micron resolution in a static radiograph. Current limits on the system resolution may be due, perhaps, to the counting statistics in the data and the motion of the relatively small, non-uniform beam spot. The beam spot motion could be due to either mechanical vibrations of the beamline components or motion of the electron beam in the storage ring. These are problems which can be readily addressed and will be dealt with in future versions of the device.

Improper calibration procedures can result in the ring-like artifacts in the reconstruction. These can clearly be seen in Fig. 1. These result from persistent, systematic errors localized on certain elements in the detector; they are probably exacerbated by the motions of the beam spot as mentioned above. Efforts are under way to reduce the effects of these errors on the reconstructions.

The presence of the Cu in the solution is manifest by the enhanced absorption by the solution at the energy above the K-edge. The solution was mixed so as to give a jump of about 1.0 in the optical density for a 0.75 mm path length upon crossing the edge. We emphasize that the data obtained are quantitative, in that each radiograph is a measure of the X-ray attenuation of the sample and this information is preserved in the reconstruction. Thus each pixel in the reconstructed image contains the value of the X-ray attenuation coefficient for the sample at that point. Indeed, the absorption coefficient determined in the reconstruction of the cupric sulfate sample agrees with the value expected, from the known concentration, to within 1%. This information can be used to quantitatively analyze the sample with respect to its density or chemical composition.

4. CONCLUSIONS AND FUTURE PROSPECTS

We have shown that microtomography is capable of producing maps of the internal structure and chemical composition of millimeter-size objects with micron-scale resolution. Our technique also yields 3-D information on a sample as rapidly as other microtomography techniques yield information about a single cross-sectional plane. We have developed reconstruction algorithms for handling the 300-fold greater quantity of projection data acquired that allow for rapid reconstruction into a 3-D image.

The potential applications of the technique are numerous. The ability to look at 3-D features on this length scale is unique and will allow for the study of such phenomena as fracture in materials, void structure in porous materials, chemical segregation in composite materials or flow of fluids. In addition there is the possibility of doing studies of biological materials, as is being pursued by BOWEN et al.[8]

The NSLS at Brookhaven has given approval to our group to construct a beamline for the purpose of carrying out a 3-D microtomography developmental and experimental program. This beamline will be on line in the middle of 1988, and will represent the only dedicated microtomography facility in the US. This will permit the further development of this rapidly expanding field.

5. ACKNOWLEDGEMENTS

We would like to thank the staff of the Exxon beamlines at the NSLS for a great deal of help with these experiments. We also thank R. Lee as well as a number of people at Exxon for making these experiments possible, including J. Dunsmuir, P. Eisenberger, S. Gruner, J. McHenry and M. Wainger.

The NSLS at Brookhaven is supported by the U.S. Department of Energy, Office of Basic Energy Sciences, Division of Materials Sciences, under contract No. DE-AC02-76CH00016.

REFERENCES

1. W. Swindell, H. H. Barrett: Physics Today $\underline{30}$, 32 (1977)
2. G. Thuesen, A. Lindegaard-Andersen: Phys. Med. Biol. $\underline{25}$, 1049 (1980)
3. A.M. Cormack: J. Appl. Phys. $\underline{34}$, 2722 (1963)
4. G. Hounsfield: Br. J. Radiol. $\underline{46}$, 148 (1973)
5. See: Applied Optics $\underline{24}$, 3948-4134 (1985) for the proceedings of a conference on the subject
6. L. Grodzins, Nucl. Inst. Meth. $\underline{206}$, 541 (1983)
7. L. Grodzins, Nucl. Inst. Meth. $\underline{206}$, 547 (1983)
8. D. K. Bowen, J.S. Elliot, S.R. Stock, S.D. Dover: Proc. SPIE $\underline{691}$, 94 (1986)
9. P. Boisseau, Ph.D. thesis, Mass. Inst. of Tech. (1987)
10. K. Usami, K. Sakamoto, H. Kozaka, T. Hirano, Y. Suzuki, K. Hayakawa, H. Shiono, H. Koono: Photon Factory Report, 162 (1986)
11. J.H. Kinney, Q. Johnson, U. Bonse, R. Nusshardt, M. Nichols: Proc. SPIE $\underline{691}$, 43 (1986)
12. B.P. Flannery, H.W. Deckman, W.G. Roberge, K.L. D'Amico: Science, $\underline{237}$, 1439 (1987)
13. H.W. Deckman, S. Gruner,J. Dunsmuir: to be published
14. W.G. Roberge, B.P. Flannery: to be published

X-Ray Microscopy by Holography at LURE

D. Joyeux[1], S. Lowenthal[1], F. Polack[2], and A. Bernstein[2]

[1]Institut d'optique, CNRS, BP 43, F-91406 Orsay Cedex, France
[2]LURE, CNRS, Université de Paris XI, F-91406 Orsay Cedex, France

1. INTRODUCTION

Reporting the first x-ray holography experiment performed in Europe, we should acknowledge previous works in the field [1-4]: their results make it no longer necessary to demonstrate that x-ray holography can be achieved with available means. However, in our opinion, there is presently no evidence that this method can become a practical, routine technique of x-ray microscopy, i.e. a reliable technique, producing reconstructed images of realistic objects, with the expected resolution and a good signal-to-noise ratio. As a matter of fact, the strategy for obtaining this result is not well established, even regarding fundamental choices such as the holographic recording geometry, and the reconstruction method. Working toward that goal is certainly a long term project. Our experiment has been designed in this context for a limited goal: realization of the whole process, including optical reconstruction, to obtain images of realistic objects, with resolution in the range 0.5-1 µm. This implies a careful analysis of the whole process, taking into account the available means (x-ray source and optics, recording medium), which is presented hereafter. Sample holograms and a reconstruction are shown. To end, a short discussion of the future of x-ray holography is given.

2. DESIGN OF THE EXPERIMENT

2.1 The Recording Geometry

There are various possible recording set-ups extensively studied in optical holography; however, most of them make use of optical components for the reference wavefront production. These components have to be made with a good accuracy, which in fact cannot be presently obtained for the soft x-ray domain (except for grazing incidence plane mirrors). This is why we chose (as others [1,3]), to work with the Gabor in-line geometry.

This set-up has advantages. Since it needs no x-ray component, the Gabor set-up is mechanically and optically very simple; in addition it has minimum requirements in terms of spatial and temporal coherence; it is also quite tolerant in mechanical stability; finally (Sect.2.2), the useful recorded field is *not* limited by the spatial coherence of the x-ray beam, but only by its energy repartition (we recorded 4 mm dia. fields with only ≃100 µm coherence area in the object plane).

However, we have to keep in mind the two major drawbacks of Gabor holography. First, in order to record correctly the phase of the object wavefront, the reference amplitude has to be everywhere greater than the amplitude diffracted from the object. In Gabor holography this ratio is determined by the object absorption; therefore, when the object has a high transmittance contrast, the condition cannot be fulfilled in the shadow of the low transmittance parts of the object. This generates reconstruction noise. The second problem arises from the well-known high sensitivity of Gabor hologram reconstruction to optical noises, in particular: diffusion of the optical reconstruction beam by optical surfaces and media, including the recording medium; nonlinearity of the recording medium, which produces higher order images *and* random noise in the useful order; and, as mentioned above, the lack of reference amplitude in some parts of the recorded pattern.

2.2 The X-Ray Source and the Coherence Conditions

We were allowed to use the undulator of the ACO ring in Orsay, with its associated monochromator (toroïdal, holographic grating, aberration corrected and blazed at 10 nm, 2400 gr/mm, grazing incidence 11'). We chose a working wavelength of 10 nm, which is optimum for the undulator and monochromator.

To perform the holographic recording, we only need *some* coherence (temporal and spatial) *in the hologram plane,* and enough flux to have a reasonable exposure time. There is no problem adjusting the temporal coherence, i.e. the spectral bandpass, through the monochromator exit slit width. This suggests therefore that one could use the exit slit itself as the holographic x-ray source. However, the measured emittance parameters of the x-ray beam (at 10 nm: 500 μm $_x$ 2 mrad) show it is spatially *incoherent*. But, as well known, the radiated field of an incoherent source exhibits *partial coherence*: we should therefore determine the *coherence degree* in the hologram plane (using the Zernicke-Van Cittert theorem [5]), and, if necessary, restrict the angular size of the source. The precise coherence condition and, from it, the source size and distance are calculated from the recording geometry parameters, which in turn are essentially fixed by the desired resolution.

Figure 1 depicts schematically the recording geometry. Space around ACO forbids source-object distances D between 1.5 m and 9 m. The latter value has been chosen, because it provides a much more uniform illumination of the object field (it can be easily shown that the illuminance in the hologram plane remains constant if D is varied *keeping the coherence in the hologram plane constant*). Let r_T be the desired object resolution, then the wavefront must be recorded with an aperture $\alpha=\lambda/2r_T$. The coherence area radius is therefore $\Delta x=\lambda d/2r_T$. In the same way the temporal coherence condition is determined by the maximum path difference Δl that rays will undergo, for a specified resolution. Δl is $\sqrt{d^2-(\Delta x)^2} - d$, which yields $\Delta l/\lambda=\lambda d/8r_T^2$. We have now to maintain over this area and this length a minimum degree of coherence, high enough to record without significant attenuation the interference frin-

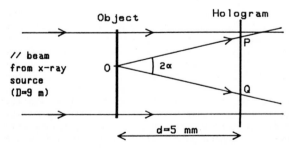

Object   Hologram

// beam
from x-ray
source
(D=9 m)

O   2α   P

Q

d=5 mm

Fig.1 Recording with a resolution r_T means an aperture $\alpha=\lambda/2r_T$; the coherence area and length are resp. PQ (dia.) and (OP-d).

ges which carry the holographic information. We used 0.75 as the minimum usable degree of coherence. In order to compute the source size, a last parameter has to be fixed, namely the object-hologram distance d of Fig. 1. The larger d, the higher the coherence constraints; but for small d, the recorded pattern looks more like a near field Fresnel pattern than a Fresnel hologram, and this yields theoretical and practical difficulties in the optical reconstruction step. We used d≃5 mm as a trade-off: this means that the hologram of a single point has 12 zones (i.e. the maximum path difference is 12 λ) at 0.7 μm resolution.

It is now simple to compute the source size, assuming for simplicity a uniform source and a square aperture [5]. With a quoted resolution of 0.7 μm, and taking into account the monochromator dispersion, it is found that a square source with a 0.6 mm side yields at least a coherence degree of 0.95 (temporal) and 0.81 (spatial), i.e. 0.76 as maximum contrast attenuation.

2.3 The Mechanical Stability Conditions

The holographic set-up consists of a "holographic camera", i.e. a vacuum chamber containing the sample holder and holographic plate; the camera is placed on a granite table (140 kg), supported by 4 air jacks for ground vibration isolation. All this is connected to the source by a vacuum line. With such a design, the stability problem is twofold: the internal stability of the camera, and the stability of the whole set-up with respect to the x-ray source. We will not go into this problem here; the main point is that these two aspects are scaled with very different values, namely the smallest fringe period for the first problem, and the x-ray source size for the second. Although it was not done in our experiment, we have come to the conclusion that air suspension, which is very efficient for vibration isolation, needs temperature control in order to avoid shifts due to air dilatation.

2.4 The Recording Medium

Ideally, the recording medium should have a high resolving power and a good sensitivity. Regarding the reconstruction step, it should also have a low intrinsic optical noise. From many former studies in optical holography, we think that a very low noise level is very important for high resolution holography, as important as the resolving power in the case of in-line Gabor geometry. It was from this argument that we rejected the use of photographic plates, even holographic ones, as too noisy.

Photoresists are then the only available class of photorecorder having a high resolving power and very low noise. We used, as a first try, an electrosensitive negative resist (OLIN-HUNT 320F), because of the mechanism of energy transfer from absorbed soft x-ray photons to matter; we chose a negative resist because of its higher sensitivity; the generally lower resolving power was not here a limitation, since a desired imaging resolution of 0.5 μm only needs to have in the resist a good modulation transfer at 1000 mm^{-1}: this is certainly obtained with any commercial resist.

As pointed out in Sect.2.1, reconstruction noise is not limited to optical diffusion noise: we have also to consider the nonlinearity noise, especially with resists, which are often said to be "binary", In fact, examination of different resists' energy response [6] never shows binary responses, but only a nonlinear one, including in some cases (essentially for negative resists) a deposited energy threshold. Such a problem is easily handled by using a classical technique of optical holography, called "preexposure". It consists in illuminating the resist coated plate with just the needed amount of energy to move to the foot of the linear part of the response curve; this can of course be done in any spectral domain where the resist is sensitive, UV for instance. In the case of a lumination threshold, the process allows high contrast patterns to be recorded without saturation problems. In addition, the higher slope of the linear part produces a gain in sensitivity. To our knowledge, the preexposure idea has been applied to resists only once [7], in a very different context (and not for x-ray exposure). We think it should be very interesting to test it on various resists, in particular for high resolution contact microscopy. More details will be published elsewhere.

Although preexposure makes the engraved height-illuminance relationship linear, holography needs merely linearity between the recorded illuminance and *the hologram transmittance*. Strictly speaking, this is impossible to obtain with a phase recording medium. However, a good approximation is obtained when the engraved heights are much lower than the reconstruction wavelength; therefore, there is a trade-off between the nonlinearity level and the diffraction efficiency of the hologram. In practice, phase holograms with heights not greater than about $\lambda/8$ are reasonably linear, and have a sufficient diffraction efficiency, provided the noise level is kept low. In other words, it is necessary to expose holograms in such a way that the engraved heights are much smaller than the coating thickness (5000 Å).

3. EXPERIMENTS

Testing a microscopy system requires "real" objects (not grids or wires), having well-identified structures, fitted to the expected properties of the process. This is why we used, along with generations of optical microscopists, the silica skeleton of diatoms as tests: they show a few classes of shapes, with known sizes; each class presents quasi periodic structures of increasing fineness and well-known dimensions, well suited to test resolution in the range 1 μm to less than 0.1 μm.

Diatoms were placed on a polyimide membrane, 1600 Å thick; its absorption is 45% at 10 nm. (A Si_3N_4 membrane would be too absorbent). Typical exposure times were from 10 to 50 min, with 100 mA in the ring. From fluorescence measurements on a P20 screen we estimate the x-ray illuminance in the hologram plane to about 45 nJ/(cm^2.s.A) per 100 mA.

Each exposure yielded a recorded field of $\simeq$4 mm diameter, containing dozens of patterns. Figure 2 shows typical aspects, viewed through a differential interference microscope. As desired, we obtained very "clean" records, free of random patterns (the broad black dots are not in the phase records, but were introduced by the microphotography device); fringes are widely spread, indicating good coherence conditions: Fig. 2c, for instance, is produced by an ovoïd diatom (Fig. 3a), about 13 μm x 26 μm. Fringe periods smaller than 1 μm are clearly visible. Examination of Fig. 2a (and of other patterns) indicates that there was probably no reference wavefront in the region of the geometrical shadow of the long diatom body. As mentioned in Sect. 2.1, this is one of the major problems of Gabor holography.

4. RECONSTRUCTION

Since we are now studying optical reconstruction, we only discuss here the key points of the problem, and present our very first reconstruction.

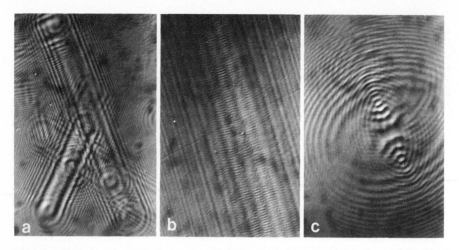

Fig.2. Typical recorded patterns, viewed through an interference microscope. Field is 130 μm×200 μm in a, 65 μm×100 μm in b-c.

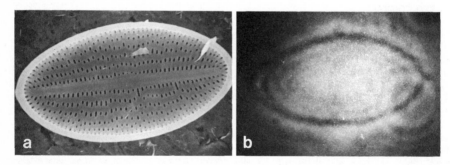

Fig.3. a) SEM image of a diatom (size 13×25 μm). Figure 2c was produced by a similar diatom. b) Image reconstructed from the hologram Fig. 2c; the original engraved pattern was directly illuminated by a spherical wavefront (primary magnification: $\simeq 60$, overall: $\simeq 3000$; $\lambda_c = 633$ nm).

Optical reconstruction is not the only possible method: numerical reconstructions have now been attempted with some success [8]. But each method certainly presents its own limitations: it is therefore useful for the future of holographic microscopy to improve both at the same time. To our opinion, *direct reconstruction* (without intermediate copy or enlargement of the hologram) has the great advantage of introducing no noise at all (through any intermediate step) between the recording process and the reconstruction process.

4.1 The Reconstruction Strategy

To begin, it should be clear that the object resolution is limited by the reconstruction wavelength. This is obvious from the general principles of optics, and does not depend on the geometry of reconstruction.

The first question to answer is: how much magnification should be obtained from the reconstruction step ? In fact, we think that the magnification which is necessary to get a comfortable observation has to be gained by the holographic reconstruction step, because this step is less demanding in the quality of optical components (no field correction). As the reference wavefront of the recording step was plane, reconstructing with a plane wavefront yields no magnification whatever the wavelength is. A spherical reconstruction wavefront is necessary to have a magnified image. However, as well known, the reconstruction process induces image aberrations; it is necessary to consider the variation of the main aberration (spherical) with the magnification. Applying the general aberration expressions [9], a minimum is found for $M=2$, but this value is not high enough. When $M \gg 1$ the aberration varies slowly, and is about 3 times greater than the minimum value. Therefore, we use a high magnification geometry, the exact value of M having no particular importance.

With this geometry, the aberration can be expressed in terms of the recording and reconstruction wavelengths λ_0 and λ_c, the recorded object resolution R (limited to $\lambda_c/2$) and the object hologram distance z_0. The aberrant phase W is found to be equal to $(2\pi/128) \times z_0 \lambda_0 \lambda_c^2/R^4$. Therefore, a smaller reconstruction wavelength improves both the aberration level and the best readable object resolution. However, the specific solution to spherical aberration correction lies in the design of special optics, compensating the aberration of the holographic process for one pair of points. This optics would be very similar (although having lower specifications) to optics used by SCHMAHL et al. [10] for holographic recording of Fresnel zone plates.

4.2 First Reconstructed Image

Figure 3b shows the very first reconstruction we obtained from the hologram shown in Fig. 2c. It includes no image processing for noise elimination or photometric equalization (however, a simple spherical aberration correction was attempted, which we no longer use, because it introduces very high field aberrations). As mentioned in Sect.3. the hologram was produced by a diatom similar to Fig. 3a (13 $\mu m \times 25$ μm, wall thickness: $\simeq 0.8$ μm). The reconstructed image shows clearly the wall shape and the measured size and thickness are in agreement with the former values: this indicates that the resolution goal has been reached.

5. DISCUSSION

The short-term target of our holography experiment is to improve the image quality, from the point of view of both noise and aberrations. This can be accomplished by improving the recording process (phase modulation optimization, reference to amplitude ratio problem: off-line geometry ?) and by introducing both image processing (for noise) and an efficient aberration correction in the reconstruction step.

More generally, optical reconstruction limits the resolution of the whole process to $\lambda_c/2$, which in turn is limited by the lowest available UV laser line ($\lambda \simeq 220$ nm). For lower wavelengths, the synchrotron source itself might be used, but vacuum operation would be needed below $\simeq 200$ nm. This would make the instrumentation much more difficult, not to speak of the optics fabrication problems. However, we consider that such a medium resolution is still interesting in many practical cases.

To conclude, we feel it necessary to discuss the supposed 3-D character of holographic imaging, often presented as its main interest, compared to the 2-D classical imaging. As a matter of fact, a careful comparison between classical and holographic imaging leads to a different insight. To that end, one can consider the holographic imaging process as the combination of two different processes, namely *imaging* and *recording*. As an imaging process, holography should be compared to classical (coherent) imaging, *with equal aperture parameter in both cases*, excluding the recording of the image.

Obviously, both classical and holographic imaging give from a 3-D object a 3-D image, which are characterized by a longitudinal and a transverse resolution. These are exactly the same in both cases for equal imaging aperture α, namely $r_t = \lambda/2\alpha$ and $r_1 = \lambda/2\alpha^2$. The difference between the two imaging processes occurs in the recording part. In the classical case, only one resolved plane can be recorded at a time from the image; by contrast, the holographic process first makes from the object a global record of the whole set of resolved planes, allowing later observation of each resolved plane . In other words, the holographic process is 3-D in its recording part, but finally, the overall 3-D resolution characteristics of both types of imaging are the same.

What are therefore the potential advantages of holographic imaging ? We think that its main advantage is its very high instrumental simplicity, in contrast to other microscopy methods (except contact microscopy). The needs for stability are basically the same for all methods, related to resolution and integration time. But holography needs no fine mechanical adjustment; this should be compared with instrumental imaging, for which focussing and alignment accuracy are scaled by the imaging resolution (it is also possible to discuss the relative difficulty of having high aperture holographic or instrumental imaging systems). The particular requirement of coherence (for holography) is presently a limitation; but this constraint will become less severe with new high emittance sources such as super-ACO, in Orsay. Finally, holography has a very specific property: a complex amplitude is recorded from the object, instead of an illuminance. This could be useful for studying x-ray phase objects.

Financial support of DRET is gratefully acknowledged.

6. REFERENCES

1. M.R Howells: in X-Ray Microscopy, ed. by G. Schmahl and D. Rudolph, Springer Ser. Optical Sci., Vol 43, (Springer, Berlin, 1984) p.318
2. E.S. Gluskin, G.N. Kulipanov, G.Y. Kezerashvili, V.F. Pindyurin, A.N. Skrinski, A.S. Sokolov, P.P. Ilyinski: in X-Ray Microscopy, ed. by G. Schmahl and D. Rudolph, Springer Ser. Optical Sci., Vol 43, (Springer, Berlin, 1984) p.336
3. S. Aoki, S. Kikuta: Jap. J. Appl. Phys., 13, 9, 1385 (1974)
4. B. Reuter, H. Mahr: J. Phys E: Scient. Inst, 9, 746 (1976)
5. M. Born, E. Wolf: Principles of Optics, 3rd ed., (Pergamon, Oxford, 1965) p.508
6. See for example X-Ray Optics, ed. by H.J. Queisser, Topics Appl. Phys., Vol 22, (Springer, Berlin, 1977)
7. L. Damé, Thèse, (Toulouse, France, 1983)
8. C. Jacobsen, J. Kirz, M. R. Howells, R. Feder, D. Sayre: this conference
9. R.W. Meier: J. Opt. Soc. Am., 57, 895, (1967)
10. G. Schmahl, D. Rudolph, P. Guttmann, O. Christ: in X-Ray Microscopy, ed. by G. Schmahl and D. Rudolph, Springer Ser. Optical Sci., Vol 43, (Springer, Berlin, 1984) p.63

Progress in High-Resolution
X-Ray Holographic Microscopy

C. Jacobsen[1], *J. Kirz*[1], *M. Howells*[2], *K. McQuaid*[3], *S. Rothman*[3], *R. Feder*[4], *and D. Sayre*[4]

[1]Department of Physics, State University of New York at
Stony Brook, Stony Brook, NY 11794 USA
[2]Center for X-Ray Optics, Lawrence Berkeley Laboratory,
Berkeley, CA 94720, USA
[3]Center for X-Ray Optics, Lawrence Berkeley Laboratory,
Berkeley, CA 94720, USA and School of Medicine, University of
California – San Francisco, Berkeley, CA 94720, USA
[4]IBM T.J. Watson Research Center, Yorktown Heights,
NY 10598, USA

Among the various types of x-ray microscopes that have been demonstrated, the holographic microscope has had the largest gap between promise and performance. The difficulties of fabricating x-ray optical elements have led some to view holography as the most attractive method for obtaining the ultimate in high resolution x-ray micrographs; however, we know of no investigations prior to 1987 that clearly demonstrated submicron resolution in reconstructed images. Previous efforts [1] suffered from problems such as limited resolution and dynamic range in the recording media, low coherent x-ray flux, and aberrations and diffraction limits in visible light reconstruction. We have addressed the recording limitations through the use of an undulator x-ray source and high-resolution photoresist recording media. For improved results in the readout and reconstruction steps, we have employed metal shadowing and transmission electron microscopy, along with numerical reconstruction techniques. We believe that this approach will allow holography to emerge as a practical method of high-resolution x-ray microscopy.

All earlier work in x-ray holography in which reconstructions were obtained made use of x-ray film as the recording medium. It has been known since the earliest days of thinking about x-ray holographic microscopy that the resolution in the Gabor geometry can be no better than the film grain size [2], which precludes sub-100 nm resolution at soft x-ray wavelengths [3]. This has led some [4,5] to turn to x-ray photoresists for recording holograms. The ultimate resolution of these photoresists is said to approach 5 nm, and they have good detective quantum efficiency [6]. However, they also suffer from low sensitivity (a necessary concomitant of small "pixel"-size), and making full use of their

253

high resolution for holography has proven to be challenging. Recently, we [7] and others [8] have taken separate approaches to overcoming these challenges, and have obtained submicron resolution in reconstructed resist holograms.

The geometry used for the recording of the holograms has been described elsewhere [7]. We used the NSLS mini-undulator beamline X17t [9] as a source of 2.5 nm x-rays, with a toroidal grating monochromator providing temporal coherence and a pinhole to provide spatial coherence. Our resulting coherent flux of typically 10^8 photons/sec [7,9] was more than 100 times larger than that obtained at the NSLS bending magnet beamline U15 [10]. Besides greatly improving both exposure time and quality, this dramatically increased illumination greatly simplified the photometry (the coherent flux at X17t was sufficient to produce a photoyield of typically 10^{-10} A on an absolutely calibrated aluminum photodiode [11]) and alignment (the coherent x-ray spot was visible when viewed on a phosphor, even with room illumination) of the experiment. The holography chamber and collimating pinhole sat on a $10 \times 3 \times 1$ foot^3 granite table, supported by vibration-isolation air pistons; consequently, any vibrations between the pinhole and the specimen-recorder package (shown schematically in Fig. 1 of Ref. [7]) were negligible compared to the 50 μm pinhole size.

Most of the holograms are of fixed and dried zymogen granules from rat pancreas acinar cells. Not only is there considerable interest in the ultrastructure of zymogen granules [12], but their small size (roughly 1 μm across) allowed us to satisfy the desirable condition of having a largely empty object plane to minimize corruption of the reference beam. The use of thin silicon nitride windows ($\sim$60% transmittive) and thin layers of resist ($\sim$80% transmittive) means that each hologram recording layer removes slightly more than half of the 2.5 nm photons from the beam [13], so that the downstream holograms still have quite good illumination. This permits us to record several holograms simultaneously; each hologram contains essentially the same information on the specimen, and this redundancy may prove useful for averaging out noise and speckles in the reconstructed image. The thin resists and windows also allow for direct examination of the developed and coated resist images in a transmission electron microscope.

Fig. 1. **A**: Hologram of a diatom fragment taken at the NSLS bending magnet beamline U15 and examined with an SEM. **B**: Hologram of a zymogen granule taken at the NSLS undulator beamline X17t and examined with a TEM. Both holograms were taken at a working distance of about 400 μm, and are sub-fields of $(200 \ \mu\text{m})^2$ total hologram areas.

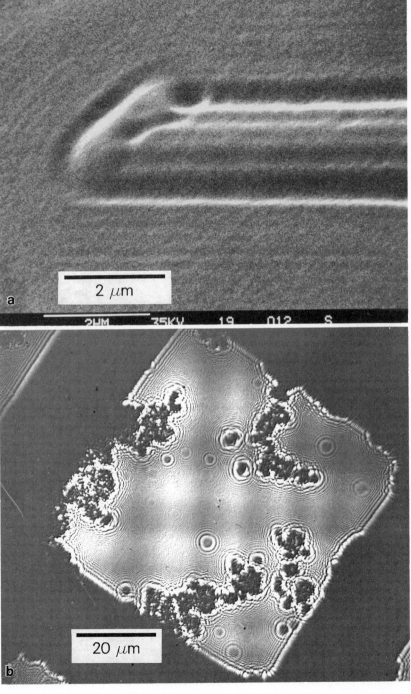

Fig. 1.

255

We had previously attempted to record holograms in this manner at the bending magnet beamline U15, and had obtained no better results than the hologram of a diatom fragment shown in Fig. 1A. We now believe that the poor fringe count and contrast on the hologram were due to inadequate flux (which we estimate was in the range of 10^4 photons/μm^2 at 3.2 nm, acquired over 10 hours) and non-optimal hologram readout. There are indications that wet development side-cutting of step-function exposure variations on polymethyl methacrylate (PMMA) resists is quite severe for absorbed doses less than roughly 100 megarads, or 10^7 photons/μm^2 at 2.5 nm wavelength [14]. This implies that extremely high photon fluxes are required to record high spatial frequency information in photoresists. (Others have come to a similar conclusion by considering the shot noise of photons illuminating (5 nm)^2 "pixels" of PMMA [15]. Finally, the U15 holograms were examined by the standard method of SEM imaging of the developed and normal-incidence-metallized resist surface, which we now feel is inappropriate for detecting the shallow height modulations of a few tens of nanometers expected for hologram fringes. Considerable improvements in the imaging of fine height variations on resists have been demonstrated by specialized SEM techniques [16] and by various replica methods for TEM examination [17,18], and we feel that this problem can benefit greatly from further study.

Because of the increased coherent flux of the X17t undulator, we were able to increase the exposure of our more recent holograms to typically 10^7 photons/μm^2 at 2.5 nm, acquired in about an hour. This exposure was high enough to see a light image of the specimen support grid bars on the undeveloped resist, and the resist required only light development (immersion for 0.5–3 minutes in 17% methyl isobutyl ketone–83% isopropyl alcohol was typical). In order to avoid the lateral distortions of the hologram that could potentially arise in replica methods, we evaporated a metal coating of ~20 nm of 60% Au–40% Pd onto the developed resist surface at typically 7° grazing angle for direct TEM examination. When processed in this manner, developed resists of nominally 200 nm initial thickness on silicon nitride substrates remained stable in a low-dose 80–100 keV TEM beam, although resist mass loss effects were invariably observed [18]. As can be seen in Fig. 1B, these holograms show a much greater information content than the lower flux, SEM-examined hologram of Fig. 1A.

A known drawback of using photoresists as a holographic recording material is that the resist thickness after development in the appropriate solvent is a non-linear function of incident illumination [19]. We have attempted to correct for this with an approximate model of the resist imaging process [20]. X-ray photoresists have been shown to respond to the absorbed x-ray dose independent of photon energy [21], so the first step is to calculate the dose absorbed by the resist as a function of incident x-ray intensity. Using published data on resist dissolution rate as a function of absorbed dose [6,14,22], one can then estimate the thickness variations of the developed resist. Finally, the TEM image contrast of thick, low-Z specimens has been modeled [23], and the response of electron microscope image films to incident electron illumination is well understood [24]. An example calculation of normalized electron film density as a function of incident x-ray intensity for typical resist parameters is shown in Fig. 2. The model indicates that an incident x-ray flux of 3×10^7 photons$/\mu$m^2 yields maximum electron film density (the resist has been fully developed away); this agrees nicely with the patches in Fig. 1B where the resist has been completely developed away with a measured peak incident flux of 2.7×10^7 photons$/\mu$m^2. The model as it now stands is certainly incomplete, however, since it estimates the response of the photoresist to uniform illumination, while what in fact is desired is an estimate of the resist modulation transfer function (MTF) [25]. Once the resist MTF is estimated and we are able to more accurately calculate developed resist surface relief, we will be able to follow a previously outlined method [19] for correcting for the effects of metal shadowing.

When holograms are to be reconstructed at a wavelength significantly different from the recording wavelength, aberrations will severely degrade the image resolution unless corrective optics are used [8] or the hologram is appropriately scaled. Because of the need to correct for resist non-linearities as well as the desire to implement hologram processing methods that are not available optically, we have instead chosen to pursue numerical reconstructions of the x-ray holograms. Towards that end, we have had several TEM negatives of holograms digitized with a microdensitometer, a step that also makes possible quantitative evaluation of the quality of the recorded holographic data. After correcting for resist non-linearity using the model sketched above, we took a random series

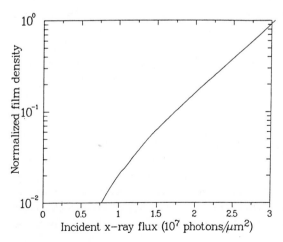

Fig. 2. Model calculation of normallized TEM film density as a function of incident x-ray intensity for a 300 nm thick PMMA resist.

257

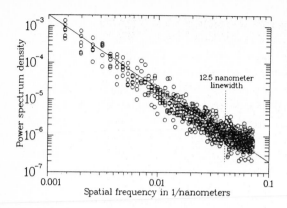

Fig. 3. Power spectra of several scan lines taken across the hologram shown in Fig. 1B. Scan lines taken both roughly parallel and perpendicular to the direction of shadowing yield similar results.

of line scans across several holograms and calculated their power spectra. The results for one hologram (which are typical for the others so far examined) are shown in Fig. 3. As can be seen, the power spectral density falls off roughly as the inverse of spatial frequency up to approximately 0.04 nm^{-1}, after which it appears to roll off to white noise. (Fringes are visible by eye on the TEM negative to a spatial frequency of about 0.01 nm^{-1}.) This suggests a finest recorded spatial period of 25 nm, or a minimum Fresnel zone width of 13 nm. A similar result is obtained if resist non-linearities are not corrected for.

Numerical reconstructions of Gabor holograms have been studied by many [26,27]. By considering the hologram to be a transparency that modulates the illuminating wave amplitude, one can use the Fresnel-Kirchhoff diffraction integral to propagate the wavefield a distance z from the hologram plane (ξ, η) to the reconstruction plane (x, y), at which point the reconstructed image intensity is obtained. In the Fresnel approximation, this can be written as

$$I(x,y) = \left| F\left\{ \tau(\xi,\eta) \exp\left(i\,\pi\,\frac{\xi^2 + \eta^2}{\lambda\,z} \right) \right\} \right|^2. \tag{1}$$

Thus, multiplying the two-dimensional hologram transmittance $\tau(\xi,\eta)$ by the quadratic phase factor

$$\exp\left(i\,\pi\,\frac{\xi^2 + \eta^2}{\lambda\,z} \right) \tag{2}$$

and then performing a Fourier transform $F\{\}$ (implemented digitally with a FFT algorithm) will produce a Fresnel transform.

We have implemented this numerical reconstruction scheme, and have used it to reconstruct the hologram shown in Fig. 1B and thus obtain the reconstructed image shown in Fig. 4. Because the hologram is at a few far-fields from the

258

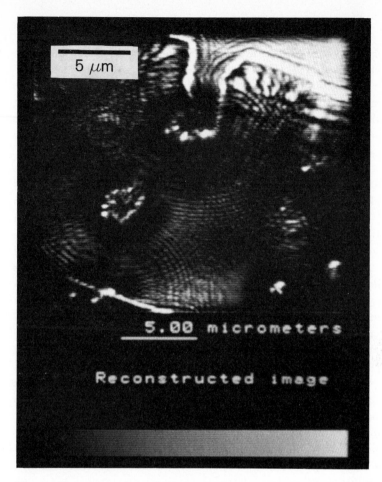

Fig. 4. Reconstruction of a hologram similar to that shown in Fig. 1B. Refer to text for discussion.

specimen, and because the "shadow" of the specimen has been apodized [28], the twin image noise inherent in Gabor holography has been eliminated (although diffraction around the Gaussian-smoothed edges of the apodizing mask itself corrupts the image to some degree). Consideration of the sampling theorem dictates that the Gabor hologram pixel size Δ_ξ be set to the diffraction-limited spot size

$$\Delta_\xi = \frac{\lambda}{2\,\text{N.A.}} = \sqrt{\lambda z/N} \tag{3}$$

(where N^2 is the number of hologram pixels), while the condition

$$(\Delta_x \Delta_\xi N / \lambda z) = 1 \qquad (4)$$

must be satisfied in order to have a discrete Fourier transform relationship in the reconstruction integral (1). This leads to the conclusion that the reconstruction pixel size Δ_x will also be equal to the diffraction-limited spot size of the numerical aperture of the hologram. For the case of a 512×512 pixel sampling of the hologram reconstructed in Fig. 4, this leads to a pixel size of 42 nm; the amount of computer time needed to perform such a reconstruction is roughly five minutes on a MicroVAX II minicomputer.

We would expect that the image quality of the reconstruction would be somewhat degraded by complications such as speckle and recording non-linearites that were not completely corrected for. The spot size of any speckle pattern should be on the same order as the diffraction-limited spot size, and since we do not see large pixel-to-pixel noise fluctuations, we conclude that the observed variations in image intensity in Fig. 4 are not due to speckle. Non-linearities in the recording of hologram intensity are more problematic. If one imagines a sine wave fringe pattern distorted into a square wave, the "extra" high spatial frequencies will manifest themselves as higher-order images (analogous to higher-order zone plate foci) and an artificial enhancement of high-spatial-frequency information on the specimen. This may be the explanation behind the bright edges and dim center of the grid bar shown in the upper right hand corner of Fig. 4. While this artifact leads us to regard the reconstructed image shown as a preliminary one, the image is reminiscent of scanning transmission x-ray micrographs taken of hydrated specimens of the same type of sample [12]. Finally, line scans taken across the grid bar edge (which may not be perfectly sharp when viewed with soft x-rays) demonstrate a knife edge transition occurring over one to two pixels, indicating that our current reconstructions have sub-100 nm resolution [29].

There are a variety of ways in which we hope to improve upon these preliminary investigations [20]. With the NSLS X1 undulator that is to be commissioned in 1988, we hope to reduce the exposure time from about one hour to a few minutes. We have used a wet cell to record holograms of hydrated specimens; examination of some of these holograms suggests that x-ray absorption in water may have reduced the exposure below the level needed to record high-resolution information. Transmission electron micrographs of a carbon replica of a crossed grating have shown us that the TEM used did not suffer from image field distortions at the micron level, although such distortions may become an issue at finer scales. We feel that there is much to be learned about the handling of photoresists, both in terms of finding ways to decrease the exposure (perhaps by adding dopants to increase the x-ray absorption of the resist) and in obtaining linear resist image readout without having to resort to metal shadowing. We have only begun to explore numerical hologram reconstruction and techniques such as phase retrieval [27] for further improving the image quality, and we are, of course, keenly interested in developments in flash sources like x-ray lasers [30]. It is our hope and expectation that x-ray holography will soon have advanced from the point of demonstration to become a useful x-ray imaging technique.

ACKNOWLEDGEMENTS

We are grateful for the help of many people. Jim Boland has provided instruction in and Marilyn Caldarolo has provided assistance with electron microscopy. P.C. Cheng, Doug Shinozaki, and Lorena Beese have shared their insights in using photoresists in x-ray microscopy. Roy Rosser, Nasif Iskander, Nadine Wang, and David Attwood participated in the recording of holograms at the U15 beamline, and Harvey Rarback and the staff of the NSLS assisted us in the use of the X17t beamline. Claude Dittmore provided us with beautiful digital data sets from the microdensitometry of the TEM negatives. Jim Grendell and Thomas Ermak have shared their insight concerning microscopy of zymogen granules, and K. Conkling and D. Joel have graciously made available their biological laboratory facilities at Brookhaven National Laboratory. We are grateful for the support of the National Science Foundation under grant BBS-8618066 (J.K., C.J.) and the Department of Energy under contract DE-AC03-76SF00098 (M.H.). This work was carried out in part at the NSLS, which is supported by the Department of Energy under contract DE-AC02-76CH0016.

1. See e.g., S. Aoki and S. Kikuta: Japan J. Appl. Phys. 13, 1385–1392 (1974); M.R. Howells, M.A. Iarocci, and J. Kirz: J. Opt. Soc. Am. A 13, 2171–2178 (1986); and references therein.
2. A. Baez: J. Opt. Soc. Am. 42, 756–762 (1952).
3. M.R. Howells, M.A. Iarocci, J. Kenney, J. Kirz, and H. Rarback: Proc. SPIE 447, 193–203 (1984).
4. G.C. Bjorklund, S.E. Harris, and J.F. Young: Appl. Phys. Lett. 25, 451–452 (1974).
5. S. Aoki and S. Kikuta: In Short wavelength coherent radiation: generation and applications, ed. by D.T. Attwood and J. Bokor (Am. Inst. Phys. conf. proc. 147, New York, 1986), pp. 49–56.
6. E. Spiller and R. Feder: In X-ray optics: applications to solids, ed. by H.-J. Queisser, Topics Appl. Phys. 22 (Springer-Verlag, Berlin, 1977), pp. 35–92.
7. M. Howells, C. Jacobsen, J. Kirz, R. Feder, K. McQuaid, and S. Rothman: Science 238, 514–517 (1987).
8. D. Joyeux, S. Lowenthal, F. Polack, and A. Bernstein: this volume.
9. H. Rarback, C. Jacobsen, J. Kirz, and I. McNulty: this volume.

10. J.M. Kenney, J. Kirz, H. Rarback, M. Howells, P. Chang, P.J. Coane, R. Feder, P.J. Houzego, D.P. Kern, and D. Sayre: Nucl. Inst. Meth. 222, 37–41 (1984).

11. R. Day, P. Lee, E.B. Saloman, and D.J. Nagel: J. Appl. Phys. 52, 6965–6973 (1981).

12. S.S. Rothman, N. Iskander, K. McQuaid, D.T. Attwood, T.H.P. Chang, J.H. Grendell, D.P. Kern, H. Ade, J. Kirz, I. McNulty, H. Rarback, D. Shu, and Y. Vladimirsky: this volume.

13. See e.g. Appendix A of J. Kirz and H. Rarback: Rev. Sci. Inst. 56, 1–13 (1985).

14. R.P. Haelbich, J.P. Silverman, and J.M. Warlaumont: Nucl. Inst. Meth. 222, 291–301 (1984).

15. D.M. Shinozaki: this volume.

16. O.C. Wells and P.C. Cheng: this volume.

17. K. Shinohara, S. Aoki, M. Yanagihara, A. Yagishita, Y. Iguchi, and A. Tanaka: Photochem. Photobio. 44, pp. 401–403 (1986).

18. P.C. Cheng, J.Wm. McGowan, K.H. Tan, R. Feder, and D.M. Shinozaki: In Examining the submicron world, ed. by R. Feder, J.Wm. McGowan, and D. Shinozaki (Plenum Press, New York, 1986) pp. 299–350.

19. G.C. Bjorklund: PhD dissertation (Microwave Laboratory report 2339, Stanford University, 1974).

20. C. Jacobsen: PhD dissertation (Department of Physics, State University of New York at Stony Brook, in preparation).

21. D. Seligson, L. Pan, P. Kink, and P. Pianetta: In Proc. Synchrotron Radiation Instrumentation Conference, Madison, 1987 (to be published in Nucl. Inst. Meth.).

22. R.J. Hawryluk, H.I. Smith, A. Soares, and A.M. Hawryluk: J. Appl. Phys. 46, 2528–2537 (1975).

23. A.V. Crewe and T. Groves: J. Appl. Phys. 45, 3662–3672 (1974).

24. See e.g., R. Valentin: In Advances in Optical and Electron Microscopy, Vol. I, ed. by R. Barer and V. Cosslett (Academic Press, New York, 1986).

25. See e.g., J.W. Goodman: Introduction to Fourier Optics (McGraw-Hill, San Francisco, 1968).

26. J. W. Goodman and R. W. Lawrence: Appl. Phys. Lett. 11, 77–79 (1967).

27. G. Liu and P.D. Scott: J. Opt. Soc. Am. A 4, 159–165 (1987).

28. M. Howells: this volume.

29. C.J. Buckley: this volume.

30. See e.g., J. Trebes *et al.*: this volume.

Fundamental Limits in X-Ray Holography

M. Howells

Center for X-Ray Optics, Lawrence Berkeley Laboratory,
University of California, Berkeley, CA 94720, USA

1. INTRODUCTION

At the present time, soft x-ray holographic microscopy is beginning to reach the point where it can be considered for use in scientific investigations [1].It has a long history which is reviewed elsewhere [2-4]. The general theme has been that improvements were brought about by advances in technology. It is not our purpose to discuss this here except. to say that we believe that there will be further advances in technology and this raises the issue of more fundamental limits to the performance of a holographic soft x-ray microscope. In this paper we consider two important classes of performance limitation: that due to sample damage and that due to the twin image problem [5]. In the spirit of the title, both are discussed for an ideal experiment involving perfect detectors, etc.

The problem of damage is treated by first estimating the number (P_{ij}) of scattered, detected x-rays that are needed to encode the position of the (i,j) sample pixel with a given resolution and a sufficient number of gray levels. This is a problem in information theory. We then argue that the sample pixel is really the unit from which scattered waves are superposed coherently at a typical point in the detector. This understanding allows us to calculate the ratio of the absorption cross section to the scattering cross section for a sample pixel and hence to obtain the sample dose associated with P_{ij} scattered x-rays. The doses needed in certain biological experiments are given as examples. This general problem has received some attention in the literature of x-ray holography [23] and other forms of x-ray imaging [21-23], in most cases with a primary focus on numerical results. The main purpose of the present treatment is to provide an explanation of the results in the language of Fourier optics.

The twin image problem is defined and is shown to have a solution in cases where the hologram is much larger than the sample. Part of the hologram is discarded during the reconstruction process such that no twin image information is mixed with the reconstructed image. The price is loss of some of the lowest sample frequencies. The scheme is analogous to the rejection of unwanted diffracted orders in zone plates by use of a central stop [6].

2. DEFINITION OF A SAMPLE PIXEL

Consider the situation shown in Figure 1. An optical system with a numerical aperture sinA is being used to image an object with light of wavelength λ. According to the Rayleigh Criterion, the smallest resolvable spacing between features (d') is given by [7]

$$d' = 0.61 \; \lambda/\sin A \tag{1}$$

This is directly related to the spatial frequency bandwidth, $K=\sin A/\lambda$, of the system where the aperture is taken to be well defined and circular and $K^2=K_x^2+K_y^2$.

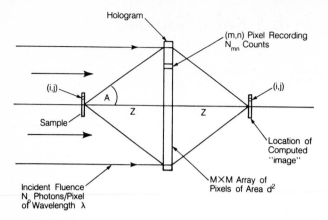

Fig.1. Notation for discussion of Gabor In-Line holography with a two-dimensional sample

The sample frequencies of a three-dimensional object that can be probed with an optical system of bandwidth K are limited by the requirement that the Cartesian components K_x, K_y and K_z of the probed frequency satisfy the Ewald Sphere [8] condition, namely

$$K_x^2 + K_y^2 + (K_z + K_o)^2 = K_o^2, \tag{2}$$

where $K_o = 1/\lambda$. In the small angle approximation, this leads to

$$K_z = K^2/K_o, \tag{3}$$

which leads to a longitudinal resolution (d_z), analogous to d', given by (1) and (3) as

$$d_z = d'^2/0.61\lambda. \tag{4}$$

We thus tentatively picture the resolvable pixel as a uniformly dense ellipsoid of revolution with semiaxes d', d' and $d'^2/0.61\lambda$. The small-angle value of the pixel volume is thus proportional to d'^4 (See Fig. 2). This argues that, if a given number of x-rays are required to be scattered by each pixel, then the number of incident x-rays (and hence the dose) needed for an experiment should scale as the fourth power of the resolution [9]. We shall see later that this argument is incomplete and that there is another type of d' dependence which must be included in calculating the dose.

Equation (4) also provides a measure of the amount of three-dimensional information that a given experimental geometry will provide in the absence of other types of limitation. It shows that present experiments, such as those described in reference 1, have a depth resolution, d_z, which is similar to the sample thickness indicating an essentially two-dimensional measurement. However, the form of (4) shows that quite a moderate improvement in resolution, say a factor of three, would give a factor nine improvement in depth resolution. It thus appears that x-ray holographic microscopy is on the verge of achieving an interesting three-dimensional capability.

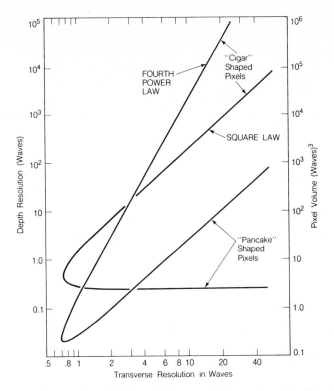

Fig. 2. Depth resolution (d_z)(square law), and pixel volume (V)(fourth power law), as a function of the transverse resolution (d) as determined by the Ewald sphere condition. The optical aperture is assumed to be circular and related to the bandwidth by $K=\sin A/\lambda$. The Ewald sphere condition then gives $K_z=-1\pm(1-2K^2)^{1/2}$, and the quantities actually plotted are defined by $d_z=1/2K_z$ and $V=d^2d_z$. Note that the region of "cigar-shaped" pixels corresponds to forward scattering by angle A and the region of "pancake-shaped" pixels to backward scattering by angle A.

The above picture of a resolution element contains a good deal of idealization. A more correct treatment would regard it as the point spread function of the system which is the Fourier Transform of the aperture function. In a real x-ray experiment the latter function is not determined by the size of a physical aperture. Instead it is a complicated entity including such effects as the decline in the detector modulation transfer function and signal-to-noise ratio with increasing signal frequency (aperture angle). These effects combine with the power spectrum of the sample and the corresponding angular distribution of scattered x-rays to give a signal which dies away gradually with increasing frequency [1,13] until it is lost in noise. It is not easy to model such a process but one might hope that the corresponding point spread function would not go negative.

3. NUMBER OF X-RAYS NEEDED TO ENCODE THE POSITION AND STRENGTH OF A PIXEL

Let us suppose that the detector shown in Fig. 1 is divided into an MxM grid of elements (detector pixels) of area $(1/2K)^2$. This is the sampling rate required by the Sampling Theorem for a signal of bandwidth K. Apart from a factor

1.22^2 it is also the same as the transverse area of the sample pixels. From now on we ignore factors of 1.22 and consider both the sample and the detector to be sampled on a square grid with spacing d=1/2K. With this definition, the small-angle pixel volume becomes d^4/2λ. Suppose an experiment consists of illuminating the system with a fluence of N_o x-rays per detector pixel. The sample, which is small compared to the detector, also receives this illumination and scatters a small fraction of it. The scattered waves interfere with the incident beam forming the hologram which is recorded as small signals modulated on to a strong reference wave. We wish to investigate the role of noise in the combined recording and reconstruction process. If we use a_{mn} and r to denote the sample and reference wave amplitudes at the (m,n) detector pixel, then the count rate N'_{mn} into that pixel is given by [10]

$$N'_{mn} = N_o \ (1 + a_{mn}^2/r^2 + (a_{mn} + a_{mn}*)/r).$$ (5)

The quantity N'_{mn} is actually a matrix of samples of a sample function of a two-dimensional random process. Further experiments would give a spread of values for the elements N'_{mn}, each of which would be a Poisson variate with variance N_o. Following Mueller [11] we subtract N_o from each N'_{mn} and neglect the second term because $|a_{mn}/r| \ll 1$ [12]. We thus arrive at

$$N_{mn}/N_o = (a_{mn} + a_{mn}*)/r \ ,$$ (6)

where N_{mn} also has variance N_o.

To compute the image, we chose to mimic the reconstruction of the real image with the original, plane reference wave. This gives us an aberration-free image [14], limited by noise. For the moment, suppose that the object is two-dimensional and let the (i,j) pixels of the object and reconstructed image be represented by the complex amplitude transparencies S_{ij} and T_{ij} respectively where $0 < |S_{ij}|,|T_{ij}| < 1$. Then, using the Fresnel approximation to the Huygens-Fresnel Principle [10], we get for the expectation of T_{ij}

$$E(T_{ij}) = F_{ij}(a_{mn}/r + a_{mn}*/r) \ ,$$ (7)

where F_{ij} is the Fresnel Transform defined as follows:

$$F_{ij}(a_{mn}) = (1/M) \sum_{m,n=0}^{M} a_{mn} \exp[(i-m)^2+(j-n)^2].$$ (8)

We have omitted a phase factor and used the fact that $d^2/\lambda z=1/M$. We now make the assumption, justified later, that the a_{mn} term in (6) and (7) can be neglected. The final result for T_{ij} then becomes, using (6),

$$E(T_{ij}) = F_{ij}(N_{mn})/N_o \ .$$ (9)

We can understand the meaning of these equations by recognizing that the intensity transmittance of the (i,j) pixel is $|T_{ij}|^2$, so that the number of x-rays P_{ij} emerging from the (i,j) pixel is given by

$$P_{ij} = N_o \ |T_{ij}|^2.$$ (10)

We also wish to calculate the variance of T_{ij}, and we can do it by recalling that the variance of N_{mn} is N_o and examining the effect of the operations carried out on a_{mn} in equations (8) and (9). Multiplication by the phase factors in (8) leaves the variance unchanged since they have modulus unity and summing over M^2 terms leads to the value $N_0 M^2$. Now, allowing for the factor $1/M$ in (8), we find that the variance of $F_{ij}(N_{mn})$ is N_0 [15]. Finally, we conclude from (9) that the variance of $|T_{ij}|$ is $1/N_0$. Summarizing, we have [16]

$$E(|T_{ij}|^2) = (P_{ij} + 1)/N_0, \tag{11}$$

which shows that the estimator of the intensity transparency consists of the x-ray signal scattered into the aperture plus a noise term which is essentially the shot noise of the reference beam [11].

The argument so far is essentially similar to that of Mueller [11], which treats the case of P_{ij} deterministic. However, in a real experiment, P_{ij} is actually a Poisson process with mean and variance equal to P_{ij}. In this context, N_0 can be regarded as exactly known. Furthermore, for any imaginable values of N_0 and P_{ij} the standard deviation of P_{ij} will be much greater than 1. We can, therefore, neglect the 1 in (11) and estimate requirements for P_{ij} on the basis of the number of gray levels required in the final image and the tolerance for errors in making the gray level assignments just as if we were doing a direct imaging experiment. We do not consider how to do this here. The problem has been considered by Rarback [17] who has found that a value $P_{ij}=10^3$ is a reasonable choice in practice.

The main conclusion of this section is that the dominant source of noise in the type of holography we are analyzing is shot noise in the number of x-rays, P_{ij}, scattered by each sample pixel; and that estimates of the number of incident x-rays and the corresponding dose to the sample needed to record a hologram can be based on a given choice for P_{ij}.

Until now we have focussed attention on a single pixel of an object that is assumed to be two dimensional. This view can also accommodate a three-dimensional object simply as a set of two-dimensional ones whose hologram records are made and read, with their appropriate z values, simultaneously with the original one. Similarly, the multiplicity of pixels of either two- or three-dimensional objects is also assumed to lead to a set of separate hologram records, each of which can be read back independently. This scenario is legitimate for a sufficiently sparse object and is similar to the practical case of holographic studies of aerosol particles [25]. We have chosen to analyze it because it allows a simple, sample-independent treatment. However, as the number of sample pixels increases, the phase delay and attenuation of the beam become significant, and the information capacity of both the wavefield and the detector become important. The analysis them becomes sample dependent, and the noise background against which each pixel record must be detected becomes more complicated. We recognize that under these conditions there will be speckle noise as well as shot noise, and this will lead to a requirement for higher doses and to limitations on the usefulness of single-exposure holography beyond those imposed by the Ewald sphere condition. However, we continue with the analysis of the more tractable "independent pixel" case with the understanding that this represents a somewhat optimistic view of what can really be accomplished.

4. THE SAMPLE PIXEL AS A COHERENT SCATTERER OF X-RAYS INTO THE SPECIFIED APERTURE

Continuing our assumption of a well-defined circular aperture (with radius h, say), the point spread function of the system is the Airy function $2J_1(\pi hK)/\pi hK$ which is obtained as the Fourier Transform of the circular aperture function [15]. We now consider the inverse problem to finding the point spread function. Consider a sample pixel whose scattering strength is distributed according to the Airy function and ask, what is the spatial coherence function for such a source? The Van Cittert-Zernike Theorem tells us immediately that it is the Fourier Transform of the Airy function, namely the uniform circular function that we started with. In other words, the sample pixel is a coherent scatterer into the aperture that defined it.

5. SCATTERING AND ABSORPTION CROSS SECTIONS

For a pixel with a scattering power of Z electrons we can write the cross section for scattering into the optical aperture as follows [18]:

$$\sigma_s = Z^2 r_o^2 \int_0^A \frac{1+\cos^2 A}{2}\, d\Omega \,, \tag{12}$$

where Ω is the solid angle and r_o the classical electron radius. For small angles, A, (12) reduces to

$$\sigma_s \simeq \pi r_o^2 A^2 (F_1^2 + F_2^2) \tag{13}$$

$$\simeq \pi r_o^2 d^6 s^2 \,, \tag{13a}$$

where F_1 and F_2 are Henke-style scattering factors [19] representing the integrated scattering power of the pixel in electrons and s is the scattering power of the material in electrons per unit volume. The small angle value for the pixel volume, $2d^4/\lambda$, is used to obtain (13a). Similarly, one can write for the absorption cross section [19]

$$\sigma_a = 2r_o \lambda F_2 \tag{14}$$

$$\simeq 4r_o d^4 a \,, \tag{14a}$$

where a is the absorbing power of the material in electrons per unit volume.

The important quantity for us is the ratio σ_a/σ_s. This is because it gives the number of absorbed x-rays (which determines the dose) per x-ray scattered into the detector. We see from (13a) and (14a) that σ_a/σ_s varies like $1/d^2$, showing that the dose problem improves with increasing d.

As a numerical example, we consider the case of protein [20] with $\lambda=30$ Å. In this case s=0.38 electrons/Å^3, and a=.15 electrons/Å^3. Taking $r_o=2.818 \times 10^{-5}$ Å, we get

$$\sigma_a/\sigma_s = 4.52 \times 10^4/d^2 \quad \text{(d in Å)} \tag{15}$$

indicating that σ_a/σ_s for a pixel equals unity when d=213 Å (which implies A=4.0°) This is a much more favorable situation than the case of a single atom scatterer for which the cross-section ratio is usually large, for example, 1.7×10^4 (C), 2.5×10^4 (N) and 2.7×10^3 (O). The single atom case is not important for experiments at 30 Å, however, since the smallest pixel volume allowed by the Ewald sphere condition (see Fig. 2) for that case is about 5700 Å^3, which corresponds to A≈45° and a transverse resolution of about 23 Å.

6. CALCULATION OF THE DOSE

It is now a simple matter to calculate the dose. The energy of an x-ray beam is deposited in the sample according to $I=I_o \exp(-\mu x)$ where x is the distance from the surface, I and I_o are the intensity at x=x and x=o respectively and μ is the absorption coefficient, given by $\mu=n\sigma_a$, where n is the atomic density.

It follows from this that the spatial rate of energy deposition at x=o is μI_o, which gives the following computational formula:

$$dose(rad) = 10^5 \mu(cm^2/gm) I_o(J/cm^2). \tag{16}$$

We now consider the case of protein with $\lambda=30$ Å and $d=200$ Å. We know that we need $P_{ij}=10^3$ scattered x-rays; and in view of the cross section ratio of 1, we need $2x10^3$ incident x-rays, which is 0.033 J/cm^2. Now, setting $\mu=1.8x10^4$ cm^2/gm, we get a dose of 58 Mrads.

We can also see how the dose scales with the resolution, d. We have seen that in the small angle case for given P_{ij}, the pixel volume scales like $1/d^4$. In addition we saw that the cross-section ratio scales like $1/d^2$. This gives the formidable conclusion that the dose scales like the sixth power of the resolution. As an example, the experiments in reference 1 required a dose of 200 Mrads with $\lambda=25$ Å, for a resolution of about 400 Å, which is roughly consistent with the above estimations for an idealized experiment.

7. THE TWIN IMAGE PROBLEM IN GABOR HOLOGRAPHY

The twin image problem in Gabor holography arises because the hologram is recorded according to (5) as

$$N'_{mn} = N_o(1 + 2|a_{mn}| \cos(\phi_{mn})/r) \tag{17}$$

where we have neglected the intermodulation term and taken $a_{mn}=|a_{mn}| \exp \phi_{mn}$ and r real. From this interference pattern we can, in principle, determine $|a_{mn}|$ and $\cos(\phi_{mn})$ and hence ϕ_{mn} up to an ambiguity in sign. This ambiguity corresponds to the physical fact that the sinusoidal fringe pattern of frequency K, say, could be formed either by two interfering beams of frequency zero (reference beam) and K (object beam) or two beams of frequency zero and -K. In other words, the object beam might be arriving "uphill" at angle A or "downhill" at angle A. In the type of holograms made by Gabor and shown in textbook illustrations of Gabor holography, the object is of similar size to the hologram so that both of these possibilities exist. In this case the reconstructed image contains the desired signal inextricably mingled with light from the twin image. There is no way to separate these signals and the conventional solution is to switch to an off-axis reference wave.

The use of off-axis reference waves is difficult in x-ray holography, and so we propose the following alternative approach. The key condition needed for the scheme to work is that the hologram be much larger than the object. We divide the hologram into two areas, first the region in the shadow of the object and second the remainder; that is, the great majority of the hologram area. In the first area the ambiguity described above prevails. Furthermore, the reference wave in this part of the hologram has passed through the sample and is, therefore, not a plane wave, as desired. We thus choose to reject this flawed information by suitable masking of the hologram (Fig. 2) after recording and use only the second area for reconstruction. This area has the advantage of an unperturbed reference wave and no ambiguity about the direction of the object beam. We can consider this approach as a somewhat distorted form of the off-axis geometry. Alternatively, we can regard it as the direct analogue of the use of a central stop to eliminate unwanted diffracted orders from zone plate lenses [6].

Either way, we need to understand what is lost in discarding the central part of the hologram. One can see from Fig. 3 that the central region contains information about sample Fourier components of frequency zero plus or minus a small spread in frequency given by $w/\lambda z$, where w is the half-width of the sample. This represents

269

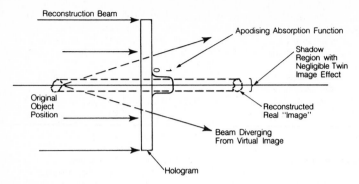

Fig. 3. Geometry of the scheme for addressing the twin-image effect of in-line holography. The diagram shows the illumination of the hologram by the original reference beam and the reconstruction of the two images, a virtual image at the original object position and a real image an equal distance downstream from the hologram. In the normal case, both the reconstruction beam and light from the virtual image would corrupt the signal at the real-image position. However, since the whole process is implemented numerically, it is possible to use a chosen absorption function to reduce this effect.

rejection of a fraction w/H of the frequency range where H is the half-width of the hologram. An initial idea of the impact on the image can be gained by examination of the point spread function for a centrally obscured aperture [20]. One can see that the point spread function is very close to the ideal (unobscured) one if the obscuration is less than about 10-20%. This is encouraging but is not the only issue. We naturally expect that the central stop will lead to some diffraction, a central (Poisson) spot, for example; and since the stop is applied by numerical processing, we are free to choose its size and transparency function to minimize the harmful effects of diffraction. The choice of an aperture function to optimize some aspect of the imaging properties of an optical system is known as the "Apodisation Problem" and has an extensive literature [26]. Studies continue regarding the best solution for this application.

Acknowledgements   The author wishes to acknowledge many helpful discussions with C. Jacobsen, J. Kirz, H. Rarback and D. Sayre. This work was supported by the U.S. Department of Energy under Contract No. DE-AC03-76SF00098.

References

1. M.R. Howells, C. Jacobsen, J. Kirz, R. Feder, K. McQuaid, S. Rothman: Science 238, 514 (1987).
2. M.R. Howells, M. Iarocci, J. Kirz: J. Opt. Soc. Am. A 3, 2171 (1986).
3. J.C. Solem, G.C. Baldwin: Science 218, 229 (1982).
4. S. Aoki and S. Kikuta: Jpn. J. Appl. Phys. 13, 1385 (1974).
5. R.J. Collier, C.B. Burckhardt, L.H. Lin: "Optical Holography," Academic, New York, 1971.
6. G. Elwert and J. Feitzinger: Optik 31, 600 (1970).
7. O. E. Myers, Jr.: Am. J. Phys. 19, 359 (1951).
8. See, for example, J. Cowley: "Diffraction Physics," North Holland, Amsterdam, 1981.

9. M.R. Howells, in "X-ray Microscopy," G. Schmahl, D. Rudolf, Eds., Springer-Verlag, Berlin, 1984. This paper contains a discussion of the scaling law for the number of x-rays as a function of resolution. It now appears that the scaling law derived there is incorrect.

10. See, for example, J.W. Goodman: "Introduction to Fourier Optics," McGraw Hill, San Francisco, 1968.

11. R. K. Mueller in "Applications of X-Ray Lasers," Physical Dynamics Rep. PD-LJ-76 132 (Physical Dynamics, La Jolla, California, 1976). A probability analysis of the problem of extracting information from photon limited interference fringes is provided by J. W. Goodman in "Statistical Optics," Wiley, New York, 1985. When the Goodman treatment using the Fourier Transform in one dimension and the Mueller treatment using the Fresnel Transform in two dimensions are properly compared, they agree apart from a factor of two arising from differences in the way negative frequencies enter in the two different linear transforms. This gives independent confirmation of (11).

12. This term is sometimes called the speckle term, although $|a_{mn}/r| \ll 1$ is not a sufficient condition for speckle to be negligible.

13. C. Jacobsen, J. Kirz, M. Howells, R. Feder, D. Sayre, K. McQuaid, S. Rothman: this volume.

14. The real image in this context is the downstream one, which, as explained in [10], is pseudoscopic but otherwise free of geometrical optical aberrations.

15. We are using the results that variances add algebraically for sums of random variables and that if $x=ky$ then $var(x)=k^2 var(y)$, where x and y are random variables and k a constant.

16. We are using the result $E(x^2)=(E(x))^2+var(x)$ where x is a random variable.

17. H. Rarback: private communication.

18. See, for example, J. D. Jackson: "Classical Electrodynamics," Wiley, New York, 1967.

19. The composition of protein is assumed to be $H_{50}C_{30}N_9O_{10}S_1$ as given in J. C. Solem: Los Alamos National Laboratory Report LA-9508-MS.

20. M. Born and E. Wolf: "Principles of Optics," Pergamon, Oxford, 1965, section 8.6.2.

21. D. Sayre et al.: Ultramicroscopy, 2, 337 (1977).

22. J. Kirz, D. Sayre, J. Dilger: Ann. N.Y. Acad. Sci. 306, 291 (1978).

23. J.C. Solem: J. Opt. Soc. Am. B, 3, 1551 (1986).

24. J. Cazaux: Appl. Surf. Sci. 20, 457 (1985).

25. See, for example, S.L. Cartright, P. Dunn, B.J. Thompson: Opt. Eng. 19, 727 (1980).

26. See, for example, P. Jacquinot and B. Roizen-Dossier: Prog. Opt. (ed. E. Wolf), Vol. III, Wiley, New York, 1964; and T. Wilson, C.J.R. Sheppard, Optik, 59, 19 (1981).

Experimental Observation of Diffraction Patterns from Micro-Specimens

D. Sayre[1], W.B. Yun[2], and J. Kirz[3]

[1]IBM Research Center, Yorktown Heights, NY 10598, USA
[2]Argonne National Laboratory, Argonne, IL 60439, USA
[3]Department of Physics, State University of New York,
  Stony Brook, NY 11794, USA

1. INTRODUCTION

In an earlier paper [1] we calculated the photon scattering cross-section of a small specimen, averaged over the elementary Shannon volumes in diffraction space, as $r_0^2 \lambda^3 m Z^2$ for the Shannon volumes in the total Ewald sphere (case of multiple orientations of the specimen in the beam, and full 3-dimensional imaging), and a/λ times that for the Shannon volumes on a single Ewald sphere (case of a single orientation in the beam, and partial 3-dimensional imaging). Here r_0 is the classical electron radius, λ is the photon wavelength, m is the mean number of atoms per unit volume of the specimen, Z is the rms atomic number of the atoms of the specimen, and a is the specimen diameter. By a small specimen is here meant one for which a is less than an absorption length for the photons. The elementary Shannon volumes pick out the informationally distinct scattering directions for the general (i.e. non-periodic) small specimen considered here. In the soft x-ray region nearly all scattered photons are scattered without energy loss, and the scattered photon pattern is conveniently spoken of as a diffraction pattern.

These formulae, assuming they are correct, are basic to x-ray imaging microscopy and holography as well as diffraction imaging, as all three methods are scattered-photon imaging methods. Part of the purpose of this paper is to cite evidence that the formulae are at least generally correct.

One piece of evidence was supplied by YUN [2]. The diffraction pattern from a uniform sphere is known analytically [3]; Yun constructed (in the computer) a model specimen consisting of uniform protein spheres of radius 1.25nm placed at random in a water sphere of radius 40nm, and for a given orientation of the specimen numerically calculated its diffraction pattern, on an absolute scale, over a fairly large range of diffraction space (Fig. 1). Yun showed that the average number of photons per Shannon volume for this diffraction pattern is in good agreement with the second of the formulae above.

Based on these formulae, one may calculate expected exposure times for the observation of diffraction patterns. For a mean 5% error in observing

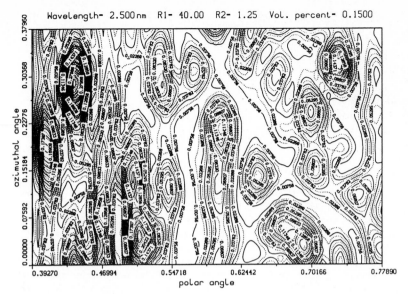

Wavelength- 2.500nm   R1- 40.00   R2- 1.25   Vol. percent- 0.1500

Fig. 1. Part of the calculated diffraction pattern of a model protein-water specimen

diffraction amplitudes (mean of 100 photon counts per Shannon volume), one obtains an estimate of 2 minutes for a single orientation of a $1\mu m$ biological specimen on a soft x-ray undulator source like that being installed on the X1 beamline at NSLS, and 1/2 day for collecting the total Ewald sphere. Note that the latter figure is independent of specimen size; as the specimen size decreases, the exposure per orientation increases, but the number of orientations required decreases. These estimates strike us as reasonably favorable for the future of scattered-photon imaging.

2. OBSERVATION OF PATTERNS USING THE U15 BEAMLINE

In the present paper, the experimental results reported are from a lower-brightness source, the bending-magnet source on the U15 beamline at NSLS. For this source the exposure-time estimate for a single orientation for a $1\mu m$ biological specimen was 10 days, which was deemed impractical. For this reason the specimens used in this paper were larger (a = 10 to $20\mu m$). Assuming the formula for the single orientation case to be correct, we expected to be able to observe diffraction from these specimens with 1/2- to 1-day exposures.

This proved to be correct. An apparatus was built [4] in which 3.2nm photons from the U15 monochromator were passed through a carefully aligned 3-pinhole collimator to form a $20\mu m$ beam with minimal stray radiation outside the beam. The specimen was placed in this beam and the scattered photons allowed to fall on a Kodak type 101 soft x-ray silver halide plate placed several centimeters downstream from the specimen. Fig. 2 shows the pattern

273

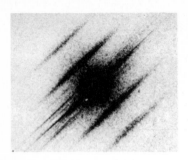

Fig. 2. Diffraction pattern of a diatom in 3.2nm x-rays

observed after a few hours exposure with a diatom with a diameter of $20\mu m$ as specimen. Evidence that the pattern is a diffraction pattern is that it (a) rotates with the specimen, (b) disappears with removal of the specimen, (c) disappears with the insertion of a soft x-ray opaque filter ($1\mu m$ mylar), (d) is unchanged by the insertion of a soft x-ray transparent filter ($0.1\mu m$ aluminum), and (e) contains prominent maxima which correspond to a spacing of 203nm, corresponding to prominent features in SEM images of the specimen with average spacings of 201nm. Patterns were also recorded with similar exposure times with muscle and bone as specimens. These patterns are believed to be the first such patterns recorded from microscopic, non-crystalline specimens.

3. COHERENCY CONSIDERATIONS

A second feature of the U15 patterns also supports the belief that our understanding of the diffraction signal intensity is essentially correct. For the diffraction pattern to have good visibility at diffraction angle θ, the uncertainty in position in diffraction space at angle θ caused by small variations $\Delta\lambda$ in the wavelength and $\Delta\theta_0$ in the direction of arriving photons must be less than the size of a Shannon volume. Quantitatively [2]

$$2(1 - \cos\theta)(\Delta\lambda/\lambda^2)^2 + (\Delta\theta_0/\lambda)^2 \leq (1/a)^2 , \qquad (1)$$

or for small θ,

$$\theta^2 \lesssim (\lambda/\Delta\lambda)^2 [(\lambda/a)^2 - (\Delta\theta_0)^2].$$

Inserting values of $\Delta\lambda$, $\Delta\theta_0$, λ, a for the U15 experiment, we conclude that the diffraction pattern should not be visible in that experiment at diffraction angles θ greater than about 5°. The results of the U15 experiment confirm this conclusion.

This point, confirmatory as it is of the theory, raises the question whether, in an X1 experiment, we may expect diffraction to be obervable to the full diffraction angle of 180°. For $\lambda = 3.2$nm, here are two sets of parameter values satisfying (1):

	(A)	(B)
Beam divergence $\Delta\theta_0$ (mr)	.1	1
Monochromaticity $\lambda/\Delta\lambda$	2000	600
Specimen size a (μm)	3.2	1
Maximum θ	180°	180°

Set (B) is well within the design range of the X1 beamline, and set (A) is about at the limit of the design range. Taken with the earlier definition of a small specimen and estimates of exposure time, X1 should thus permit observation of soft x-ray diffraction patterns from small specimens to $\theta = 180°$.

4. SUMMARY

The findings of this paper, partly experimental and partly theory, are that even small non-periodic specimens scatter soft x-rays strongly enough to permit scattered-photon imaging. For the type of imaging principally studied here (diffraction-pattern imaging), the requirements on divergence and monochromaticity of the illumination are also moderately well understood, and appear to be capable of being sufficiently well met by undulator beamlines to permit imaging to the diffraction-limited resolution of $\lambda/2$, provided of course that a method of phasing the diffraction pattern is available. It may be remarked that for holography the illumination requirements, while varying with the form of holography, are more severe than for diffraction-pattern imaging, while for imaging microscopy they may be considerably less severe, depending on the angular and frequency bandwidths of the imaging element. It would appear to be for this reason that imaging microscopes are currently achieving exposure times of a few seconds even on bending-magnet beamlines [5].

REFERENCES

1. D. Sayre, R.P. Haelbich, J. Kirz, W.B. Yun: In X-Ray Microscopy, ed. by G. Schmahl and D. Rudolph, Springer Ser. Opt. Sci., Vol.43 (Springer, Berlin, Heidelberg 1984), p.314
2. W.B. Yun: Ph.D. thesis (State University of New York at Stony Brook 1987)
3. B.L. Henke: In Low Energy X-Ray Diagnostics, ed. by D.T. Attwood and B.L. Henke, AIP Conf. Proc., Vol.75 (American Institute of Physics, New York 1981) p.146
4. W.B. Yun, J. Kirz, D. Sayre: Acta Cryst. A43, 131 (1987)
5. J. Thieme: This volume

An X-Ray Microprobe Beam Line for Trace Element Analysis

B.M. Gordon[1], A.L. Hanson[1], K.W. Jones[1], W.M. Kwiatek[1], G.J. Long[1], J.G. Pounds[1], G. Schidlovsky[1], P. Spanne[1], M.L. Rivers[2], and S.R. Sutton[2]

[1]Brookhaven National Laboratory, Upton, NY 11973, USA
[2]University of Chicago, Chicago, IL 60637, USA

1. INTRODUCTION

The application of synchrotron radiation to a x-ray microprobe for trace element analysis is a complementary and natural extension of existing microprobe techniques using electrons, protons, and heavier ions as excitation sources for x-ray fluorescence. This was first recognized by HOROWITZ and HOWELL [1] in their development of the first synchrotron radiation microprobe at the Cambridge Electron Accelerator. SPARKS, et al. [2] used a miniprobe beam at the Stanford Synchrotron Radiation Laboratory in an attempt to find natural occurring superheavy elements by x-ray fluorescence of characteristic L-lines. The ability to focus charged particles leads to electron microprobes with spatial resolutions in the sub-micrometer range and down to 100 ppm detection limits and proton microprobes with micrometer resolution and ppm detection limits. The characteristics of synchrotron radiation that prove useful for microprobe analysis include a broad and continuous energy spectrum, a relatively small amount of radiation damage compared to that deposited by charged particles, a highly polarized source which reduces background scattered radiation in an appropriate counting geometry, and a small vertical divergence angle of ~ 0.2 mrad which allows for focussing of the light beam into a small spot with high flux. The features of a dedicated x-ray microprobe beam line developed at the National Synchrotron Light Source (NSLS) are described.

2. EXPERIMENTAL

In the first stage of development described here, the continuous spectrum was used with spot sizes produced by a set of remotely-controlled collimator slits. Appropriate absorbers were used to alter the excitation spectrum to maximize the sensitivity for a desired element or range of elements. The photon excitation beam, after traversing the collimator system, passes through a helium-filled ionization chamber for beam monitoring after which it falls on the target, held on a computer-controlled X-Y-Z-θ translational stage with a 1-μm reproducibility. The target, held at 45° to the beam, was observed by a Si(Li) detector positioned at 90° to the excitation beam in order to minimize the scattered radiation background. The target was continuously viewed by a microscope and video camera attachment. The slit system was capable of producing beams as small as 20 x 20 μm². GIAUQUE, et al [3] used a Lawrence Berkeley Laboratory (LBL) designed synthetic multilayer focussing system to achieve a monochromated 10-keV beam with 10-μm resolution and 1-keV energy spread.

The sensitivity for beam spot analyses in thin (5 mg/cm^2) low Z targets was approximately 0.5 ppm for first row transition elements using an unfocussed continuous spectrum at 75 μm resolution in a 300-sec run. Using the LBL focussing system, a 10 × 10 μm^2 beam spot using 10 keV radiation yielded a range of 2 ppm for Cu to 5 ppm for Mn with the absolute detectable weight being in the 10- to 20-fg range.

The present scanning technique uses a point microbeam and the target is scanned both vertically and horizontally across the beam. Regions on interest (ROI) are set up for each elemental x-ray fluorescence peak to be determined as shown in fig. 1 by the hatched areas. The integrated counts for each ROI, totally or with background subtracted, is determined on-line and recorded in the computer for off-line analysis. The spectrum is of a fetal rat bone explant after being cultured in a medium containing 25 μM Ga(NO$_3$)$_3$, for a study in the use of gallium in cancer therapy. A 30-μm section was made of the bone and an area of 1 × 3 mm^2 was scanned with a 40 × 40 μm^2 pixel size and data collected at the rate of 2 sec/pixel. The right side of the section contained the bony tissue whereas the left side was made up of cartilage. The results can be illustrated by three-dimensional histograms, color contour pictures and, as illustrated in fig. 2, contour maps where the tick marks indicate the direction of lesser counts, i.e. downhill. The calcium, zinc, and gallium are seen to be held within the bony structure as well as most of the iron. The copper, whose origin is not known at present, resides principally on the surface of the cartilage. The highest integrated values for counts/pixel ranged from 15,000 for Ca, 5000 for Ga to 400 for both Zn and Cu.

Other special applications of the x-ray microprobe beam line include the trace element analysis of living cells, made possible by the unique characteristics of synchrotron radiation enumerated above. A prototype wet cell was designed, constructed, and tested using cat cardiac myocytes,

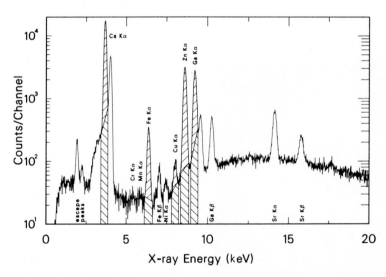

Fig. 1. X-ray fluorescence spectrum of fetal rat bone section after administration of gallium. The shaded areas are regions of interest for the elements being determined.

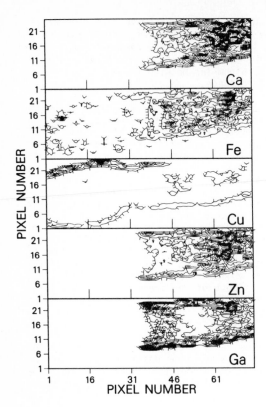

Fig. 2. Contour plots for the distribution of counts/pixel for determined elements. Tick marks point in direction of lesser counts.

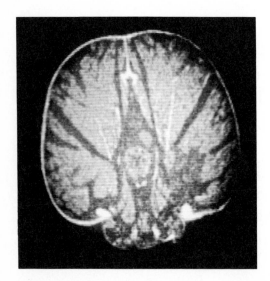

Fig. 3. Transmission tomograph of freeze-dired caterpillar head. The pixel size is 30 x 30 μm² and the scan dimension is 177 x 177 pixels.

and major trace elements such as iron could be quantitated in single myocytes. Nutrient solution can be passed through the irradiation cell, which essentially is made up of two Kapton foils separated by a Kapton spacer foil with the perimeter of the cell cut out. Thusly, cells of about 10-μm depth can be used.

Figure 3 illustrates a transmission tomogram of a freeze-dried caterpillar head with a 30-μm pixel size. This example of microtomography was made in 50 min using both a translational and a rotational stage. In the future, mappings of trace elements will be made in cross section (depth) by detection of fluorescence x rays with a point beam or in thin samples using a line source, as illustrated by KNOCHEL et al. [4] using the light source at DESY.

3. ACKNOWLEDGMENTS

Work supported in part by Processes and Techniques Branches, Division of Chemical Sciences, Office of Basic Energy Sciences, US Department of Energy, under Contract No. DE-AC02-76CH00016; applications to biomedical problems by the National Institutes of Health as a Biotechnology Research Resource under Grant No. P41RR01838; applications in geochemistry by National Science Foundation Grant No. EAR-8618346; and applications in cosmochemistry by NASA Grant No. NAG 9-106.

4. LITERATURE REFERENCES

1. P. Horowitz and J. A. Howell: Science 178, 608 (1972)
2. C. J. Sparks, Jr., S. Raman, H. L. Yakel, R. V. Gentry, M. O. Krause: Phys. Rev. Lett. 38, 205 (1977)
3. R. D. Giauque, A. C. Thompson, J. H. Underwood, Y. Wu, K. W. Jones, M. L. Rivers: Anal. Chem., submitted
4. A. Knochel, M. Bavdaz, N. Gurker, P. Ketelsen, P. Petersen, M. H. Salehi, T. Dietrich: Proceedings of Symposium on Accuracy in Trace Analysis, National Bureau of Standards, Gaithersburg, Maryland, September 28, 1987, to be published

Possibilities for a Scanning Photoemission Microscope at the NSLS

H. Ade[1], J. Kirz[1], H. Rarback[2], S. Hulbert[2], E. Johnson[2], D. Kern[3], P. Chang[3], and Y. Vladimirsky[4]

[1]Department of Physics, State University of New York,
Stony Brook, NY 11794, USA
[2]National Synchrotron Light Source, Brookhaven National Laboratory,
Upton, NY 11973, USA
[3]IBM T.J. Watson Research Center, Yorktown Heights,
NY 10598, USA
[4]IBM T.J. Watson Research Center, Yorktown Heights,
NY 10598, USA
and Center for X-Ray Optics, LBL, Berkeley, CA 94720, USA

1. INTRODUCTION

Of the three main surface analysis techniques: SIMS (Secondary Ion Mass Spectrometry), XPS (X-ray Photoelectron Spectroscopy) and e^-AES (electron induced Auger Electron Spectroscopy) , both SIMS and e^-AES have lateral resolution in the 50 nm range [1]. Yet only XPS makes use of the wealth of chemical shift information. Recent publications on radiation damage [2,3] estimated (considering only microscopic causes of the damage) that for the same lateral resolution and induced damage, XPS is one to two orders of magnitude more sensitive then e^-AES. This is mainly due to the intrinsic difference in the signal/background ratio of the two techniques, which is 10 to 100 for XPS and 0.1 to 1 for e^-AES. Taking into account macroscopic damage, such as local heating and charging, the advantages of XPS are even greater, particularly for polymers and other radiation sensitive materials. Experimental investigation is required, however, to fully assess the advantages of an XPS microscope over other techniques.

It is clearly desirable to improve the lateral resolution of XPS, which for practical instruments is now several microns. In the microprobe approach there are three different types of focusing schemes which are in principle able to overcome this practical limit on resolution: Schwarzschild Objectives (SO), Wolter Objectives (WO) and Zone Plates (ZP). SPILLER [4] is building a microscope with a multilayer SO. TRAIL [5] is in the process of setting up an x-ray laser based multilayer Schwarschild microscope, F. CERRINA [6] is planning to build a microscope at Aladdin (SRC Wisconsin) with a SO specifically for XPS/UPS. FRANKS [7] is building a Wolter type microscope. Both the multilayer based Schwarzschild microscope and the grazing incidence Wolter microscope might be particularly attractive with smaller, cheaper, laboratory sources. Present day SO can only work efficiently with photon-energies up to 280 eV, and it is unclear if the Wolter type can go to very high resolution. ZPs, however, have already demonstrated resolutions of 50-75 nm at 400 eV [8] and present Au ZPs will work efficiently with photon energies up to 800 eV. We will therefore concentrate on the energy region between 400 and 800 eV, which gives us access to spectral peaks from most of the elements, and in particular from the K-shell of oxygen, nitrogen and carbon. With applications to different energy regions (multilayer microscopes up to 280eV, Wolter microscopes up to several keV) these different x-ray focusing instruments tend to complement each other. ZPs together with a dedicated source such as an undulator should therefore make a useful instrument for photoemission microscopy and are the basis of our instrument.

2. SOURCE AND BEAMLINE

The proposed instrument is designed for the X1A undulator beamline at the NSLS. The coherent output of the X1 undulator in the energy range from 400 to 800eV is calculated [9] to be about 4×10^{11} photons/s/0.1%BW/200mA. Including expected losses in the beamline (2 reflections and monochromator $\simeq 5\%$, ZP efficiency $\simeq 4\%$) we expect to have 8×10^8 photons/s/0.1%BW/200mA in a diffraction limited spot on the specimen. The spot size sets the resolution and depends critically on the ZP. Some of the proposed ZP parameters are listed in table I. In a first stage instrument the expected resolution will be $\sim$120 nm. Anticipating continuing progress in ZP fabrication, resolution below 50 nm may become possible in the future. The flux on the sample can be increased by trading off energy and/or spatial resolution, but should be in most cases sufficient to allow for reasonable data aquisition times.

TABLE I: ZP characteristics

	ZP50	ZP100
Diameter	125 μm	125 μm
Central stop diameter	50 μm	50 μm
Central stop thickness	1000 nm	1000 nm
Zone thickness	120 nm	150 nm
Number of zones	625	312
$\lambda/\Delta\lambda$ required	625	312
Focal length (λ=1.6 nm/2.4 nm)	3.9 mm/2.6 mm	7.8 mm/5.2 mm
Depth of field (λ=1.6 nm/2.4 nm)	4.0 μm/2.7 μm	15.6 μm/10.4 μm
Δr	50 nm	100 nm
Resolution (theor.)	61 nm	122 nm

3. XPS MICROPROBE

The device will reside inside a 6-way cross, with five 8" ports and one 10" port, with the scanning stage mounted off the mating 10" flange. The upstream port will hold a Cylindrical Mirror Analyser (CMA) with the x-rays passing through its center. The focusing ZP and the Order Selecting Aperture (OSA) are mounted from the CMA. The remaining ports are used for the ion pump and other useful auxiliary equipment, such as Ar-ion sputtering, e^-gun, partial yield detector for ZP/OSA alignment, etc.

In order to remove the unwanted orders of the ZP, one places the OSA (a pinhole) in a strategic place, as shown in Fig.1. This arrangement increases significantly the signal

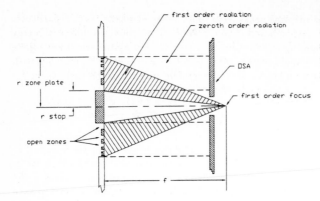

Fig.1 Side view of the ZP and OSA, showing central stop and placement of OSA to eliminate background radiation

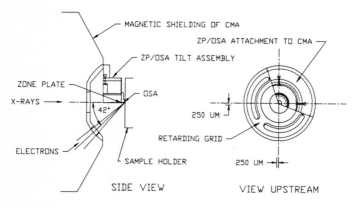

Fig.2 Side view and view along optical axes (upstream) of the CMA and ZP/OSA tilt mechanism. 2/3 of the CMA acceptance cone is unobstructed

to background ratio, but it can also block the electrons emerging from the surface. In our design the mount for the pinhole and the pinhole itself extend only over 130° in the plane perpendicular to the x-ray beam, as shown in Fig.2. The dimensions in the other 230° are kept small enough (250 μm from the center of the pinhole, which has 50 μm diameter) to keep 2 of the 3 grid holes of the CMA free from obstruction. For proper operation the ZP and OSA have to be carefully aligned. To align them with respect to each other, they are mounted on a 130° section of a tube, which is connected by the corresponding section of a flexible diaphragm to a larger tube, thus allowing them to be tilted.

The specimen will be scanned in discrete steps with a stepping motor driven demagnifying stage. The proposed instrument will use a micro stepping motor with 1 μm cardinal step size and up to 30 micro steps depending on the interchangeable PROMs of the micro stepper drive used. The demagnifying stage is a flexible pivot stage with a 1/20 reduction. This results in 50 nm steps of the specimen, further subdivided by the microsteps. The stepping motor based design allows for a very clean vacuum. With the motion being fed through bellows to the aluminum alloy stage inside, there are only well established

UHV-compatible materials in vacuum. We rejected alternative actuators, such as piezos and magnetic coil drives mounted directly on the stage, as they, besides other shortcomings, produce electric and magnetic fields, which complicate the electron energy analysis. Data aquisition and stepping is done via a Camac crate and a slave computer, which is networked to the host MicroVAX II. The instrument will have two modes of operation. In the imaging mode one tunes an energy window of the single pass CMA to the spectral peak of an element to map the spatial distribution of this element. In the analysis mode the stage remains fixed and the electron kinetic energy is swept (EDC) to get detailed information on the composition and the chemical state at this particular point.

4. SUMMARY

The proposed instrument will scan the specimen across a small X-ray probe formed by a ZP. Photoelectrons will be analyzed by a CMA to yield information on surface structure with a resolution of 120 nm or better. Good signal/noise and low damage should make this a powerful and versatile instrument.

5. ACKNOWLEDGEMENTS

The authors would like to acknowledge stimulating discussions with Mike Browne from Kings College, London. The work at Stony Brook is supported by the National Science Foundation, the NSLS and the Center for X-Ray Optics are supported by the Department of Energy.

REFERENCES
1. for SIMS see for example: R. Levi-Setti, G. Crow, Y. L. Wang: Scanning Electron Microsco. (1985) II , p.535-551; and A. R. Bayly, A. R. Waugh and P. Vohlralik: Spectrochimica Acta, 40B (1985), p.717
   for e^-AES see for example: J.Cazaux: J.Micros. 145 (1987) p.257
2. J. Cazaux: In Application of Surface Science 20 (North-Holland 1985) p.457
3. J. Kirschner: In X-Ray Microscopy , ed. by G.Schmahl and D.Rudolph, Springer Ser. Optical Sci., Vol.43 (Springer, Berlin, Heidelberg, 1984) p.308
4. Spiller: In X-Ray Microscopy , op.cit., p.226
5. J. A. Trail and R. L. Byer: In Applications of Thin-Film Multilayer Structures to Figured X-Ray Optics, SPIE 563 (1985) p.90 and this volume
6. F. Cerrina, G. Magaritondo, J. H. Underwood, M. Hettrick, M. Green, L. J. Brillson, A. Franciosi, H. Hoechst, P. M. Deluca and M. N. Gould: Proc. of the 5th National Conference on Synchrotron Radiation Instrumentation, Madison, Wisconsin (1987) (to be published in Nucl. Instr. Meth.)
7. Franks: In X-Ray Microscopy , op.cit., p.129
8. H.Rarback, D.Shu, S.C.Feng, H.Ade, J.Kirz, I. McNulty, D. P. Kern, T. H. P. Chang, Y. Vladimirsky, N. Iskander, D. Attwood, K. Quaid and S. Rothman: Rev. Sci. Instr. (to be published about Jan. 1988)
   D. Rudolph, B. Niemann, G. Schmahl and O. Christ: In X-Ray Microscopy , op.cit., p.192
9. C.Jacobsen and H.Rarback: International Conference on Insertion Devices for Synchrotron Sources, SPIE 582 (1986) p.201

Design for a Fourier-Transform Holographic Microscope

W.S. Haddad[1], *D. Cullen*[1], *K. Boyer*[1], *C.K. Rhodes*[1], *J.C. Solem*[1], *and R.S. Weinstein*[2]

[1]MCR Technology Corporation,
P.O. Box 10084, Chicago, IL 60610, USA
[2]Corabi International Telemetrics, Inc.,
2201 Campbell Park Drive, Chicago, IL 60612, USA

1. Abstract

A design for a Fourier-transform holographic microscope to operate toward the lower end of the water window is presented. The x-ray beam illuminates both the reference scatterer and the specimen, which can be simultaneously dropped into the path of the beam or statically mounted on a very thin foil. Charge-coupled devices are used for recording, their output being digitized and linked directly to a computer.

A critical element of this design is the reference scatterer, which is a grazing incidence reflecting sphere. A survey of the periodic table shows that nickel is the optimum reflector near the carbon K-edge while osmium is optimum near the nitrogen K-edge. Both elements are sufficiently reflective to make this form of microholography feasible. It has been found that nickel spheres in the 2–10 μm range can be fabricated to a very high degree of surface smoothness and roundness, sufficient for this application.

A unique computer algorithm for reconstructing Fourier-transform holograms is described. The results of a numerical experiment demonstrate the efficacy of this algorithm and clearly indicate that the extension to practical x-ray holography is feasible.

2. Introduction

In recent years, many researchers have been engaged in the development of laboratory x-ray lasers[1]. The impetus for much of this effort has been to create a source of coherent x-rays for holographic imaging of hydrated biological specimens. The characteristics of a laser source, as opposed to pulsed plasmas or synchrotrons, dictate a very special holographic apparatus. X-ray lasers will produce exceedingly high intensity beams with very narrow divergence. High intensity makes "snapshot" imaging practical[2]. Narrow divergence mitigates against Fresnel transform holography[3] because this form of holography requires a large area to be illuminated with planar reference radiation, which would require impractical distances between laser and recording surface.

A means has been devised to advantageously utilize the narrow divergence by exploiting the unique features of Fourier transform holography[4]. In this approach, diverging reference waves are generated by reflecting the laser beam from a tiny sphere. Both the sphere and specimen could be dropped into the beam, which is envisioned as the ultimate realization of this technology, or they could be statically mounted on a very thin foil, which could provide a near-term realization. The static mounting could also be used in prototype experiments with synchrotron sources. Since Fourier–transform holography does not require a high–resolution recording medium, we could use a mosaic of charge coupled devices (CCD's), which are efficient, linear, and could be linked directly to a computer. Development of computer algorithms to translate the CCD–recorded hologram into images and data useful to the life scientist is an important part of our research program.

In this paper, we discuss the status of research development of each element of this holographic apparatus.

3. Reference Scattering Sphere

The reference scatterer must be a sphere because its spatial orientation cannot be controlled. Unlike the parabolic reference scatterer usually discussed in connection with spherical Fourier transform[5] holography, the spherical reference scatterer does not produce true spherical waves –– they emanate from a line rather than a point and are spherical only at large distance.

Most life scientists believe that the interesting wavelength region for x–ray microscopy (in the near term) is the "water window", the band between the $K\alpha$ edges of carbon (44.7 Å) and oxygen (23.6 Å). Within the water window the greatest contrast appears at the $K\alpha$ edge of nitrogen (31.6 Å) for both nucleic acid and protein[5]. Specular reflectivity at moderate angles deteriorates with decreasing wavelength, therefore only two wavelengths are of interest: 44.7 Å, water–window threshold; and 31.6 Å, maximum contrast. The penalty of reduced reflectivity and consequent reduction in area of reference illumination would not be adequately compensated by higher resolution at the short–wavelength end of the water window.

Figure (1a) shows the λ = 44.7 Å reflectivity[6] at grazing angles of $10°$ and $14°$ for 30 elements. The elements were chosen to be representative of regions of the periodic table and clustered around atomic numbers where local maxima were expected. Nickel (Z = 28) is clearly the best reflector for λ = 44.7 Å at grazing angles up to about $14°$ and is surpassed only by rhenium (Z = 75) and osmium (Z = 76) at greater angles. The reference illumination scattered from nickel is probably adequate for good

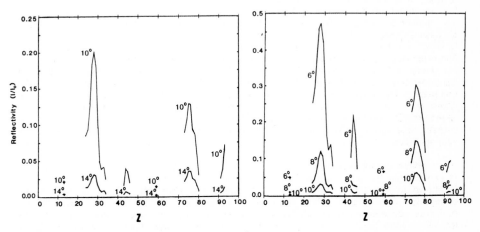

Figure 1: (a) Survey of reflectivities at 44.7 Å: $10°$ and $14°$. (b) Survey of reflectivities at 31.6Å: $6°$, $8°$, and $10°$.

285

fringe visibility up to a grazing angle of about 15° where it is 3% of the incident intensity. X-rays scattered from the specimen will generally be less intense at greater angles, so the mismatch with reference illumination will not be great. A grazing angle of 15° corresponds to a scattering angle of 30°, thus about $\pi/9$ steradians will receive adequate reference illumination. Fig. (1b) shows the 30-element survey at λ = 31.6 Å, the wavelength of maximum contrast. Osmium is superior to nickel at 10° where its reflectivity is about 6%. The maximum usable scattering angle is about 22°, corresponding to about $\pi/17$ steradians.

Transverse resolution is limited to about $\lambda/2\theta$, where θ is the scattering angle, while longitudinal resolution[7] is limited to about $\lambda/4\theta^2$. The usable scattering angles for both the 44.7 Å and 31.6 Å case are sufficient to not be a critical factor in limiting either transverse or longitudinal resolution. Barring other factors, resolution should approach the diffraction limit.

A further requirement of the reference sphere is that it be round and smooth. Lack of roundness will result in geometric distortions of the reconstructed image, which will affect quantitative interpretation but will probably not affect qualitative interpretation. Surface roughness, however, could introduce speckle that would severely inhibit accurate reconstruction. We have examined a variety of microspheres fabricated by differing techniques. Scanning electron micrographs reveal that some commercially available spheres already have the requisite roundness and are smooth to less than 20 Å. More advanced fabrication techniques may reach smoothness of 3 Å.

4. Data Acquisition and Analysis

CCDs[8] capable of quantum efficiencies as high as 80% in the soft x-ray regime can be prepared by removing most of the backing layer of a conventional CCD. Such CCDs are available now with 2000 x 2000 pixels and 10^4 x 10^4 seems feasible, especially if they can be assembled in a mosaic. Spatial resolution is important in Fourier transform holography only to constrain the overall size of the instrument. Pixel size of about 10 µm is available now and 1 µm seems achievable in the near future, thus compact "table-top" instruments should be possible.

We have developed and demonstrated a computer algorithm for reconstructing Fourier transform holograms. This algorithm does not utilize a Fourier transform of the numerical data. Contrary to its name, Fourier transform holography maps physical distances uniformly into spatial frequencies only when the recording surface is very far from the specimen and reference source, or when the paraxial[9] approximation applies. Our approach is to calculate a set of "basis functions", which are holograms of individual points in the reconstruction volume. The image is reconstructed by "projecting" out from the hologram function, the component of each of these basis functions. The necessary calculations are carried out with the help of a fast numeric coprocessor, and precalculated basis function data is stored on high-volume high-speed media to further improve the efficiency of the reconstruction process. Of course, distortions associated with numerical operations of this type will be a limiting factor.

We have demonstrated the efficacy of this algorithm in computer experiments and in experiments using a CCD and a visible laser. The essence of a computer experiment is shown in Fig. (2). In the interest of speed and simplicity, one of the transverse dimensions was ignored. The specimen was the "happy face" shown in Fig. (2a), its horizontal dimension in the longitudinal direction and its vertical dimension in the transverse direction. This two-dimensional specimen was then numerically mapped into the one-dimensional hologram shown in Fig. (2b). The reconstruction algorithm was then applied resulting in the picture shown in Fig. (2c). This successful demonstra-

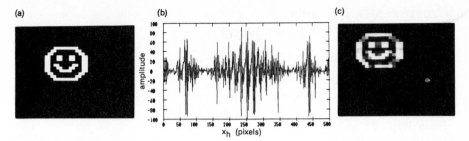

Figure 2: Numerical reconstruction of Fourier-transform hologram: (a) original, (b) hologram, (c) reconstruction.

tion is especially satisfying considering that it was performed with the limited computing capacity of an IBM PC/AT. Can you see the happy face in Fig. (2b)?

5. Conclusion

We have demonstrated the essential elements for a laser driven x-ray holographic microscope appropriate for imaging hydrated biological specimens. All of the elements we have discussed are presently available or within the scope of near future technology. The element most crucial to its ultimate realization is the x-ray laser itself.

6. Acknowledgements

This work has been supported by the Department of Energy, Office of Basic Energy Sciences, Division of Advanced Energy Projects, under Grant No. DE-FG02-86ER13610.

7. References

1.   For a summary of recent progress, see 1987 Annual Meeting, Optical Society of America, Technical Digest, p. 52, Session TuC. OSA, 1816 Jefferson Place, N.W., Washington, D.C. 20036.

2.   J. Solem and G. Baldwin, Science 218, 229 (1982); J. Solem, J. Opt. Soc. Am. B 3, 1551 (1986).

3.   D. Gabor, Nature 161, 777 (1948); D. Gabor, Proc. R. Soc. London Ser. A197, 454 (1949).

4.   J. Winthrop and C. Worthington, Phys. Lett. 15, 124 (1965); G. Stroke and R. Restrick, Appl. Phys. Lett. 7, 229 (1966).

5.   J. C. Solem and G. F. Chapline, Opt. Eng. 23, 193 (1984).

6.   A. H. Compton and S. K. Allison, X-Rays in Theory and Experiment, Van Nostrand (1935). Also see U. Fano and J. Cooper, Rev. Mod. Phys. 40, 441 (1968).

7.   A. Kondratenko and A. Skrinsky, Optical Information Processing (Plenum, New York, 1978), Vol. 2, p. 1.

8.   J. Janesick et. al., Rev. Sci. Instrum. 56, 796 (1985); K. Marsh et. al., Rev. Sci. Instrum. 56, 837 (1985); D. Lumb, G. Hopkinson, and A. Wells, Advances in Electronics and Electron Physics (Academic, London, 1985) Vol. 64B, p. 467.

9.   R. Collier, C. Burckhardt, and L. Lin, Optical Holography (Academic Press, New York, 1971).

Photo-resist Studies
in X-Ray Contact Microscopy

G.J. Jan[1], *L.F. Chen*[1], *and D.C. Hung*[2]

[1]Synchrotron Radiation Research Center, Taipei, Taiwan 10757,
Peop.Rep.of China and Dept. of Electrical Engineering,
National Taiwan University, Taipei, Taiwan 10764,
Peop.Rep.of China
[2]Synchrotron Radiation Research Center, Taipei, Taiwan 10757,
Peop.Rep.of China

ABSTRACT

A stationary anode soft x-ray generator system has been set up since last year. We use home-made soft x-ray contact microscopic technique to investigate the temperature dependence of the dissolution rate of PMMA resists with respect to exposure time.

The experimental results show that development temperature affects the dissolution rate of exposed resist (PMMA) and the dissolution rate increases as temperature increases. The thickness of the dissolved PMMA resist is measured by an instrument called Taly-Step. The detailed results and interpretations are described in this paper.

1. INTRODUCTION

The stationary anode x-ray generator can provide a simple, inexpensive and easily-used x-ray source. This x-ray source is used to study the photo-resist (PMMA) characteristics and soft x-ray contact imaging of biological specimens of adult sea urchins' tooth [1].

The PMMA photo-resist is a considerably important material to develop the submicron photo-lithography technique [2]. Synchrotron radiation and laser-induced plasma x-ray sources are considerably powerful x-ray sources in this research subject. However, they are not so popular from the point of view of cost-effectiveness and location. Therefore, the stationary anode x-ray source can provide a convenient, low cost and in-laboratory facility to investigate photo-lithography and contact imaging in biology. It is a nice facility to practice technique in those research topics before synchrotron and/or laser-induced plasma x-ray sources are available.

The key parameters of well-resolved image on PMMA photo-resist are energy peak of the source, exposure dose, PMMA film quality, and development temperature as well as development time [3-5]. A carbon target is used on the stationary anode, so characteristic x-ray peak is fixed. The x-ray intensity is almost fixed, due to constant current of high voltage power supply. Two factors are studied in this photo-resist investigation; one is exposure time and the other is development temperature. The dissolution rate with respect to x-ray exposure time (similar to exposure dose) has been studied.

The relative step thickness of exposed and unexposed areas of a photoresist film has been measured by Taly-Step. The result shows that the development temperature considerably affects the dissolution rate. The interpretation of the experimental results will be described in the discussion section.

The biological contact image of adult sea-urchin tooth has been carried out on this system. The structure of pure calcite and the ultra-fine structure in calcium carbonate tissue are shown in the contact image. The image is viewed by conventional phase contrast microscope and transmission electron microscope (TEM). The results show the clear fine structure of calcite structure on adult sea urchin tooth and calcite crystal structure on the TEM image.

2. EXPERIMENTAL SYSTEM AND SAMPLE PREPARATION

The stationary anode x-ray generator and soft x-ray contact microscopy system has been set up. A carbon target has been used in this series of studies. The parameters of the x-ray source are set at 4KV and 15mA. The vapor pressure of the working chamber is about 2×10^{-6} torr. The exposure is made on the PMMA photo-resist and the resist developed at 0°C and 25°C temperature. The relative step thickness was measured by Taly-Step.

Polymethylmethacrylate (PMMA) x-ray resist was dissolved in chlorobenzene (6% solution). A drop of the PMMA solution was then placed on the center of a microglass substrate and spun at high speed (~5000 RPM) for 60 seconds. A thin PMMA film (~1 μm in thickness) was formed on the surface of the glass substrate, then it was placed in an oven at 150°C for 1 hour to dry the residual solvent of the resist.

Metal mask or biological specimens were placed in intimate contact with the PMMA coated glass and irradiated with a suitable dosage of x-rays. After x-ray exposure, the exposed PMMA resist was placed in the developer (methyl isobutyl ketone/isopropyl alcohol, 1:1 v/v) for 60 seconds, and the substrate was carefully removed from the developer and blown dry with dry N_2 gas. After development, a "contact print" of the biological specimens was revealed and viewed by phase contrast optical microscopy and TEM.

3. RESULTS AND DISCUSSION

Fig. 1 shows the relative step thickness of exposed and unexposed areas of PMMA resist film. The step thickness (sensitivity) with respect to exposure time is changed exponentially. However, the sensitivity of the resist below

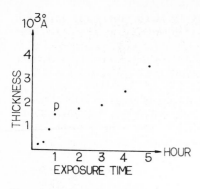

Fig. 1. Step film thickness
vs exposure time

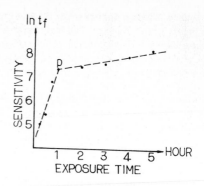

Fig. 2. Step thickness t_f
plotted logarithmically
against exposure time

the 1 hr exposure time is higher than that above it. The turn point P is
clearly shown in the figure. The higher sensitivity of the exposed resist
is due to the irradiation of the soft x-ray on the PMMA surface. The bond
chain is broken by the soft x-ray photon. The PMMA resist in the deep layer
has less sensitivity due to bond rechaining of the bulk-effect.

Fig. 2 shows the natural logarithm of the step thickness vs exposure time.
The dual slope curve in Fig. 2 represents the different sensitivity which
depends upon x-ray dosage. The low dosage has higher sensitivity than high
dosage. Two interaction mechanisms may occur in the interaction of the PMMA
resist with the soft x-ray photon.

The temperature of the PMMA resist development process considerably affects
the dissolution rate with respect to exposure time. The 5 hours exposure time
at 0°C development temperature and the 1 hour exposure time at 25°C development temperature give the same dissolved thickness, about 3000Å. However,
high temperature in the development process will also reduce the resolution
and increase the sensitivity of the dissolution rate.

In order to study the adult sea urchin tooth structure in biological
specimens using home-made x-ray source, two samples of adult sea urchin tooth
were exposed to soft x-ray microscopic technique. The rough image is viewed
by phase contrast optical microscopy shown in Fig. 3. The micro-image of the
specimen is viewed by transmission electron microscope shown in Fig. 4. In
Fig. 3, the picture shows the x-ray image of the adult sea urchin tooth on
the copper mesh of the EM grid. This image shows the calcite structure on
the specimens. The fine structure of the pure calcite in the specimen can
be viewed by TEM. Fig. 4 shows the needle sharp crystal structure of pure
calcite in the specimens. The magnification of TEM image is about 10^4. The
background of the TEM image shows that the PMMA thin film coated on the glass
substrate is not perfectly smooth. However, the fine structure of calcite
crystal is clearly presented in the picture.

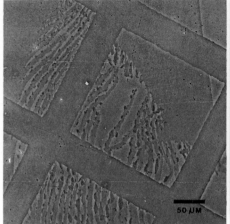

Fig. 3. Contact image of the
adult sea urchin tooth viewed
by optical microscope.

Fig. 4. Contact image of the
adult sea urchin tooth viewed
by TEM.

ACKNOWLEDGEMENT

One of the authors (G.J. Jan) would like to thank Miss Mei-Yu Tsai who is a research assistant in the users' training division of SRRC. I also appreciate deeply Professor M.C. Wang's support of this research. I wish to thank Professor P.K. Tseng and Miss Tou for providing much technical support in the experiments. This research work is supported by the National Science Council under grant contracts NSC-75-0608-E002-05 and NSC-77-0608-E002-02.

REFERENCES

1. G.J. Jan, L.F. Chen, Y.J. Twu, P.K. Tseng, and C.P. Chen: In X-Ray Microscopy: Instrumentation and Biological Applications, ed. by P.C. Cheng, G.J. Jan (Springer, Berlin, Heidelberg, 1987)
2. P.C. Cheng: Ph.D. thesis (University of Illinois at Chicago, Health Science Center, 1985)
3. D.L. Spears, H.I. Smith: Solid State Technology 15, 21 (1972)
4. W. Schnakel, H. Sotobayaski: In Progress Polymer Science, Vol.9, (Pergamon Press, 1983), p.297
5. J.M. Warlaumont, J.R. Maldonado: J. Vac. Sci. Technol. 19, 1200 (1981)

Imaging X-Ray Microscopy with Extended Depth of Focus by Use of a Digital Image Processing System

L. Jochum

Universität Göttingen, Forschungsgruppe Röntgenmikroskopie,
Geiststraße 11, D-3400 Göttingen, Fed.Rep.of Germany

1. Introduction

Observing a thick object by a x-ray microscope with a high spatial resolution, only a thin plane of this object will be in focus; the other object planes are defocused. In-creasing the depth of field by decreasing the aperture results in a reduced spatial resolution. The conflicting requirements for high spatial resolution and large depth of field are a fundamental problem in imaging microscopy. Therefore, the extension of depth of focus by digital image processing is of interest, especially in high resolution microscopy. In this paper, the local variance weighted averaging (LVWA) method, developed by S. A. Sugimoto and Y. Ichioka /1/ and its application at the Göttingen x-ray microscope is presented.

2. The Optical System

The images shown in this paper have been taken at the Göttingen x-ray microscope. For details of the schematic outline see /2/. The optical elements of the x-ray microscpe used for these experiments are:

- the condenser zone plate KZP 3
- the micro zone plate MZP 3

With the zone plate parameters of MZP 3 which are listed in /3/ the depth of focus can be calculated by /4/

$$\Delta f = \frac{\lambda \; f(\lambda)^2}{4 \; r_n^2} \; . \tag{1}$$

which results in $\Delta f = 0.68 \; \mu m$.

3. Data Acquisition

The images were first stored on a film emulsion, then they were magnified through a Zeiss microscope and digitalized by a CCD camera to be ready for image processing. This intermediate step using a film emulsion before digitalizing the images results of course in a loss of image quality, compared to the direct image recording by the CCD camera. The main problem at this point was to overlay the images congruently. This problem has been solved satisfactorily, but a slight loss of contrast in the processed image was unavoidable. In the near future, a direct image recording at the Göttingen x-ray microscope with the efficient CCD camera system designed by W. Meyer-Ilse /5/ will be possible, and thus, basic problems in image recording for further processing will be overcome.

4. The LVWA Method

Let us assume an object which is thick compared to the depth of focus of the optical system. The object plane is moved through the object by moving the MZP in steps of one depth of focus. The result is a set of focal-series images $I_k(m,n)$ with different in-focus and out-of-focus segments (Fig. 1a-g). $k = 1,...,K$ is the number of the image slice, (m,n) are the pixel coordinates in the image plane. The local variance $V_k(m,n)$ of the image $I_k(m,n)$ is defined by

$$V_k(m,n) = \frac{1}{W} \sum_{p=-W_x}^{W_x} \sum_{q=-W_y}^{W_y} [I_k(m+p,n+q) - \bar{I}_k(m,n)]^2 \ , \quad (2)$$

where $\bar{I}_k(m,n)$ is given by

$$\bar{I}_k(m,n) = \frac{1}{W} \sum_{p=-W_x}^{W_x} \sum_{q=-W_y}^{W_y} I_k(m+p,n+q) \qquad (3)$$

and $\qquad W = (2W_x + 1)(2W_y + 1) \qquad\qquad\qquad (4)$

is the number of pixels within the local area around the central pixel (m,n). At these pixels, where the local area around the pixel is in focus, the local variance becomes maximum and it decreases when the image region within the local area is out of focus. The image with increased depth of focus (Fig. 2) is obtained by calculating and adding up the weighted average of each individual focal series image:

$$I(m,n) = \sum_{k=1}^{K} V_k(m,n) \, I_k(m,n) \qquad\qquad (5)$$

and normalizing the result by

$$\| V(m,n) \| = \sum_{k=1}^{K} V_k(m,n) \qquad\qquad\qquad (6)$$

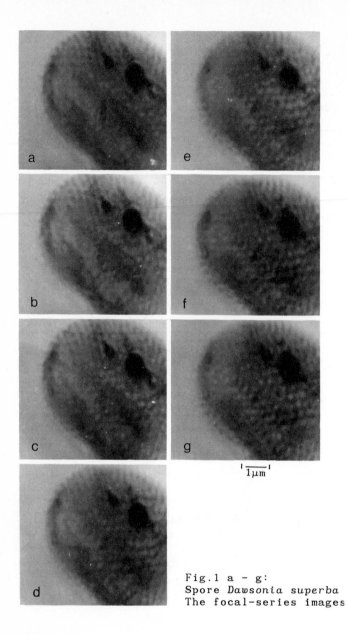

Fig.1 a - g:
Spore *Dawsonia superba*
The focal-series images

It is obvious from eq.(5) and (6) that image regions with a
high local variance (i.e. the in-focus segments) of each
individual focal-series image mainly contribute to the image
with increased depth of focus.

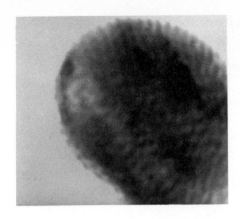

Fig.2:
The image with increased
depth of focus

Acknowledgements

This work has been funded by the Federal Ministry for Research
and Technology (BMFT) under the contract number 05266SL.
I thank the members of the Göttingen x-ray microscopy group
for helpful discussions and V. Sarafis, who collected the
spores.

References

1. S. A. Sugimoto, Y. Ichioka: Appl. Optics, Vol. 24, No.
   14, p.2076 (1985)
2. D. Rudolph, B. Niemann, G.Schmahl, O. Christ: in
   X-Ray Microscopy, eds.: G. Schmahl   and   D.Rudolph,
   Springer Series in Optical Sciences, Vol 43, Springer
   Verlag, 1984, p. 192-202
3. G.   Schmahl, D. Rudolph, P. Guttmann, O. Christ: in
   X-Ray -Microscopy, eds.: G. Schmahl   and   D. Rudolph,
   Springer Series in Optical Sciences, Vol 43, Springer
   Verlag, 1984, p. 63-74
4. P.Guttmann: in X-Ray Microscopy, eds.: G. Schmahl and
   D. Rudolph, Springer   Series   in   Optical   Sciences,
   Vol 43, Springer Verlag, 1984, p. 75-90
5. W. Meyer-Ilse: Application of charge coupled detectors
   in x-ray microscopy, this volume

A Zone Plate Soft X-Ray Microscope Using Undulator Radiation at the Photon Factory

Y. Kagoshima[1;3], S. Aoki[1], M. Kakuchi[2], M. Sekimoto[2], H. Maezawa[3], K. Hyodo[3], and M. Ando[3]

[1]Institute of Applied Physics, University of Tsukuba,
  Tsukuba Ibaraki 305, Japan
[2]NTT LSI Laboratories, Atsugi, Kanagawa, 243–01, Japan
[3]Photon Factory, National Laboratory for High Energy Physics,
  Tsukuba, Ibaraki 305, Japan

1. Introduction

A soft X-ray microscope using Fresnel zone plates as optical imaging elements has been developed in the last few years at an undulator beamline BL-2 of the Photon Factory. As was already reported, /1/ magnified images of copper #1000 and #2000 meshes were obtained preliminarily. In this paper, we describe the new version of the soft X-ray microscope with the improved optical and mechanical arrangements, and report some recent results newly obtained with the improved system.

2. Beamline BL-2B$_1$ for A Soft X-ray Microscope

A 60-period permanent magnet soft X-ray undulator has been in operation at the Photon Factory./2,3/ The first harmonic of the radiation can be tuned over the wavelength range from 1.3 nm to 3.0 nm by varying the gap width of the undulator. The fractional band width of the quasi-monochromatic first harmonic was observed as 1/20-1/14.

As shown schematically in Fig. 1, the beamline for the soft X-ray undulator has two branches, a straight branch and a deflection branch. As a main part of the soft X-ray microscope, an optical bench with a high precision linear translator is installed in the deflection branch 25 m

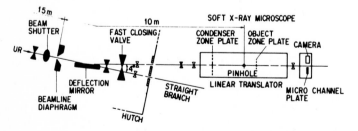

Fig. 1 A schematic drawing of the undulator beamline at the Photon Factory

distant from the center of the undulator. A plane deflection mirror is inserted upstream of the beamline for filtering out undesired harmonics of undulator radiation. A water-cooled copper diaphragm of 1 mm diameter is also inserted upstream of the mirror for rejecting highly off-axis components of the undulator radiation.

3. Optical System and Mechanical Arrangement of the Soft X-ray Microscope

The optical system of the zone plate soft X-ray microscope is shown schematically in Fig. 2. The optical system is composed of a condenser zone plate, a pinhole, an object zone plate and a screen. For determining numerical parameters of the optical elements, the following three practical conditions had to be considered; 1) the wavelength range should be included in the tunable range of the 1st harmonic of the undulator radiation, 2) a source point is supposed not to be the pinhole inserted in the beamline, but the undulator itself, /4/ and 3) an optical matching must be satisfied under the practical boundary conditions of the beamline. Taking account of these conditions, numerical parameters of the condenser and the object zone plates were determined, as shown in table 1, aiming at the resolution of about 0.3 μm and the magnification of about 150-200.

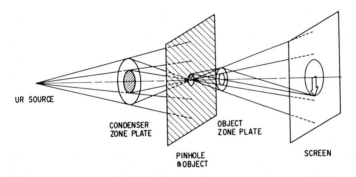

Fig. 2 Optical system of the zone plate soft X-ray microscope

Table 1 Numerical parameters of zone plates

	condenser zone plate	object zone plate
radius of innermost zone	15.8 μm	5.0 μm
number of zones	1000	100
diameter	1.0 mm	0.1 mm
width of outermost zone	0.25 μm	0.25 μm

As a stopping material of the zone plate, tantalum was chosen because of its high absorption coefficient approximately the same as that of gold in the soft X-ray region /5/ and its high residual stress four times higher than that of gold. Reactive ion-etching(RIE) technique can easily be applied to patterning process of tantalum./6/

Figure 3 shows the schematic mechanical arrangement of the soft X-ray microscope. The mechanical system of the microscope consists of four parts,

297

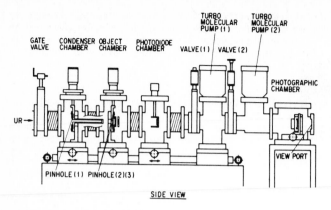

SIDE VIEW

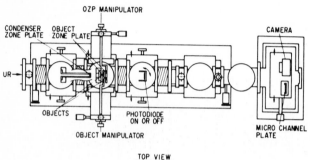

TOP VIEW

Fig. 3 Schematic machanical arrangement of the soft X-ray microscope

i.e., a condenser chamber, an object chamber, a photodiode chamber and a photographic chamber. Each of three chambers except for the photographic one has a laterally adjustable stage. The condenser chamber can be scanned along the beam axis with a high precision linear translator constructed on the optical bench. The object chamber has two manipulators on both sides, one is for the object zone plate and the other for samples. The optical system can be aligned by these two manipulators in the object chamber.

An MCP(micro channel plate) is mounted in the photographic chamber so that the magnified image can be monitored visibly through a view port equipped at the end of the chamber. A camera is mounted on a movable stage in the same chamber next to the MCP. After completing the alignment of the optical system, the camera can be inserted on the beam axis in place of MCP and the magnified images can be taken on the film.

4. Experimental Results

To examine the imaging performance of the soft X-ray microscope, a copper #1000 mesh was used as a model sample. Figure 4 shows a magnified image of the mesh at the wavelength of the first harmonic of 2.4 nm. A film used was FUJI MINICOPY FILM. The exposure time was 4 seconds at the ring current of 157 mA and the magnification was 180. Black thick crossed lines correspond to those of the mesh, with the line width of 7.5 μm. A white blurry circle

Fig. 4 Magnified image of copper mesh #1000

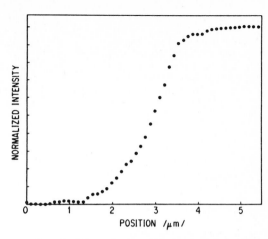

Fig. 5 Microphotometer trace of a wire of the mesh

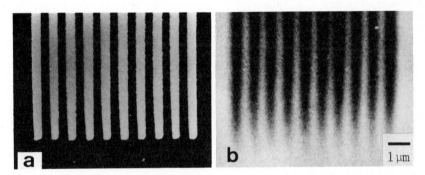

Fig. 6 Magnified images of a 0.8 μm pitch tantalum grating obtained by (a) scanning electron microscope and (b) soft X-ray microscope

covering over the image corresponds to the defocused image of a pinhole of 20 μm in diameter. A microphotometer trace of the image is shown in Fig. 5. Assuming that a piece of a wire of the mesh has a knife edge, the resolving power of the soft X-ray microscope is estimated about 0.8 μm.

Next experiment was performed by using tantalum grating patterns with various pitches. Figure 6 shows magnified images of a 0.8 μm pitch grating pattern obtained by both a scanning electron microscope and the soft X-ray microscope. The wavelength of the first harmonic of the undulator radiation was 2.98 nm. The film used was FUJI MINICOPY FILM. The exposure time was 4 seconds at the ring current of 163 mA and the magnification was 140. The picture clearly shows that the soft X-ray microscope can resolve the grating pattern with a pitch of less than 0.8 μm.

5. Summary

An improved optical and mechanical arrangements of the soft X-ray microscope were described. Magnified images of copper #1000 mesh and a grating pattern of 0.8 μm pitch were obtained. Considering the knife-edge profile of the mesh wire, the resolution of the microscope was estimated better than 0.8 μm. Furthermore, the magnified image of the grating pattern showed that structure of 0.4 μm was able to be resolved by the improved system.

Reference

1. S. Aoki, Y. Kagoshima, H. Yamaji, M. Kakuchi, T. Tamamura, M. Sekimoto, A. Ozawa, H. Yoshihara, M. Ando, H. Maezawa K. Hyodo: to be published in Proceedings of X-Ray Microscopy 86, National Taiwan University, Taiwan, 1986.
2. H. Maezawa, Y. Suzuki, H. Kitamura and T. Sasaki: Nucl. Instr. and Meth. A246, 82 (1986)
3. H. Maezawa, Y. Suzuki, H. Kitamura and T. Sasaki: Appl. Opt. 25, 3260 (1986)
4. H. Maezawa, A. Mikuni, M. Ando and T. Sasaki: Jpn. J. Appl. Phys. 26, L3 (1987)
5. B.L. Henke, P. Lee, T.J. Tanaka, R.L. Shimabukuro and B.K. Fujikawa: "Low Energy X-Ray Interaction Coefficients: Photoabsorption, Scattering, and Reflection." At. Data Nucl. Data Tables. 27, 1 (1982)
6. M. Sekimoto et al.: "X-Ray Zone Plate with Tantalum Film for X-Ray Microscope." (this volume)

Scanning X-Ray Microradiography and Related Analysis in a SEM

D. Mouze, J. Cazaux, and X. Thomas

Laboratoire de Spectroscopie des Electrons, UFR Sciences,
F-51062 Reims Cedex, France

1 Introduction

Current scanning x-ray microscopes or microprobes require x-ray optics such as Fresnel zone plates and mirror systems. Because of different loss factors /1/ these systems need high photon flux sources, such as synchrotron radiation (SR). Despite this, the resulting x-ray photon density still remains in the $10^7 - 10^9$ph/(s μm^2) range. As shown by WITTRY and GOILIJANIN /2/, such values can be obtained from a classical source combined with a toroidal curved crystal, a system which affords all the advantages of a laboratory source in comparison with SR.

Another step towards simplification is direct use of the characteristic x-radiation issued from a foil target, with the specimen kept close to the back of the foil. With such a system, using a low beam current (10-100 nA), the x-ray source size can be reduced to the μm^2 range (which is much lower than that obtained with a rotating anode). The photon density attainable with a characteristic line is in the $10^6 - 10^8$ ph/(s μm^2) range.

A simple system of this kind has been tested in a non-modified SEM. We report here some results obtained in the field of x-ray analysis, namely x-ray fluorescence analysis (XRF) and x-ray absorption spectroscopy (XAS), associated with Scanning Electron X-ray Imaging (SEXI). The next section summarizes the principle of SEXI and related x-ray analysis and the following section gives some practical applications.

2 Principles

As described in a previous study /3/, this system uses a conventional scanning electron microscope with a lateral SiLi detector, unmodified except for a slight change in the sample-holder : the target-specimen system is tilted in such a way that it is the transmitted x-rays which are detected instead of the emitted x-rays as in conventional electron probe microanalysis. Thus the angle between the normal to the object plane and the direction of the detector is set to about 45 ° (Fig.1a).

In SEXI, the electron beam is rastered on the anode-foil and the detected signal is fed to a multichannel analyser. The single channel output allows the desired x-ray band to be selected in order to obtain the corresponding absorption image. All the scanning and synchronous recordings are monitored by a microcomputer and the images are displayed on a CRT and recorded onto a Winchester disk.

301

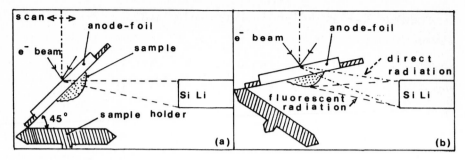

Fig. 1   Geometry used for x-ray analysis (a) XAS and imaging (b) XRF

In the static probe mode, one can perform point elemental analysis, i.e XAS and XRF. Whereas XAS requires the same configuration as in imaging, for x-ray microfluorescence analysis a slight modification of the geometry is necessary : the modified sample holder is slanted toward the detector so that the detected x-rays emerge at a glancing angle of a few degrees above the object plane (Fig.1b). The reason for such a modification is quite obvious : this tilting increases the path length of the direct radiation (characteristic and continuous) across the anode and consequently decreases the detected direct signal. As regards the fluorescent signal which originates from the sample, the path length to travel is much shorter and, consequently, so is the absorption. This leads to an improvement of the signal-to-background ratio in XRF.

This is the first time that XAS has been applied in a SEM. It requires a heavy metal to be used as anode to generate intense bremsstrahlung. Then the continuous radiation undergoes characteristic absorptions when it travels through the specimen towards the detector. A reference spectrum is obtained with the anode alone, without the sample (this is most often done by focussing the probe outside the specimen region).

Finally such a system allows characteristic imaging : by means of a single channel analyser we can select for imaging any region of the transmitted spectrum. A comparison between these images gives information about the elemental composition of the sample.

Having established the principles, we now go on to review the experimental results.

3 Results and discussion

Several experiments were made to test the capabilities of this system. Some of the results have been reported elsewhere /4/. This paper gives some new results regarding the analytical techniques mentioned above.

i)   X-ray imaging of a biological sample and the associated XRF point analysis.

ii)   X-ray absorption spectroscopy (XAS).

iii)   Characteristic imaging versus thickness imaging.

An example illustrating the first point is radiographs of pollen grains at two magnifications (Fig.2a-b). These images were recorded in about 25 minutes using a Ni anode (1 µm thick) and an 8.5 keV/10 nA electron beam. The fluorescence spectrum obtained when the probe was located over a single pollen grain is shown Fig.2c. This spectrum was obtained in 45 minutes (8.5 keV/70 nA). The analysed volume, in relation to the whole region lit by the x-ray beam, is a few 10^{-5} mm^3 which is much less than in conventional XRF. As for the lateral resolution in imaging, it was shown to be in the range of a few microns /3/.

The second point, i.e. XAS (ii), is illustrated in Fig. 3. In Fig. 3a two spectra are shown : the continuous line spectrum is obtained from a pure Ni foil anode (1 µm thick) while the dotted line spectrum shows the same spectra after absorption by a silver foil (0.5 µm - thick). The acquisition time was 20 minutes. The silver absorption curve $\mu(E)$ was then deduced (after normalization and Log. plot) - Fig. 3b - and the results are comparable with the published tables. Similar experiments were performed on biological samples, such as a human hair fragment (Fig. 3c). This spectrum exhibits an unambiguous absorption feature in the sulfur K-edge region.

The last point, i.e. Z-contrast versus thickness contrast, is illustrated in Fig.4. Two images of an integrated circuit are shown. The Si substrate was used as the anode to generate characteristic lines and bremsstrahlung. The first image (Fig. 4a) was recorded by selecting the SiK line (1740 eV) to show up the aluminium structures (K_{abs} =1559 eV) of the circuit. This image also reveals some defects attributable to the roughness of the substrate (resulting from the mechanical thinning). So a

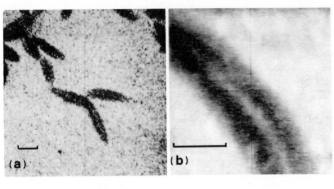

(a) (b)

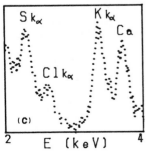

(c)

S k_α K k_α Ca

Cl k_α

2   E (keV)   4

Fig.2 (a) (b) Radiographs of pollen grains.
(c) XRF spectrum of a single pollen grain.
Bar = 50 µm

303

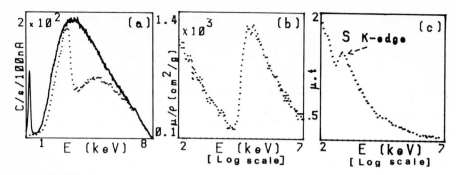

Fig.3 (a) Reference —— and absorption spectra ... of a silver foil (Ni foil anode). (b) Silver absorption curve. (c) Hair fragment absorption curve

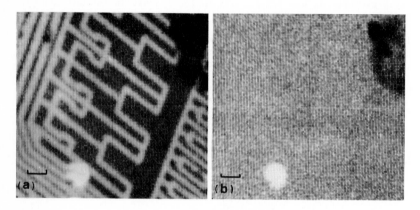

Fig.4 Scanning x-ray image of an integrated circuit recorded with (a) SiK α (b) bremsstrahlung. Bar = 60 μm

second image was recorded in exactly the same conditions, except for the energy window which was set in the high energy region over the Si K α line (3 - 8 keV). A broad band was used to compensate for the low count rate in the bremsstrahlung. In the corresponding image (Fig.4b), the Al structures do not appear but one can again see the defects of the substrate, i.e. the image is a pure thickness contrast one. The next advance will be to correct the first image for thickness contrast by numerical processing in order to obtain a pure chemical contrast image.

4 Conclusion

We have demonstrated that SEXI is a simple method which is very easy to implement in a SEM. Moreover, x-ray microanalyis (XAS and XRF) can be performed in the same apparatus. Obviously the simplicity of the method must be traded off against some disadvantages : lateral resolution limited by the range of x-ray generation in the anode material (in the μm range);

large divergence of the emitted x-ray beam (and consequent loss of resolution in XRF) ; low count rate per % BW when bremsstrahlung is used. Some of these drawbacks could be overcome with substantial modifications making the system not directly usable in a SEM. But even in the present form it seems to us that such a simple system will be useful for people involved in SEM and electron probe microanalysis, as well as for those concerned with SR for preliminary experiments and instrumental alignment.

References

1 P. Guttmann, B. Niemann : "A Detector System for High Photon Rates for a Scanning X-Ray Microscope", this volume,
2 D.B. Wittry, D.M. Goilijanin : 11th Int. Congr. on X-ray Optics and Microanalysis (IXCOM) London, Canada, 51-55 (1986).
3 D. Mouze, J. Cazaux, X. Thomas: Ultramicroscopy $\underline{17}$, 269-272 (1985).
4 D. Mouze, X. Thomas, J. Cazaux : 11th Int. Congr. on X-ray Optics and Microanalysis (IXCOM) London, Canada, 63-66 (1986).

Image Capture in the Projection Shadow X-Ray Microscope

S.P. Newberry

CBI Labs, Box 11 S. Westcott Road RD 6, Schenectady, NY 12306, USA

1. Introduction:

In any form of X-Ray microscopy the problem of capturing the image and converting it to a visible form is a central issue because the x-ray flux is invisible. The classical method of recording directly on photographic emulsion suffers from the long delay of film processing and printing but more seriously from inability to provide for interactive specimen and illumination adjustment prior to exposure. The Shadow Projection Method some what reduces these problems because of its fixed focus and its versatile image capture possibilities. This method employs a simple point source of x-rays created at or in the window of a very fine focus X-Ray tube. The specimen is placed beyond the window and thus may be very close to the source while remaining in air or external crude vacuum. Likewise the photographic plate is in air or crude vacuum and thus readily accessible. Since shadow projection can provide most or all of the image magnification, the photographic requirements are greatly reduced. For set-up and less critical work the shadow projection image can be captured on Polaroid instant film, either the plain or the positive negative variety. Ordinary lantern slide emulsion is adequate for even the most critical recording. The fixed focus of the microscope permits one to simply set specimens in place and photograph them, sometimes for hours on end, without refocusing the X-Ray Source.

While much of the work is taken care of by "point and shoot" techniques it is often desirable to align the specimen and or illumination critically in an interactive mode rather than tedious cut and try through a succession of photographs. The source, which contains complex electron-microscope-like optics in a demountable vacuum environment, becomes unstable as its apertures contaminate. One can appreciably extend the useful period between column clean up by stigmation and or by touch up of electron beam alignment, if one can only work with the x-ray image interactively. Also the specimen may require critical alignment or preselection to show regions of interest or avoid uninteresting samples. Up to the present work, there has been no convenient way to accomplish these tasks in real time using the x-ray image. The nearest approach required the operator to dark adapt for half an hour, and work in total darkness, in order to adequately view the image on a fluorescent screen. This paper is a report of work in progress to capture the image electronically in real time with consumer-priced components wherever possible.

Rapid advances in personal TV recording and in personal computing power lend considerable encouragement to these efforts.

2. Test Set-Up

The test set-up is shown in Fig. 1. It consists of a series of components mounted vertically by means of laboratory support rods from the camera housing of the General Electric version of the Projection Shadow Microscope [1]. The components in Fig. 1 are: reading from top to bottom, a Panasonic model Pk-452S color TV camera (a) with a Newvicon type S4400 tube; a three-stage fiber optic coupled image intensifier tube (b) Ernest F. Fullam inc. Cat. # 1729; an optical microscope body (c) carrying a fluorescent screen and a lens to project the screen image on the input fiber optic plate of the intensifier. One can also see a B&W monitor sitting on top of the X-Ray microscope and an AT compatible computer behind the monitor. In the background is a color TV set to test false color concepts. Figure 2 is a closer view with the test components removed to show the test sample location and the optical microscope body on its side to show the Willemite granular fluorescent screen. Note the heavy brass rings around sample and screen which give an expandable x-ray shield which permits the screen to sample distance to be varied even while viewing. The rod which supports the optical microscope body has been added.

Fig. 1 Test set-up on X-Ray Microscope showing (a) TV camera (b) image intensifier (c) optical package

Fig. 2 View moving closer to show fluorescent screen and sample

307

3. Results To Date

Preliminary tests proved, as expected, that direct TV camera viewing of the fluorescent screen by a consumer product camera yields no image. An industrial low light level camera possibly could view the screen directly but has not yet been available for test. Using the intensifier as described above, the Newvicon tube proved to be superior to standard vidicon tube cameras despite having to interface through a video cassette recorder (JVC model HR-6700U). The standard photoconductive vidicon tubes have equal sensitivity to the Newvicon but suffer from blooming, lag and noise. All subsequent tests were conducted using the image intensifier and Newvicon camera tube.

In analog display mode, without the computer, it has been possible to see simple grid images on the display monitor or the color TV with sufficient resolution and contrast to readily analyze, correct and optimize the x-ray source including arrival at best focus. These convenient, normal TV presentations are qualitatively the same as visual dark adaption images.

Figure 3 shows the monitor screen with an analog image of a 200 mesh copper grid. The same grid at higher magnification is shown in Fig. 4 by a direct exposure on Polaroid film. The X-Ray microscope had considerable astigmatism at the time which made it a better test vehicle. The photograph in Fig. 3 of the TV monitor screen does not do justice to the visual monitor image because of scan jitter and mechanical vibration in the very tall, laboratory clamp mounted system as shown in Fig. 1.

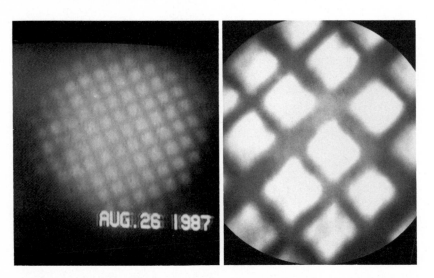

Fig. 3 Photograph of the monitor screen with analog display of 200 mesh grid image

Fig. 4 Direct exposure (at higher magnification) for comparison

Using a digital frame store (eg MaxVision AT-1 version 2.46 by Data Cube 4 Dearborn Road Peabody, MA 01960) and by frame averaging for about one second the TV display surpasses the dark adaption image yet is still sufficiently real-time to not hinder interactive adjustment of focus, astigmatism etc. Sufficient experience has not been gained to determine whether online processing for times comparable with photographic exposure times can equal or even approach the quality of the photographic images. Certainly simple frame averaging will not remove all the noise introduced by the TV capture system. Many of the standard image programs such as edge enhancement look promising and are easy to implement in real time. At present the best procedure is to align by real-time TV, expose photographically to obtain a low noise picture, and then process the photo image off line.

4. Discussion of Results

These results clearly establish one point, namely that despite the problems with currently available hardware, TV image capture is well worth the cost and added complexity. Since only one system has been investigated, and it has been limited to specific hardware available to the author, it would be imprudent to suggest that the demonstrated performance is near its limit or to seriously compare alternate systems. Each alternative system which has passed preliminary scrutiny currently contains one major cost item. In the system described above it is the Image intensifier which is both expensive and demanding in vertical space although a right angle prism could alleviate the last problem. A current revival of flat screen image intensifiers could possibly reduce both cost and space problems. Systems based on industrial low light level cameras are about an even cost break with the image intensifier plus ordinary camera but are more compact vertically and give larger pixel fields. Systems based on liquid nitrogen cooled CCD cameras are the most expensive but may in long term be the best choice. They hold the most promise for bypassing photography altogether. For time lapse studies this would be a great boon.

5. Future Work

For the immediate future flat screen or other space conserving image intensifiers should be explored. Longer term, optimized versions of alternative systems should be constructed and compared.

6. Acknowledgements

The Author is grateful to Ernest F. Fullam Inc. for loan of the image intensifier, and to Electro-Scan Inc. of Danvers MA and Data Cube Inc. of Peabody MA for loan of AT compatible image processor systems.

7. References

1. S.P. Newberry, S.E. Summers: In Proc. Int. Conf. Electron Mic., London July 1954 (Royal Microscope Soc., London, 1956) p.305

The Stanford Tabletop Scanning X-Ray Microscope

J.A. Trail[1], R.L. Byer[1], and J.B. Kortright[2]

[1]Applied Physics Department, Stanford University,
Stanford CA 94305, USA
[2]Center for X-Ray Optics, Lawrence Berkeley Laboratory,
University of California, Berkeley, CA 94720, USA

1 Introduction

The recent success of X-ray microscopy has brought the field to the point where biologists are beginning to use X-ray microscopes as a valuable tool. However, these present microscopes require large scale facilities such as synchrotrons or large laser systems where, for the biologist, there is often a lack of accessibility, control, and ease of use. A complete X-ray microscope on the size scale of an electron microscope could have significant appeal to the biological research community.

At Stanford we have constructed a scanning X-ray microscope which is not much larger than an electron microscope. The microscope uses Laser-Produced Plasmas (LPP) as the source of soft X-rays and a reflecting Schwarzschild objective coated with multilayers as the focusing optics. Although our ultimate goal is to develop a compact microscope that operates in the water window between 23Å and 44Å our first microscope has been constructed to operate at 140Å. At this longer wavelength, mirror performance is high and relatively reliable, and our existing plasma source has greater brightness, both of which improve the expected flux levels in our first microscope.

In this paper we outline the design and construction of our microscope. Component parts are discussed first, followed by description and a photograph of the complete system. Extension of the present system to operation at 40Å is also discussed.

2 The Laser-Produced Plasma Source

By focusing a high peak power laser pulse onto a sufficiently dense material a plasma is formed which emits radiation across the soft X-ray region (10Å - 300Å) and into the X-ray region (<10Å) [1]. The conversion efficiency from laser radiation to broadband soft X-rays can be high, routinely from 10% to 50% [2], and for many target materials the emission spectrum is a continuum with minimal or no line structure [3].

Historically there has been the perception that the high peak laser power necessary for a LPP soft X-ray source required large, low repetition rate, low average power laser systems. This has led to the impression that the LPP is suitable only for 'single shot exposure' X-ray microscopy, and not scanning type microscopes where a high average brightness source is required. In fact LPP's having high brightness in the soft X-ray regime can be produced with millijoules or less of laser energy from systems occupying no more than a few square meters of space [4]. These systems can also have high repetition rates resulting in sources of high average brightness in addition to their high peak brightness.

Our present LPP source is produced by focusing pulses of 1064nm radiation from a Q-switched Quanta Ray DCR II Nd:YAG laser onto a solid copper cylinder located inside a vacuum chamber. The laser operates at 10Hz with a pulse length of 8ns and pulse energies of up to 800 mJ. The copper target is rotated on a 40 pitch threaded rod to produce a clean area of target for each shot, aiding in the shot to shot reproducibility of the plasma and the soft X-ray emission. Our LPP source has yet to be used in the microscope, however we have used the source extensively in the past several months in measurements of the normal incidence reflectance of multilayer mirrors [5]. We have observed that using a laser pulse energy of only 10mJ focused to a 30μ diameter spot on the target we produce a plasma with an average spectral brightness at 150Å of 2 x 10^7 photons sec^{-1} mm^{-2} mrad^{-2} in a 1Å bandwidth. The low pulse energy of 10 mJ was used in order to avoid saturation of the microchannel plate detector in our monochromator. We expect to obtain increased brightness at the higher laser power levels of our laser.

Although our present LPP source is sufficient for our first demonstration microscope it is not an optimal source for microscopy. A preferred laser system would produce 100ps pulses at kilohertz repetition rates. The short pulse length would allow high laser intensities at the target with only moderate pulse energies, which in turn would produce hotter plasmas with more emission at 40Å. The short pulse length would also result in a smaller plasma expansion with a smaller emitting area and hence further increase the brightness. Such laser systems are being developed and their predicted average brightness values are summarized in table 1.

Table1   Comparison of Average Spectral Brightness at 40Å (photons s^{-1} mm^{-2} mrad^{-2} 1Å Bandwidth)

Synchrotron [6]	NSLS Bending Magnet	5 x 10^{14}
	NSLS Undulator	5 x 10^{17}
Laser-Produced Plasma [a]	30 W 100Hz 200ps 30μ [7]	2 x 10^{10}
	5 W 5 KHz 100ps 10μ [8]	3 x 10^{10}
	200 W 1 KHz 100ps 20μ [9,10]	3 x 10^{11}

(a) Assumptions: plasma is a black body with >1% of the radiated energy in a 1% bandwidth at 40Å (valid if 50eV < temperature < 130eV), 10% conversion from laser energy to black body emission, spherical plasma with diameter equal to laser focal diameter , plasma radiates into 4π steradian.

3 The Schwarzschild Objective

The design of our Schwarzschild microscope objective has been presented previously [11]. The zerodur mirror blanks were custom fabricated by Zygo Corporation and the parameters are given in table 2. The small size of the objective makes for a compact microscope and also reduces the effect of surface figure error on the focal spot diameter. The objective has a numerical aperture of 0.3, a compromise between efficiency of the optic and ease of both alignment and fabrication.

In fabricating the mirror blanks, great effort was made to achieve good surface figure and low surface roughness. By figure error we refer to deviations from the ideal surface on a large lateral scale i.e., on the order of 100μ or greater, whereas roughness is a measure of deviations on a lateral scale of a few microns or less. In a practical sense figure errors produce minute deviations in the directions of the reflected rays which act to increase the size of the focal spot of the objective, while surface roughness produces large angle scatter which reduces the specular reflectivity of the multilayer mirrors.

Determining the effect of surface figure error on the focal spot size is a difficult problem requiring detailed knowledge of the lateral scale of the deviations[12,13]. As an estimate of the effect of figure error on spot size we take a simple geometric approach. If we take the peak-valley error on the concave mirror to have a lateral scale of 10mm then the slope error is near 10^{-6} radians. With a distance from mirror to focus of 36mm this slope error would produce a ray deviation at the focal plane of 700Å.

Table 2   Schwarzschild Objective

	Convex Mirror	Concave Mirror
Radius of Curvature	13.4 mm	23.7 mm
Mirror Diameter	8.0 mm	22.2 mm
Surface Figure Error [a] (Peak-Valley)	160 Å	120 Å
Surface Figure Error [a] (rms)	12 Å	24 Å
Surface Roughness [b] (rms)	7 Å	7 Å

a) Surface figure measured using a Zygo Mark III Interferometer.
b) Surface roughness measured using a Wyco Profilometer

The effect of surface roughness in reducing the specularly reflected intensity for the case of normal incidence reflection from materials of refractive index near 1.0 is often described by the expression

$$\frac{I}{I_o} = \exp\left[-\left(\frac{4\pi\sigma}{\lambda}\right)^2\right] ,$$

where σ is the rms roughness and λ is the wavelength [14,15]. By this criterion our surface roughness of 7Å rms will reduce the reflectivity at 140Å by the tolerable factor of 0.67. For the wavelengths near 40Å the surface roughness will need to be further reduced before good mirror performance can be achieved.

The mirrors of the objective were coated with molybdenum / silicon multilayers designed to give a peak objective throughput at 140Å. These coatings were deposited using the facilities at the Center for X-ray Optics in the Lawrence Berkeley Laboratory. We have measured over 50% peak reflectivity on similar coatings deposited on polished glass substrates [5]. After including the roughness factor above we expect a single Schwarzschild mirror to have a normal incidence reflectivity near 40% with a 7Å bandwidth.

An important consideration in constructing a two mirror objective with narrow bandpass mirrors is the overlap of reflectance of the two mirrors for the various rays passing through the objective. We were careful to ensure that our coating procedure would result in optimum overlap of the reflectance bandwidths of the two mirrors in the Schwarzschild objective. With the vertices of the two mirrors placed at the same distance from the sputtering source in the coating chamber the overlap is virtually perfect across the full aperture of our objective. The expected throughput of the coated microscope objective is 6% across a 10Å bandwidth at 140Å.

Fabricating multilayers with reasonable normal incidence performance at wavelengths shorter than the carbon K edge at 44Å requires further development of the multilayer technology. Tungsten / carbon multilayers have been fabricated for normal incidence reflection at 45Å with an estimated peak reflectivity of 5% to 10%, however very little work has been done to date on multilayers for wavelengths shorter than this.

4 Mirror Alignment

Proper alignment of the two Schwarzschild mirrors is extremely important . For the present objective, in order to have a misalignment contribution to the rms focal spot diameter of less than 300Å, the mirror separation must be correct to within ±1μ and the centers of curvatures of the mirrors must be collinear with the source to better than ±0.2μ [11]. Given the difficulty of obtaining and keeping such alignment we have chosen to use an automated alignment system in which the mirrors are positioned by five computer controlled, encoded DC motors. A simple optimization program is used to minimize the focal spot diameter by adjusting the positions of the mirrors. Automating in this manner means that the final alignment can be performed using X-rays rather than visible light, an important consideration as the alignment tolerances for a visible light diffraction limited spot are much less stringent than those for the desired X-ray focal spot. Additionally the automation allows us to routinely check that our alignment is optimal, again using X-rays.

5 Scanning Stage, Detector, and Instrumentation

The sample holder is mounted on a three axis piezoelectric stage. The X and Y axes use a folded lever flexure mount arrangement to amplify the motion of the piezoelectrics giving a range of 50μ. The range in the Z direction is 15μ. The piezoelectric stage is mounted on a three axis Newport translation stage driven by encoded DC motors. The coarse stage has a range of 1.25cm in X and Y, 2.5cm in Z , and a resolution of 0.1μ.

To detect the photons that are transmitted by the sample we have chosen to use a microchannel plate (MCP) detector rather than a proportional flow counter. The MCP is better suited to the high peak flux of our pulsed system, although it has the disadvantage of requiring a vacuum of 10^{-5} torr.

In keeping with the spirit of the microscope the instrumentation is simple, compact, and easy to use. All control of the microscope is performed by an IBM AT personal computer with the software written in Pascal. A single rack of electronics interfaced to the computer is used to drive the eight encoded DC motors and the three low voltage piezoelectrics. The signal from the MCP detector is processed in a boxcar integrator before being sent to the computer via an RS232 link.

6 The Microscope

A schematic of the full microscope configuration is shown in fig. 1. Figure 2 shows a photograph of the complete microscope. All components of the microscope, including the laser system, are shown, except for the IBM AT. At the present operating wavelength of 140Å it is necessary that all parts of the microscope be in vacuum. The system is evacuated to 10^{-5} torr using a 330 l/s turbo pump. A summary of the microscope parameters is given in table 3.

Table 3   Microscope Parameters

Source to Sample Distance	135.6 cm
Demagnification	100
Focal Length	13.3 mm
Numerical Aperture	0.3
Expected Resolution	1500Å - 3000Å
Expected Flux at Sample (at 140Å with 5μ source pinhole)	> 5,000 photons per second

The resolution of the microscope is given by the rms diameter of the focal spot. The four factors which influence this diameter and their estimated contributions are given in table 4. The most uncertain of the factors is the effect of surface figure as discussed in section 3.

The expected flux through the microscope is obtained by multiplying the estimated brightness of our present source at 140Å by the etendue and efficiency of our coated Schwarzschild objective. In section 2 the LPP brightness with 10mJ laser pulses was given as 2×10^7 photons sec^{-1} mm^{-2} mrad^{-2} in a 1Å bandwidth. The Schwarzschild etendue for a 5μ source

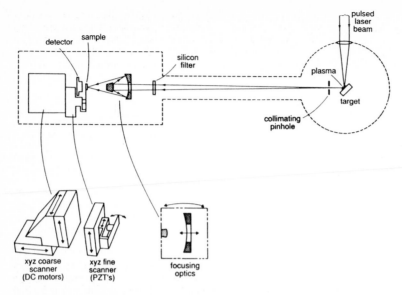

Figure 1 A schematic showing the arrangement of the microscope components

Figure 2 A photograph showing the complete microscope. The target chamber for the laser-produced plasma is in the lower right corner. The laser is just visible at the top left of the optical table. The boxcar integrator, alignment and stage control electronics, and MCP power supply sit on the instrumentation tables above the optical table.

size, including obscuration by the convex mirror is 4 x 10^{-4} mm^2 mrad^2. The efficiency is given by the transmission of the Schwarzschild which is 6% across a 10Å bandwidth. The expected flux at the sample with 100mW of laser power is therefore 5,000 photons per second. At the higher laser power levels available from our laser we expect to obtain higher flux. With sufficient flux to take a single pixel of data at every laser shot we could take a 100 x 100 pixel picture in just over 16 minutes using our present laser.

Table 4   Contributions to Focal Spot Diameter

Mirror Figure	500 - 1500 Å
Mirror Alignment	300 Å
Diffraction (1.22 fλ)	300 Å
Demagnified Source Size (5µ pinhole)	<u>500 Å</u>
Total	1500 - 3000Å

7 Summary

We have constructed a scanning soft X-ray microscope which fits entirely on a single optical table. Our approach is to use compact ,high average power, pulsed lasers to produce an X-ray source of high average spectral brightness, and to combine this source with high efficiency X-ray optics. The present microscope operates at a wavelength of 140Å; however, the approach should scale to 40Å given the predicted performance of short pulse, high repetition rate LPP sources, and with further progress in multilayer coatings and mirror surface quality.

All that remains before testing of the microscope is alignment of the mirrors. The ease of implementing our automated alignment process is uncertain, however we expect to be taking our first images at 140Å within the next several weeks. The first images will be of fine mesh, fine lithography patterns in thin free-standing metal films, and biological specimens which have been prepared for electron microscopy.

The work at Stanford is supported by the National Science Foundation under grant ECS-8611875. The Center for X-ray Optics is supported by the Department of Energy under grant DE-AC03-76SF00098.

References

1   B. Yaakobi, P. Bourke, Y. Conturie, J. Delettrez, J.M. Forsythe, R.D. Frankel, L.M. Goldman, R.L. McCrory, W.Seka and J.M. Soures, A.J. Burek and R.E. Deslattes, Opt. Comm. 38, 196, (1981).
2   R. Kodama, K. Okada, N. Ikeda, M. Mineo, K.A. Tanaka, T. Mochizuki, and C. Yamanaka, J. Appl. Phys. **59**, 3050 (1986)
3   H.C. Gerritsen, H. van Brug, R. Bijkerk, and M. J. van der Wiel, J. Appl. Phys. **59**, 2337 (1986)
4   H.C. Kapteyn, M.M. Murnane and R.W. Falcone, Optics Letters, **12**, 663, (1987)
5   J.A. Trail, R.L. Byer, T.W. Barbee Jr., Submitted to Applied Physics Letters.
6   D.T. Attwood and K.-J. Kim, Nucl. Instr. Meth. A246, 86 (1986).
7   A.L. Hoffman, G.F. Albrecht, E.A. Crawford and P.H. Rose, Proc. SPIE **537**, 198, (1985).
8   I.N. Duling III, T. Norris, T. Sizer II, P. Bado and G.A. Mourou, J. Opt. Soc Am. B, **2**, 616 (1985).
9   S. Basu, T.J. Kane, R.L. Byer, IEEE J. Quantum Electronics, **QE-22**, 2052 (1986)
10   S. Basu and R.L. Byer, Optics Letters,**11**,617,(1986)
11   J.A. Trail and R.L. Byer, Proc. SPIE, 563, 90, (1985).
12   A. Franks, B. Gale and M. Stedman, Proc. SPIE, **830**, (1987).
13   R.H. Price, Proc. AIP, **75**, 189 (1981).
14   T.W. Barbee, in _X-Ray Microscopy_, ed. G. Schmahl and D. Rudolph, 144, (Springer Verlag, 1984).
15   H. Hogrefe and C. Kunz, Applied Optics, **26**, 2851 (1987).

Examination of Soft X-Ray Contact Images in Photoresist by the Low-Loss Electron Method in the Scanning Electron Microscope

O.C. Wells[1] and Ping-chin Cheng[1];*

[1]IBM Research Center, Yorktown Heights,
   NY 10598, USA
*Present address: Department of Electrical and Computer
   Engineering, State University of New York at Buffalo, Buffalo,
   NY 14260, USA

1. INTRODUCTION

X-ray contact microradiography uses a high molecular weight polymer as the photochemical detector. A contact image is obtained by placing the specimen in close contact with photoresist and then exposing to x-rays. After chemical development of the exposed resist, an image is found on the surface of the resist in the form of surface relief. This can then be magnified by either light microscopy or electron microscopy[1]. The scanning electron microscope (SEM) operated in the secondary electron (SE) mode is commonly used to magnify the contact image.

The purpose of this work was to evaluate the low-loss electron (LLE) method[2-4] as an alternative to the SE imaging mode. The LLE signal is formed by collecting electrons that have lost less than typically 200eV in the specimen by means of an energy-filtering detector (Fig. 1). Prior work[3,4] has shown that the shallow information depth and insensitivity towards specimen charging of the LLE image makes it suitable for examining the surface of uncoated photoresist. For a 10keV electron in carbon, the average rate of energy loss is 4.02eV/nm. An

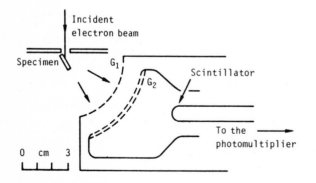

Fig. 1. Schematic diagram of LLE detector

energy loss of 200eV then corresponds to a 50nm (average) penetration path length in the specimen. If one collects the LLE that leave the specimen tangentially to the surface in the forward direction, then fine surface structures can be observed more clearly than in the SE image. The LLE imaging method has been successfully applied to both uncoated and coated photoresist and biological specimens.

For resist coated with 15nm of Au-Pd, the surface charging problem is eliminated so that the SE method can be used. However, in some cases the LLE image still shows the surface details more clearly. Since the SE and LLE detectors can operate simultaneously, it is clearly best to have both images available to obtain complementary information.

2. METHODS AND MATERIALS

Human fibroblasts were cultured in Eagle's medium on the surface of a 100nm thick Formar film supported by a gold electron microscopy index grid. The cells were fixed in 1% glutaraldehyde in phosphate buffer saline (PBS), dehydrated in ethanol, and critical point dried in CO_2. For x-ray contact imaging, polymethyl methacrylate (PMMA, MW=420,000) photoresist was used. The contact image was exposed with 3.5nm monochromatic synchrotron radiation (National Light Source at Brookhaven National Lab., VUV storage ring, beamline U15) in vacuum and developed in a 1:1 mixture (by volume) of methyl isobutyl ketone (MIBK) and isopropanol (IPA) for 30 sec.

Contact images in photoresist were observed under a Cambridge Instrument Co. S-250 Mk. III SEM with a LaB_6 single-crystal cathode. The samples were coated with 15nm of Au-Pd. The scattering of LLE occurs mainly within the surface layer, and this results in an order of magnitude increase in the collected LLE signal as compared with uncoated photoresist. The sample was mounted with a glancing angle of incidence of 30° as shown in Fig. 1. Both a SE detector and a LLE detector were used. The SEM acceleration voltage was 10kV, and the LLE detector was operated with an energy window of 200eV.

3. RESULTS AND DISCUSSION

Figures 2(a) and 2(b) show a comparison pair of SE and LLE images from the same field of view. The effective electron penetration distance in the LLE image is limited by the energy window of the LLE detector, and this gives fringes at sharp edges that are narrower than for the SE image. Thus in some cases it is possible to see finer details in the LLE image. (The major problem of the LLE method, that a rough sample shows strong shadowing effects, does not apply in this situation.) These images show that the LLE image can provide useful additional information to the more familiar SE method.

The electron bombardment needed to obtain the LLE image is typically five to ten times greater than for the SE image. A certain amount of beam damage is therefore unavoidable. For very high resolution imaging, a metal replica of the resist surface should be used[1].

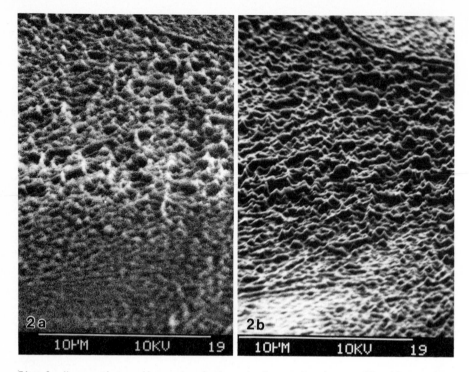

Fig. 2. X-ray microradiograph of the cytoplasm of a human fibroblast. The image was exposed with 3.5nm monochromatic x-rays on PMMA resist. The surface was coated with 15nm of Au-Pd. (a) SE image. (b) LLE image. Note the high image contrast and fine topographic details in the LLE image.

ACKNOWLEDGMENTS

We would like to thank Dr. J. Kirz for the use of U15 beamline at the Brookhaven VUV storage ring.

REFERENCES

1. P. C. Cheng, D. M. Shinozaki, K. H. Tan: In X-ray Microscopy - Instrumentation and biological applications, ed. by P. C. Cheng and G. J. Jan (Springer, Berlin, Heidelberg 1987) p.65

2. O. C. Wells: Appl. Phys. Lett. 19, 232 (1971)

3. O. C. Wells: Appl. Phys. Lett. 49, 764 (1986)

4. O. C. Wells, P. C. Cheng: J. Appl. Phys. 62, 4872 (1987)

Soft X-Ray Contact Microscopy at Hefei

Xing-shu Xie[1], Cheng-zhi Jia[1], Ji Ren[1], Tao Jin[1], Shi-xiu Kang[1], and Hen-Jia Shen[2]

[1]Center for Fundamental Physics,
 University of Science and Technology of China,
 Hefei, Peop. Rep. of China
[2]Medical School of Anhui, Hefei, Peop. Rep. of China

I. Introduction

Hefei Synchrotron Radiation Machine,a dedicated light source,is under construction [1]. This machine is planned to provide photons and will be available to the experimentalists in 1988-1989. Figure 1 shows the plane layout of Hefei machine. Table.1 is the main parameters of the storage ring. Around the ring, all together 27 ports can be set up (three from proposed 3 insertions). Four beam lines from 3 ports will be built in the construction period of the ring. One of the lines will be used for soft x-ray scanning microscopy as well as contact microscopy and other x-ray imaging studies. The design and construction of the soft x-ray beam line [2] and scanning x-ray microscope [3] are in progress. Preliminary studies of soft x-ray contact microscopy using a home-made stationary soft x-ray source have been performed.

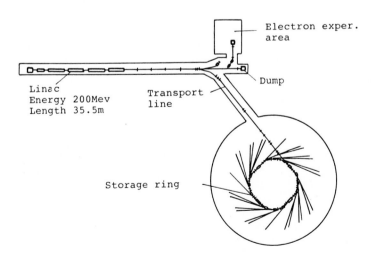

Fig.1. Plane layout of Hefei machine

Table.1 Main parameters of the storage ring

Energy	0.8 Gev
Average beam current	300 Ma
Bending magnet field	1.2 T
Radius of B-Magnet	2.222 M
Critical wave length	24 A
Long straight section length	3.36 M
Number of straight sections	4
Radiation power	4.89 Kw
Circumference of the ring	66.13 M

2. Preliminary studies

Among various x-ray imaging methods, soft x-ray contact micro-scopy is the best developed method for biological applications at the present time. The aim of our studies is to establish the routine procedure of this technique such as specimen prepa-ration, exposure, development and examination of the x-ray replica. Some kinds of biological specimens have been examined in our lab.:alga filaments, mouse myocardial tissues and mouse T lymphoma cells.

Mouse YAC-1 cells, a continuous T cell lymphoma line from a strain A mouse, were cultivated in suspension in RPMI-1640 me-dium supplemented with 10% heat-inactivated fetal calf serum, 2mM glutamine and antibiotics penicillin(100μg/ml) and strepto-mycin(100μg/ml). Appropriate aliquots of cell suspension were centrifuged at 1000 rpm for 8 min, the supernatant discarded. When washed for three times with Hank's solution (pH 7.4), the tumor cells were resuspended in 0.1 ml Hank's solution. Then a droplet of cell suspension was smeared on the surface of the resist which is the 0.6μm thick PMMA layer spun on a silicon substrate and dried in air. Sections of mouse myocardial tissue were prepared as usual and immersed in normal saline. The sec-tion was spread carefully on the PMMA resist, after dried in air, covered with copper grid to fix it.

The exposure source used in this study was a home-made sta-tionary target soft x-ray generator with carbon anode. A cali-brated proportional counter was used to measure the x-radiation flux and equivalent dosage on the specimen. The specimens fixed on PMMA were exposed with carbon K radiation. The exposure time was 8 hours for YAC-1 cells and 13 hours for section of mouse myocardial tissue. After wash off the specimens, PMMA was deve-loped in a 2:1 mixture of methyl isobutyl ketone(MIBK) and iso-propyl alcohol (IPA). The dissolution rate of the resist depended on the ratio of MIBK to IPA, exposure time and tempe-rature. The x-ray replicas then viewed under a light microscope and a scanning electron microscope. For SEM examination the resulting replicas were coated with a gold film (8 nm).

Figure 2 shows the micrograph of the replica in PMMA of mou-se YAC-1 cells. The whole cell profile, the geometric positions

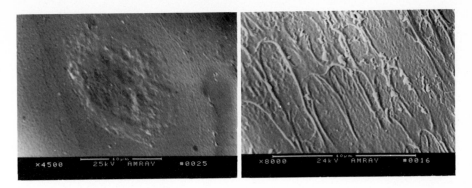

Fig. 2   Fig. 3

Fig. 2 [left] Scanning electron microgragh (at 45° tilt) of the
replica in PMMA of mouse YAC-1 cells, made with C-K radiation.
The x-ray source was operated at 4 kV and 50MA, with a target
to specimen distance of 30 cm and exposure time of 8 hours.

Fig. 3 [right] Scanning electron micrograph (at 45° tilt) of a
section of mouse myocardial tissue. The x-ray condition is same
as in Fig. 2 but exposure time of 13 hours.

of nucleus and cytoplasm of YAC-1 cells are clearly revealed.
The diameter of nucleus is about 8-9 μm, in accord with that
observed by light microscope. Both nuclear and cytoplasmic features show heterogeneous structure reflecting the complex
tissue components.   Figure 3 shows the x-ray resist images of
the mouse myocardial tissue. As compared with that of TEM, the
x-ray resist images are more clear with sharp contrast and show
the stereo features. The best resolution of the feature is less
than 80 nm. The YAC-1 cell image is less clear than that of
myocardial tissue and further work is needed to improve it.

3. Future plans

High intensity soft x-ray source is particularly important for
contact microscopy studies. An improved stationary target soft
x-ray generator which is more powerful and has a small focal
spot size will be used for exposure very soon in our lab. An exposure station which will be connected to the soft x-ray beam
line of synchrotron radiation source is also in construction.
Laser produced plasma x-ray source is ideal for imaging wet and
living biological specimens. A high power Nd: glass laser and
target system in laser lab. at USTC [4] is planned to be used
for flash contact microscopy in the near future.

    Several techniques have been developed for viewing the x-ray
resist image with a transmission electron microscope [5, 6, 7].
TEM image of x-ray resist has higher resolution and is easier
to interpret. We will use TEM examination method and compare
the TEM images of both the x-ray resist and the original biological specimen in future work.

Increasing interest has been shown in the biological applications of soft x-ray microscopy. Our studies are in the initial stage. We will do more work to gain further experience of the imaging method and of recognizing and delineating the x-ray images.

Acknowledgements

This work is supported by Hefei National Synchrotron Radiation Laboratory, USTC. The authors are grateful to Wu Xi Chemical Institute for providing the PMMA resist. We are indebted to Prof. Kirz and Dr.Cheng and our colleagues in China for the fruitful discussions and continuous help.

References

1. Z.-M. Bao: Nucl. Instr. Meth. A246, 18 (1986)
2. Y.-G. Su, S.-J. Fu, Y.-W. Zhang: Nucl. Instr. Meth. A246,655 (1986)
3. X.-S. Xie, S.-X. Kang, C.-Z. Jia, T.Jin: Nucl. Instr. Meth. A246, 698 (1986)
4. H. Wu, D. Guo, Y. Zhou, L. Zhang, W. He, Y. Dai: J. China Univ. Sci. Technol. 15-1, 30 (1985)
5. R. Feder, J.L. Costa, P. Chaudhari, D. Sayre: Science, 212 1398 (1981)
6. P.C. Cheng, H.B. Peng, R. Feder, J.W. McGowan: Electron Microscopy, 1, 461 (1982)
7. K. Shinohara, S. Aoki, M. Yanagihara, A.Yagishita, Y.Iguchi, A. Tanaka: Photochemistry and Photobiology, 44-3, 401 (1986)

A Projection X-Ray Microscope
Converted from a Scanning Electron Microscope, and Its Applications

K. Yada and S. Takahashi

Research Institute for Scientific Measurements,
Tohoku University, 2–1 Katahira, Sendai 980, Japan

1. Introduction

Projection X-ray microscopy which had been developed by a Cambridge group [1,2] and others [3] around 1955 is very simple in the operational principle and certain useful results had been obtained both in the biological and engineering fields with unique advantages of very big focal depth and capability of stereo observation of the samples held in air. However, it has not been widely used thereafter probably because of its technical difficulties of realization of sufficiently intense micro X-ray source with a desirable wavelength.

If we utilize a scanning electron microscope (SEM) with a transmitting X-ray target, it is rather easy to focus the electron beam exactly on the target by observing the SEM image of the target. The X-ray intensity will be increased by improvement of electron optical elements such as electron emitter and electron lenses. We tried to build a projection X-ray microscope which had a resolution of 0.2 µm by utilizing a commercial SEM [4,5]. Intensity of micro X-ray source in relation to electron beam spread in some typical target materials and contrast of X-ray image with target materials were studied. It was found that Ti and Sc are most suitable as the target material for biological substances such as plant cells, micro insect, sections of human bone and neuron stained with silver [4,5,6]. In this paper, recent results obtained with Ti and Ta targets for some biological materials will be shown.

2. Resolution when Ti target is used

Resolution of projection X-ray microscope which is limited by factors of X-ray source size and Fresnel diffraction is expressed as

$$\delta = (\delta_e^2 + \delta_s^2 + \delta_F^2)^{1/2},$$

where δ_e is half value width of electron beam size impinging on the target, δ_s spread of X-ray producing region due to electron scattering in the target and δ_F width of Fresnel fringes from the specimen edge to the first maximum. It is possible to realize smaller δ_e than 200-500 Å rather easily in a normal SEM, though rather long exposure time is required when such a small electron beam is used for the X-ray microscopy. Thinner target composed of high Z material is desirable so as to have a minimum spread of X-ray producing region in the target if the mechanical strength of the target is sufficient as a vacuum window. For Ti target, 1.5 -2 µm thick foil is estimated to be optimum from the viewpoints of the X-ray intensity and the spread of impinging electrons [4]. Therefore, we used a 2 μm thick Ti target most frequently in the experiments.

The limitation imposed by δ_F which is expressed as $\delta_F = (\lambda b)^{1/2}$, where λ is wavelength of X-ray used for imaging and b distance from the target to the specimen, is a rather serious factor to realize the high resolution. If K_α-line of Ti (2.75 Å) is used for imaging and resolution better than 0.1 μm is desired, b must be less than 36 μm, which is fairly difficult condition not only to the X-Y adjustable mounting but also to the tilting for stereo observation.

Some pictures were taken considering other factors such as instrumental stability, film sensitivity and contamination of the target in addition to above-mentioned factors. Fig. 1 shows a stereo pair of 200 μm thick section of Golgi-stained Parkinje neuron of a rat taken with 30 kV beam and 1 hr exposure. Fig. 2 shows X-ray images of diatomite mounted on a thin cleaved mica with 15 kV beam, b ∿ 0.6 mm and 10 min. exposure, where arrays of small holes of about 0.5 μm in diameter and 0.3 μm thick wall of honeycomb structure are clearly seen. Reduction of accelerating voltage from 30 kV to 15 kV was slightly effective to enhance the image contrast because of reduction of hard component in the continuous X-rays. From the X-ray micrographs of Au coated micro grid which contains many holes smaller than 0.1 μm in diameter, the resolution of the X-ray micrograph was estimated to be about 0.3 μm with the same instrumental condition as Fig. 2. With decrease of b, the resolution was improved to about 0.2 μm. Fig. 3 shows an example of 110 μm thick human bone, where dark spots in the light micrograph A are seen as lighter and considerably smaller spots in the X-ray micrograph B. It was found from stereo observation that these spots "lacuna" are actually oval shaped and randomly oriented flat inclusions having approximate dimensions of 18 x 6 x 1 μm.

3. Image contrast when Ta target is used

Shinohara et al.[7] showed that 10 Å X-ray gives sufficient contrast to the contact X-ray image of HeLa cells. As the Ti K_α-line (2.75 Å) was too hard for observation of HeLa cell, we selected M-line of Ta from view points of wavelength, melting point and electric conductivity so as to yield softer X-ray than Ti. Fig. 4 shows EDX reflection spectra from Ta target excited with different beam energy, 30, 15 and 10 kV. It is seen that the Ta L-lines are excited both at 30 and 15 kV but nearly disappear at 10 kV. In addition, undesirable Fe K-line and Cr K-line from stainless steel target holder also decrease at 10 kV. We can expect to form images with the Ta M_α-line of 7.25 Å if the beam energy is lowered to 10 kV, though the X-ray intensity is decreased by several times at the same time. The roughly estimated intensity ratio from total X-ray counts was $I_{30kV}/I_{10kV} = 7.6/1$ without correction of the beam current. In order to avoid too much absorption of Ta M-line in the target, its thickness should be controlled to be rather thin. Actually, Ta film of about 1000 Å thick was made by vacuum evaporation. It was found that the resolution was improved to about 0.1 μm owing to reduction of the target thickness without remarkable loss of X-ray intensity. Fig. 5 shows effect of the Ta target to image contrast of the diatomite when electron energy was changed from 15 kV (a) to 10 kV (b), where the pictures were printed under the same photographic condition. It is seen that image contrast is remarkably increased in (b). Contrast of the image at 15 kV (a) was nearly same as compared with that by Ti target at 15 kV. Resolution better than 0.2 μm was obtained even at 10 kV.

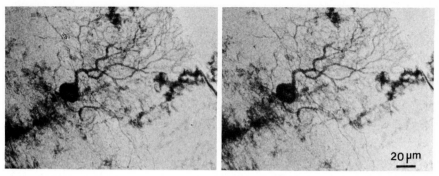

Fig. 1   Stereopair of X-ray micrographs of a Golgi-stained 200 μm thick section of neurons of rat, 30 kV, Ti target.

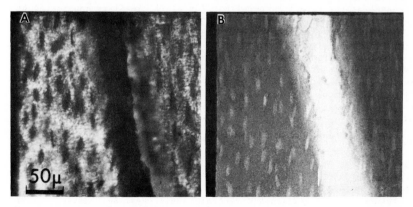

Fig. 2   X-ray micrograph of diatomite at 15 kV, 2 μm thick Ti target. 10 min.

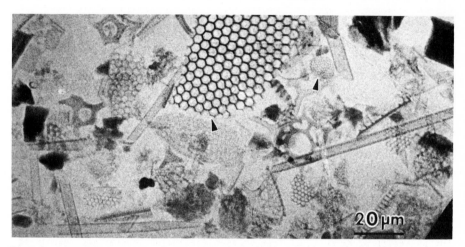

Fig. 3   Micrographs of 110 μm thick section of human bone. A: light micrograph, B: X-ray micrograph at 30 kV, 3 μm thick Ti target.

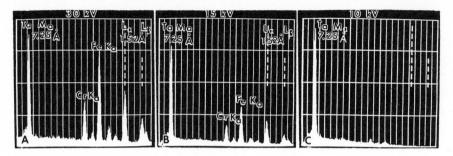

Fig. 4 EDX reflection spectra from Ta target at A: 30 kV, B:15 kV and C: 10 kV.

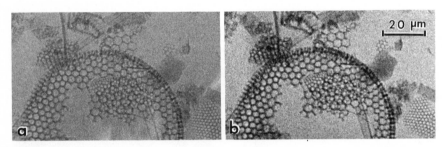

Fig. 5 X-ray micrographs of diatomite using 0.1 μm thick Ta target. (a):15 kV, 1 min., (b) 10 kV, 10 min.

4. Conclusion

A projection X-ray microscope has been developed based on a SEM with unique advantages of great depth of focus, high resolution even better than light microscope and stereo observation capability if focusing of the electron beam is monitored by a scanning system. For most biological samples, Ti was a good choice as the target material but for micro organisms consisting of lighter elements, thin Ta film target was found to be very suitable when Ta M_α-line (7.25 Å) is preferentially excited with 10 kV electrons. Image contrast was certainly improved by using Ta target, and resolution better than 0.2 μm was obtained at 10 kV.

References
1. W. C. Nixon, Proc. Roy. Soc. A232 (1955) 475.
2. V. E. Cosslett and W. C. Nixon, X-ray Microscopy, Cambridge University Press (1960).
3. S. P. Newberry and S. E. Summer, Proc. Internat. Conf. Electron Microscopy, London (1954) 305 (Royal microscopical Soc. Londi)
4. K. Yada and H. Ishikawa, Bull. Res. Meas., Tohoku Univ. 29 (1980) 25 (in Japanese).
5. M. Takahashi, S.Takahashi, M.Kagayama and K. Yada, Tohoku J. exp. Med. 141 (1983) 249.
6. S. Takahashi, K. Yada and K. Shibata, Bull. Res. Sci. Meas., Tohoku Univ, 29 (1980) 13.
7. K. Shinohara et al., Photochemistry and Photobiology 44 (1986) 401.

Part IV

Applications
of X-Ray Microscopy

Applications of Soft X-Ray Imaging to Materials Science

D.M. Shinozaki

Department of Materials Engineering,
The University of Western Ontario, London, Ontario,
Canada, N6A 5B9, Canada

1. Introduction

The traditional approach in materials science has been to relate the microstructure to the bulk properties of the material. Modern methods of electron microscopy (scanning, transmission and scanning transmission) can be used to examine extremely small structures. At the scale of small numbers of atoms, it is possible to analyze the crystal structure and chemical composition. The absolute resolution of soft X-ray imaging is about two orders of magnitude poorer. At these scales (structures less than 500 Å) the methods of soft X-ray imaging will not generally replace electron microscopy. Presently the existing soft X-ray microscopes are situated at synchrotron radiation facilities, and even contact microradiography requires these high intensity sources to reduce the exposure times to manageable proportions. For most users of soft X-ray imaging facilities, the instrument is geographically distant from the laboratory in which the specimens are prepared. For the microscopist in biology as well as in materials science, the specimen preparation is at least as important as the instrument itself, and the inconvenience of separating preparation facilities from the microscopy facilities can only be justified for particular kinds of experiments which can be performed using soft X-rays. In materials science, these unique applications are (1) in the study of thick specimens not easily examined with 120 kV electrons, (2) microchemical analysis of light elements not easily made with electrons, and (3) the imaging of dynamic systems using pulsed sources.

Examples of these applications include the study of buried interfaces in specimens which are thick compared to those visible at normal accelerating voltages in the transmission electron microscope. Solid state reactions which occur at these interfaces would be visible without sectioning the specimen. Precipitation and segregation processes would be visible over large areas of the specimen.

The potential to perform microchemical analysis of light elements such as C, N, O etc., exists because the K edges in the absorption spectrum lie in the soft X-ray region. There remains some question about the relative sensitivity and spatial resolution when compared with windowless detection of X-rays generated from electron beams.

Finally the methods used to image living, moving biological cells using pulsed sources of soft X-rays might be used in examining reactions which redistribute chemical elements in solid-solid, solid-liquid and solid-gas systems. The success of this depends on the design of the environmental cell and the spatial resolution using the pulsed sources.

2. Microstructures of Al-Cu Thin Films: Preliminary Studies

A useful metallic system to examine initially is the Al-Cu alloy. These metals are easily processed over wide ranges of composition and produce microstructures which vary from precipitates less than 200 Å in diameter (the expected resolution of the contact method) to grains of the order of 1 micron or larger. The contrast of these structures is largely generated by the redistribution of Cu in the Al matrix. Furthermore, the detailed microstructures of these alloys is well documented in the electron microscopy literature since the Al-Cu system is an extremely important commercial alloy in a wide variety of applications from structural components to microelectronics.

The experiment was designed in the following way to produce Cu concentration gradients which would result in Cu-rich precipitates or phase regions of varying scales. A layer of Al (1000 Å thickness) was vacuum evaporated through a hexagonal TEM grid onto a silicon nitride window. The windows were silicon nitride of 1000 Å thickness inside a silicon frame, and were supplied by Ralph Feder of IBM (Yorktown). They had outside dimensions suitable for direct insertion and examination in a standard TEM holder. The hexagonal TEM grid was removed and 200 Å of Cu was evaporated through a TEM grid consisting of a series of equally spaced lines. The entire surface was then overcoated with 100 Å of Al. The starting layered structure is shown in Figure 1. Other specimens examined included a simple 2 layer structure of 1000 Å of Al covered with 200 Å of Cu.

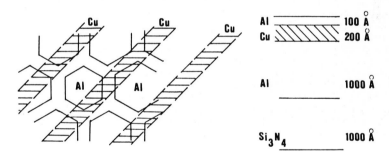

Figure 1: Starting layered structure

The thickness of Al and Cu were measured only approximately, but the phase diagram of the binary alloy has a large allowable composition range for the desired two-phase microstructure of theta plus kappa. The alloy consists of approximately 60% Al and 40% Cu, which lies well within the two-phase region. The theta phase consists of approximately 47% Al and 53% Cu while the kappa phase is 97% Al. About one half of the volume of the microstructure will be theta phase. The exact volumes depend on the exact thickness of Al and Cu laid down.

Examples of the images obtained from these Al-Cu specimens are given in Figures 2 to 4. Attempts to image Al-Cu conducting lines are shown in Figures 5 and 6.

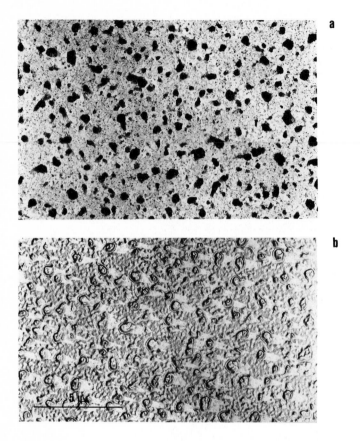

Figure 2:
Al-Cu microstructures (a) original specimen compared to (b) the
soft X-ray contact image (18 Å radiation, metal shadowed carbon
replica of the exposed and developed surface). The coarse
structure is readily replicated, and considerable fine structure
is visible.

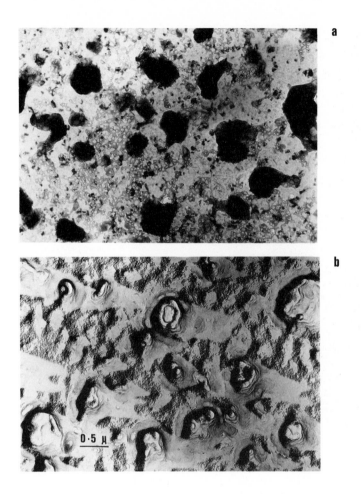

Figure 3:
Higher magnification comparison of (a) the original specimen and
(b) the contact image (18 Å, metal shadowed carbon replica). This
is typical of the detail visible over very large areas of the
specimen, typically spanning 1 mm or more. An experimental
advantage of the contact method is the large areas of specimen
which are visible at high resolutions. The minimum feature size
identifiable in these kinds of images is about 200 Å.

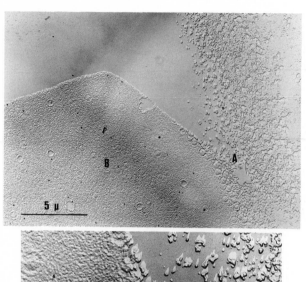

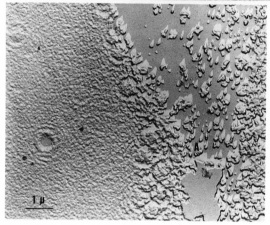

Figure 4:
Low magnification image taken at a wavelength of 100 Å (with a significant second order content at 50 Å). The precipitate distribution at the boundary of the Cu stripe and the Al region is readily visible (A). At these long wavelengths, the pure Al regions (B) show up with strong contrast, but regions high in Cu content are essentially opaque and no internal structure is visible.

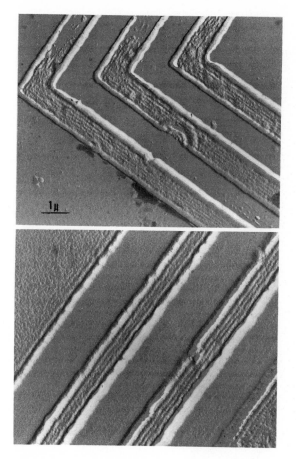

Figure 5:
Attempts to image 1 micron wide Al-Cu conducting lines with 35 Å radiation shows the strong diffraction effects which obscure any internal structure within the line itself.

Figure 6:
Using polychromatic soft X-rays (zero-order illumination)
eliminates the diffraction problems seen in Figure 5. The
internal distribution of theta phase (Cu-rich regions) is
readily observed.

The Use of Soft X-Rays to Probe Mechanisms of Radiobiological Damage

D.J. Brenner and C.R. Geard

Radiological Research Laboratories,
College of Physicians & Surgeons of Columbia University,
630 West 168 St, New York, NY 10032, USA

1. Introduction

For the past four decades, soft X-rays have been perceived as a useful tool for the investigation of the mechanisms whereby all types of ionizing radiations damage living cells. In this review we shall describe this "prima facie" unusual use for low-energy X-rays.

We will briefly describe the pertinent properties of soft X-rays, followed by a historical overview of their use in this field. Current techniques will be described, followed by a discussion of what has and can be learned from interpreting the results of radiobiological experiments with soft X-rays.

2. Soft X-Ray Properties

In the photon energy range of interest here ($\sim 0.1 \rightarrow 5$ keV) and in the range of absorbers of interest radiobiologically ($Z \leq 8$, with small amounts of trace elements), essentially all the photons interact through photoabsorption, yielding either a photoelectron, or a photoelectron and an Auger electron. As an example, for Al K X-rays (1.5 keV) passing through living

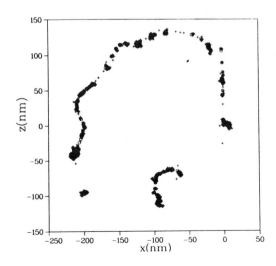

Fig. 1. Two-dimensional projection of typical simulated tracks in water of electrons emitted after the absorption of 0.27, 1.5, and 4.5 keV photons. Circles represent the positions of ionization events, and crosses the positions of excitations.

335

tissue, about 90% of the resultant dose deposited will be due to energy deposition by the ∿0.45-keV and 0.51-keV electrons emitted after absorption by an oxygen nucleus. After emission, the electrons then move in a random path through the tissue; an example of such paths, which may be simulated by Monte-Carlo techniques, is given in Fig. 1 for electrons emitted after absorption of 0.27 (C, K), 1.5 (Al, K), and 4.5-keV (Ti, K) characteristic X-rays on oxygen nuclei. It may be seen that the ranges of the energy depositions are considerably smaller than the size of typical mammalian cells (diameter ∿10 µm).

A second pertinent property of soft X-rays is their strong attenuation in any material. For example, the attenuation length of carbon K X-rays (278 eV) in living tissue is about 1.8 µm, resulting in 50% attenuation in just over 1 µm. The problems engendered by this large attenuation will be discussed at length below.

3. History

The first, pioneering experiments with soft X-rays were performed by LEA and CATCHESIDE in 1943 [1]. At that time, techniques were not available for assaying survival or genetic damage in mammalian cells, and thus they measured the yield of chromatid aberrations in the pollen tubes of the plant Tradescantia. Such experiments present considerable technical difficulties due to the rapid attenuation of the X-rays in the material through which they travel before reaching the genetic material in the nucleus. The technique used involved sowing pollen grains on an agar surface and, after germination, irradiating the thin emergent pollen tubes (diameter ∿6 µm) which contain the generative nucleus.

Some results from the LEA and CATCHESIDE experiments are shown in Fig. 2. They observed, for equal doses, a dramatic fall-off in the yield of chromosomal alterations as the photon energy was decreased from 170 kVp to 1.5 keV. A natural interpretation of this was in terms of thresholds: In other words, there must be some minimum energy deposition (or perhaps track length) necessary to produce biological damage — anything less than this threshold is highly ineffective.

These celebrated experiments were not repeated for over two decades, and the conclusion that there must be a threshold amount of energy

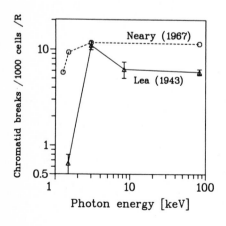

Fig. 2. Yield of single chromatid breaks per unit dose in Tradescantia pollen tubes as a function of incident photon energy. The full curve is from Ref.[1] and the dashed curve from Ref.[4].

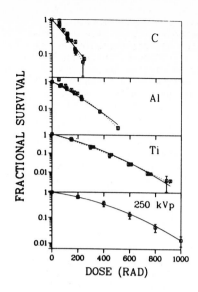

Fig. 3. Survival of Chinese hamster (V79) cells exposed to various energy X-rays. The data are derived from Ref.[5] and references therein.

necessary to elicit a biological response was accepted and elaborated during this time (e.g., HOWARD-FLANDERS [2]). In the 1970s, however, NEARY and his colleagues at Harwell did repeat the experiments [3,4], with the same endpoint, i.e., chromatid aberration in <u>Tradescantia</u> pollen tubes. Some of their results are also shown in Fig. 2. Notable is the major difference at low energies, where a sharp fall-off was <u>not</u> observed. Indeed more recent experiments (see, for example, Fig. 3, for mammalian cell inactivation) show a biological effectiveness <u>continuously increasing</u> <u>as the energy decreases</u> from 250-kVp X-rays through 4.5-keV X-rays, 1.5-keV X-rays, to 0.28-keV X-rays.

It is hard, several decades <u>post facto</u>, to assess realistically the cause of the anomalous results of the lowest energy irradiations of LEA and CATCHESIDE [1] in <u>Tradescantia</u>. A likely cause is that the generative nuclei were often partially or totally shielded by the agar in which the grains were sown, i.e., the problems associated with soft X-ray attenuation and hence accurate intracellular dosimetry were greatly underestimated. Nevertheless, the influence of the erroneous low-energy point in the data of LEA and CATCHESIDE [1] has been far reaching.

4. Current Measurements

As techniques for the handling and evaluation of biological damage to mammalian cells under the stringent conditions necessary for soft X-ray irradiation were developed, so several groups began to obtain quantitative data on the effects of soft X-rays. Groups at Harwell [5], Columbia [6], and Los Alamos [7] have all recently published data and formulated interpretations on the radiobiological effects of soft X-rays. Techniques for generating the soft X-rays have used either incident electrons or protons on a target to produce its characteristic X-rays, the latter having the considerable advantage of producing essentially no bremsstrahlung.

The considerable attenuation undergone by the soft X-rays in matter leads to two major problems: First, there must be minimal absorber between the produced X-rays and the cells, and secondly, accurate dosimetry becomes less certain. The first problem is overcome by plating the cells on very thin-bottomed (1.5→6 μm of mylar) dishes, allowing the cells to attach firmly to the mylar, and then irradiating through the mylar. This is preferable to irradiating the cells from the top, as the liquid medium in which the cells are grown tends to stick to the top of the cells, even when suctioned off. There remains, however, in all cellular systems currently under investigation, a small fraction of cells that remain shielded from the radiation, typically either because they are growing on top of another cell or because they are anomalously large. In terms of the evaluation of a dose-response curve for cellular survival, this will yield severe distortions at survival levels comparable to or smaller than the fraction of shielded cells. An example is given in Fig. 4, where it is clear that a few percent of the cells are being shielded from the radiation.

The second problem is that of dosimetry. Clearly, accurate absolute dosimetry is essential for quantitative comparisons and interpretations of the results. However, most conventional dosimeters have entrance windows far thicker than can be tolerated with soft X-rays. One of the most promising solutions is a thin-window ion chamber such as that shown in Fig. 5. It has several important features, notably: (A) A front window of the same thickness and composition as the mylar used to support the cells. (B) A variable-height collecting volume, which may be adjusted to simulate the mean height of the cells, and thus give a direct measure of the dose to the cells. (C) An accelerating electrode constructed out of thin tissue-equivalent plastic to reduce the possibility of creating detectable photoelectrons in the back wall of the chamber, and (D) a second gas volume, immediately downstream from the collecting volume. By means of a differential screw system the heights of the two regions are kept identical to reduce penetration of the accelerating grid by field lines originating at the collector; thus a sharply defined collecting volume is maintained.

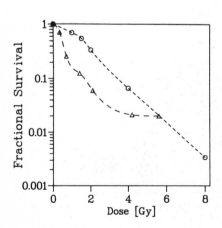

Fig. 4. Survival of Chinese hamster (EM9) cells exposed to Al X-rays (long dash) and γ-rays (short dash).

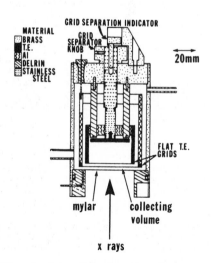

Fig. 5. Schematic drawing of an extrapolation chamber for soft X-ray dosimetry.

5. Interpretation of Results

There have been two basic approaches to understanding curved dose-response curves such as those in Fig. 3. The first — whose origins lie in the early work of Lea and Catcheside discussed earlier — suggests that the basic damage is due to a single critical energy-deposition event of a certain minimum size. The curved nature of the response curves is attributed to a different phenomenon, namely saturation of cellular repair mechanisms with increasing dose. This approach will be discussed further, below.

The second approach has its origins with obervations such as those of Sax, in the 1940's [8]. He oberved that the dose-response curves for chromosomal aberrations due to a single break were linear, whereas those for aberrations involving a pair of breaks were curved and actually quadratic. Such observations resulted in the Theory of Dual Radiation Action [9] in which it is suggested that radiation produces elementary injuries, termed sublesions, in the cell at a rate proportional to the energy deposited locally. These sublesions may then interact — perhaps in some distance-dependent manner — to produce observable damage. From such considerations, the shape of the dose-response curve for a radiation type "i" is predicted [9] to be described by

$$\alpha_i/\beta = \int t_i(x)\gamma(x)dx \ . \tag{1}$$

Here α and β refer respectively to the constants in a quadratic representation of a dose(D)-response relationship

$$\text{Effect} = \alpha_i D + \beta D^2 \ . \tag{2}$$

The function $t_i(x)dx$ is the so-called proximity function of energy depositions. It is a description of the pattern of the energy deposition events in the cell, and is defined as the expected energy deposited in a spherical shell, radius x and thickness dx, centered at a random energy deposition point; this random selection at a point being weighted by the energy deposited at that point. Clearly, it is basically a function of the radiation field, rather than the target, as indicated by the subscript "i".

The second function, $\gamma(x)$, is the probability that two energy depositions or transfers, a distance x apart, will result in observable damage. It depends on the target cellular system rather than the irradiating field. For a set of radiations, i, (1) are Fredholm equations and, given sufficient numbers of the pairs $[\alpha_i/\beta, t_i(x)]$, can be unfolded to yield $\gamma(x)$. Thus, the procedure to obtain $\gamma(x)$ involves three steps: Firstly, fit experimental dose-effect data, for a variety of radiations, to (2); secondly, calculate proximity functions, $t_i(x)$, for all the radiations used in the first step; finally, unfold $\gamma(x)$ from these data and (1).

Techniques for calculating proximity functions have been described elsewhere [10] — essentially, they require detailed Monte-Carlo electron-transport codes. Typical proximity functions are shown in Fig. 6, and from these and the data in Fig. 3 the function $\gamma(x)$ in Fig. 7 has been unfolded [11]. The nature of this function, which is sharply peaked at nanometer dimensions, makes it clear why soft X rays are of such considerable interest — in order to unfold $\gamma(x)$ from (1) it is clear that radiations whose proximity functions are peaked at nanometer dimensions (see Fig. 6) must be used: This is just the property exhibited by soft X-rays.

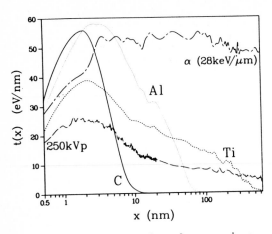

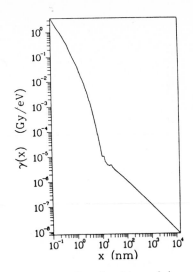

Fig. 6. Proximity functions for a various radiations [11]

Fig. 7. The function $\gamma(x)$ [see (1)] derived from the data of Figs. 3 and 6 for V79 cells [11]

In order to corroborate this theory, it is important to use $\gamma(x)$ predictively. Thus, for example, in order to predict the response of the cells to 24-MeV alpha particles, the following steps would be taken: (A) Calculate the proximity functions for this radiation; (B) integrate this function $t(x)$ [see (1)] with the function $\gamma(x)$ obtained above (from X-ray data). Then the result should be a prediction of α/β (i.e., the shape of the dose-response curve) for these alpha particles. The curve in Fig. 8 shows this predicted response for alpha particles [11], compared with some measured data [12]. The agreement is quite encouraging. There are other examples in the literature (see, for example Ref.[6]) where a gamma function obtained with soft X-rays has been used predictively for qualitatively different types of radiations (or vice versa). Thus the theory appears to be consistent and have considerable predictive power for relative biological effects.

The alternate theory discussed above, that a single critical lesion is responsible for damage, has also been quantified and tested using soft X-rays. As discussed above, the basic assumption, first suggested by LEA [13], is that the endpoint occurs with a frequency proportional to the frequency that a minimum amount of energy (or number of ionizations) is deposited within a given sensitive target. Thus in order to quantify this model it is necessary to find a particular minimum energy deposition in a particular site size, the frequency of which correlates, for different radiations, with the biological effectiveness. This was done for the data of Fig. 3 in Ref.[14] for X-rays, where it was concluded that a minimum of about three ionizations in a site diameter of ~100 nm was necessary for cell killing. Again, the small distances involved emphasize the importance of using soft X-rays for such analyses. Within this model, however, a problem arises when an attempt is made to use these numbers predictively: They fail to predict the response to alpha particles, in contrast to

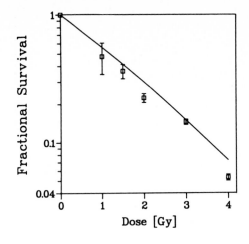

Fig. 8. Predicted (from the α-particle proximity function in Fig. 6 and (x) in Fig. 7) response of V79 cells to 24-MeV α particles (curve) compared with measured results [12] (points)

the "sublesion" model (see Fig. 8). In fact, a critical number of around 10 ionizations in a diameter of about 10 nm appears to emerge from alpha-particle data [15].

Thus if one is to require a consistent model for the damage mechanisms of all ionizing radiations, the soft X-ray data have provided a validation of the interacting "sublesion" model over the "single lesion" model.

We conclude that soft X-rays have provided radiobiologists with a clear insight into the importance of energy depositions on a nanometer (subnuclear) level. They have allowed quantification of different models of radiation action, and perhaps provided critical tests between model. On the other hand, the inherent attenuation problems will always partially obscure the true shape of the dose-response curve and introduce uncertainties into any conclusions drawn.

6. Acknowledgement

This investigation was supported under Grants CA 15307 and CA 41468 to the Radiological Research Laboratory, Department of Radiation Oncology, from the National Cancer Institute.

7. Literature

1. D.G. Catcheside, D.E. Lea: J. Genet. 45, 186 (1943)
2. P. Howard-Flanders: Advan. Biol. Med. Phys. 6, 553 (1958)
3. G.J. Neary, J.R.K. Savage, H.J. Evans: Int. J. Radiat. Biol. 8, 1 (1964)
4. G.J. Neary, R.J. Preston, J.R.K. Savage: Int. J. Radiat. Biol. 12, 317 (1967)
5. R. Cox, J. Thacker, D.T. Goodhead: Int. J. Radiat. Biol. 31, 561 (1977)
6. D.J. Brenner, R.P. Bird, M. Zaider, P. Goldhagen, P.J. Kliauga, H.H. Rossi: Radiat. Res. 110, 413 (1987)

7. M.R. Raju, S.G. Carpenter, J.J. Chmielewski, M.E. Schillaci, M.E. Wilder, J.P. Freyer, N.F. Johnson, P.L. Schor, R.J. Sekring, D.T. Goodhead: Radiat. Res. 110, 396 (1987)
8. K. Sax: Genetics 25, 41 (1940)
9. A.M. Kellerer, H.H. Rossi: Radiat. Res. 75, 471 (1978)
10. D.J. Brenner, M. Zaider: Radiat. Res. 98, 14 (1984)
11. D.J. Brenner, M. Zaider: Radiat. Res. 99, 492 (1984)
12. J. Thacker, A. Stretch, M.A. Stephens: Int. J. Radiat. Biol. 36, 137 (1979)
13. D.E. Lea: Actions of Ionizing Radiations on Living Cells (Cambridge University Press, London & New York 1955)
14. D.T. Goodhead, D.J. Brenner: Phys. Med. Biol. 28, 485 (1983)
15. D.T. Goodhead, D.E. Charlton, W.E. Wilson, H. Paretzke: In Radiation Protection, Fifth Symposium on Neutron Dosimetry, ed. by H. Schraube, G. Burger, J. Booz (Commission of the European Communities, Luxembourg 1986) p.57

X-Ray Contact Microscopy, Using Synchrotron and Laser Sources, of Ultrasectioned, Heavy Metal Contaminated Earthworm Tissue

K.S. Richards[1], A.D. Rush[2], D.T. Clarke[3], and W.J. Myring[3]

[1]Department of Biological Science, University of Keele, Keele,
Staffs.,ST5 5BG, UK
[2]Rutherford Appleton Laboratory, Didcot,
Oxon.,OX11 0QX, UK
[3]Daresbury Laboratory, Warrington, Ches.,WA4 4AD, UK

1 INTRODUCTION

The technique of contact microscopy using soft X-rays is now established /1, 2,3/ and enables a 1:1 image of the specimen to be formed in an X-ray sensitive resist which, after development, can be viewed by electron microscopy. The quality of the image so obtained depends on several things:- sensitivity of the resist; intimate contact of the specimen with the resist; exposure and development times; thickness/thinness of the sample. Because of the preferential absorption of X-rays by features within biological specimens, the degree of information gained will depend on the structural heterogeneity within the material. The technique also has the potential of imaging mineral deposits within a tissue, providing that the element has a high absorption cross section in the X-ray band used.

In order to maximize these aspects of the technique, the tissue surrounding the intestine of earthworms was chosen. This chloragogenous tissue is structurally heterogeneous and, in metal-contaminated worms, is the site of metal sequestration /4,5,6/. The ultrastructure of this tissue is known /4,5/ and a pilot study on lead-polluted tissue using synchrotron radiation /7/ enabled a comparison to be made between contact imaging (CI) and more conventional TEM techniques.

More recently, both laser and synchrotron sources have been used, different resist types have been tested, and earthworm tissue contaminated with either cadmium or lead (at 2 different levels) has been imaged.

2 MATERIALS & METHODS

2.1 Earthworm Tissue

Imaging was performed on ultrasections (≈60 nm), mounted on copper grids, of glutaraldehyde fixed chloragogenous tissue of control earthworms and worms contaminated either naturally with lead (*Dendrobaena rubida*; see /4,5/ for site details) or experimentally with cadmium (*Lumbricus rubellus*; /6/).

2.2 X-Ray Sources

The Synchrotron Radiation Source (SRS) at the SERC Daresbury Laboratory routinely operates at 2 GeV with average currents of 150 mA /8/. At Station 3.4 the beam is filtered using a premirror (cylindrical chromium-coated quartz or toroidal platinum-coated quartz) with a grazing angle of 5° and a 400 nm thick carbon foil. These conditions are calculated to give a soft X-ray beam of wavelength range 2.0-4.4 nm. Sample exposure was 20s in a chamber operated at 1×10^{-6} mbar.

The Vulcan Laser (Neodymium-glass) at the Central Laser Facility, SERC Rutherford Appleton Laboratory produces a single nanosecond pulse of infra red radiation (1.06 μm) of around 100 J /9/. Frequency doubling gives a pulse of green light (0.53 μm) of around 30 J that is focussed to give a 300 μm spot on a gold target, producing a plasma emitting X-rays strongly, but not exclusively, in the 2.3-4.4 nm band.

2.3 Resist Types, Specimen Mounting, Development

2.3.1 Philips PM15 (PM)

This resist, a methacrylic co-polymer /10/, 0.5 μm thick on 60 mm^2 glass plates (Balzers Union Ltd.) was only used with synchrotron radiation. Grid/ resist contact was maintained using adhesive tape, and after exposure and removal of the grid, the resists were etched in a 3:1 mixture of isopropyl alcohol (IPA) and methyl isobutyl ketone (MIBK) for ≈15 s.

2.3.2 P(MMA-MA) Terpolymer (TP)

This resist, 0.5 μm thick on 5 mm^2 silicon wafers /11/ (supplied by Dr. R. Feder, IBM J.Thomas Watson Research Center, New York) was used with both the synchrotron and laser sources. Resists and grids were placed in a modified spring-loaded holder /12/ together with a 100 nm thick silicon nitride window to absorb, in the laser exposures, any UV light also produced, and to protect the resist from target debris. TP resists were developed, after cleaning in an ultrasonic ethanol bath (2 min), in 1:1 IPA and MIBK mixture for 1-2 min.

Development times for all resists were assessed using a Nomarski interference microscope, and all resists were then sputter coated with 30 nm gold and viewed in a JEOL JSM-840 or a Philips 501 b SEM (5-15 kV; WD 10-15 mm).

3 RESULTS

3.1 Lead-Contaminated Tissue

3.1.1 High Lead Levels

Chloragogenous tissue from worms (body burden 6000-7000 ppm lead) living in lead-contaminated soil has high lead levels (≈17,000 ppm) /4/. The ultra-structure (Figs. 1,2 and 7) differs from control tissue (≈70 ppm lead) in that electron-dense deposits occur associated with the chloragosome granules (≈1 μm diam.) and within debris vesicles in the polluted tissue. The sections from which these micrographs were taken had previously been irradiated with synchrotron radiation (SRS) and the contact imaging (CI) shown in Figs. 4-6

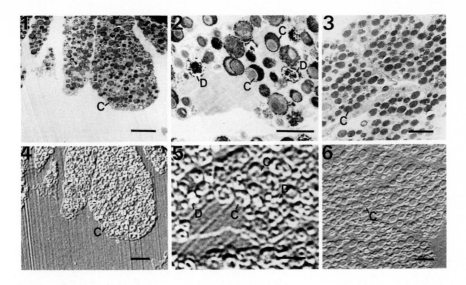

Figs. 1-3, TEM; Figs. 4-6, CI (SRS/PM)

Fig. 1. TEM, high lead tissue. Note chloragosomes (C) with peripherally associated electron-dense deposits. Bar=5 μm

Fig. 2. TEM. Higher magnification of Fig. 1 showing heterogeneous chloragosomes (C) and debris vesicles (D). Bar=2 μm

Fig. 3. TEM control tissue. The chloragosomes (C) lack peripheral deposits. Bar=5 μm

Fig. 4. CI of *same* section as Fig. 1. C, chloragosome. Bar=4 μm

Fig. 5. CI of Fig. 2 (*same* section). Note precise imaging of chloragosomes (C) and debris vesicles (D). Bar=2 μm

Fig. 6. CI of control tissue (*adjacent* section to Fig. 3). C, chloragosome. Bar=4 μm

produced using PM type resists. Resolution of cell processes ≈70 nm wide has been achieved /13/. CI of lead-contaminated tissue (Figs. 4,5 and 8:SRS/PM) also differs from that of the control tissue (Fig. 6). Absorption heterogeneity associated with the chloragosomes and debris vesicles is apparent in the polluted tissue. The 'white' areas, indicating high X-ray absorbance, have an identical distribution to the TEM electron-dense areas that gave, using energy dispersive X-ray microanalysis, prominent peaks for lead.

Imaging of this tissue using the laser source (Fig. 10:Laser/TP) resulted in images that, in terms of resolution of the heterogeneous distribution of the metal, were less informative than those produced with SRS using the PM-type resists. Since a different resist type had been used, sections of this tissue were imaged in the SRS using the TP resists (Fig. 9:SRS/TP). Again, the resolution is less satisfactory than when PM-type resists were used in the SRS (see Figs. 4,5,8), and the results more closely resemble those achieved using laser radiation (cf. Fig. 10).

TEM   CONTACT IMAGING
                      SYNCHROTRON
                PM   TP   LASER
                                                  TP

7 TEM:High Pb   **8** CI:High Pb:SRS/PM   **9** CI:High Pb:SRS/TP   **10** CI:High Pb:LAS/TP

High Pb tissue:

note Pb deposits
associated with
granules in TEM

11 TEM:Med.Pb   **12** CI:Med.Pb:SRS/PM   **13** CI:Med.Pb:LAS/TP

Medium Pb tissue:
Very little Pb
associated with
granules in TEM

14 TEM:Cd   **15** CI:Cd:SRS/PM   **16** CI:Cd:SRS/TP   **17** CI:Cd:LAS/TP

Cd tissue:

metal is not
associated with
granules in TEM,
but in vesicles
(arrow)

Figs. 7-17. Comparisons between TEM (Figs. 7, 11, 14) and CI (Figs. 8-10,
12, 13, 15-17) of lead (Pb) and cadmium (Cd) polluted tissues using synchrotron (SRS) and laser (LAS) radiation sources, and PM & TP resists.
Bars = 2 μm

3.1.2 Medium Lead Levels

In tissue from polluted worms with lower body burdens of lead (less than
4000 ppm /5/), the electron-dense deposits around the chloragosomes, in TEM,
are less marked and appear more uniform (Fig. 11). This decrease in the
heterogeneity of the heavy metal deposition is reflected in the more homogeneous imaging of this tissue using both synchrotron (Fig. 12:PM resist)
and laser sources (Fig. 13:TP resist).

3.2 Cadmium-Contaminated Tissue

The ultrastructure of cadmium-contaminated tissue /6/ differs from lead-polluted tissue in that the metal is not associated with the chloragosomes but
occurs in membrane-bound vesicles in the cytoplasm (Fig. 14). The CI of
this tissue (Fig. 15:SRS/PM; Fig. 16:SRS/TP; Fig. 17:Laser/TP) reflects
the more homogeneous nature of the chloragosomes which appear similar to
those of control tissue (see Fig. 6).

4 DISCUSSION

The results demonstrate that both synchrotron and laser sources allow imaging
of the structural heterogeneity of earthworm chloragogenous tissue and, in
the case of the SRS, a resolution of ≃70 nm has been achieved /13/. The

346

precise imaging of the lead deposits in the tissue has produced, in effect, a lead 'map', and in less polluted tissue the lower lead levels were reflected in the contact images. The study has also shown that when a tissue contaminated with cadmium, which has a different tissue distribution /6/, was imaged, the results differed from those of the lead-polluted tissue.

These observations suggest that the technique might be usefully developed to exploit the potential of preferential absorption by elements within biological tissues. To achieve this unequivocally, monochromation of the radiation is necessary, together with imaging either side the absorption edge, as has been achieved with bacterial spores /14, and this volume/. It might then be possible to determine the lowest level of metal that could be satisfactorily imaged.

The present study has allowed a comparison to be made between different radiation sources whilst using closely adjacent sections from the same tissue piece. With the high lead tissue, apparently more successful imaging was achieved using the SRS and the PM resist type than with the laser source and the TP resist type. Furthermore, the TP resist used in the SRS produced images less informative than those when the PM resist was used with this source. Several possible reasons can be advanced to explain these differences, not the least of which is the chemical nature of the resists. The TP resist is more sensitive than the PM type and therefore better results might have been expected. However, the 'shelf-life' is considered to be shorter, and deterioration of the TP resists might have occurred prior to their use. The abnormally high levels of lead in the tissue might have caused scattering, resulting in a spread image in the highly sensitive TP resists. Alternatively, the method of mounting the sample when using the small TP resists differed from that used with the PM resists. It could be that the contact with the sample was consistently less good, thus producing less precise imaging of the metal deposits. Furthermore, the different development times used with the two resist types may have contributed to the less satisfactory imaging when using the TP resists. Although a range of times was undertaken during this study, development was judged subjectively using Nomarski optics. Clearly, a more objective assessment of resist development is needed.

The similarity of the imaging of camium-contaminated tissue when using the same resist type but different radiation sources (and, to a lesser extent, the similarity of the high lead tissue imaging under these conditions) might suggest that the two sources are equally able to image heavy metal polluted earthworm chloragogenous tissue. However, at neither source was a detector present in the exposure chamber. Although the presence of a premirror and carbon foil at Station 3.4 at the SRS, and the frequency doubling and gold target at the laser source result in a predominance of X-rays in the 2.0-4.4 nm range, the possibility of other wavelengths reaching the sample has not been fully investigated. Diagnostics at the site of the exposure chamber would seem to be an essential requirement in the future development of the technique of imaging metal deposits in biological material.

5 ACKNOWLEDGEMENTS

The SERC grant to KSR (GR/D 58482) is acknowledged, and we are grateful to S.Parry and C.J.Veltkamp for assistance with the SEM photography.

6 REFERENCES

1. Springer Series in Optical Sciences: X-Ray Microscopy, ed. by G.Schmahl, D.Rudolph (Springer, Berlin, Heidelberg 1984)
2. Journal of Microscopy 138, Pt.3 (1985)
3. Examining the Submicron World, ed. by R.Feder, J.Wm.McGowan, D.M.Shinozaki, NATO ASI Series B: Physics Vol. 137 (Plenum, New York, London 1986)
4. M.P.Ireland, K.S.Richards: Histochemistry 51, 153 (1977)
5. K.S.Richards, M.P.Ireland: Histochemistry 56, 55 (1978)
6. M.P.Ireland, K.S.Richards: Environ. Pollut. Ser. A 26, 69 (1981)
7. K.S.Richards, A.D.Rush, D.T.Clarke, W.J.Myring: J. Microsc. 142, 1 (1986)
8. J.S.Worgan: Nucl. Inst. Meth. 195, 49 (1982)
9. I.N.Ross, M.F.White, J.E.Boon, D.Craddock, A.R.Damerell, R.J.Day, A.F. Gibson, P.Gottfeldt, D.J.Nicholas, C.J.Reason: IEEE J. Quant. Electr. QE-17, 1653 (1981)
10. E.D.Roberts: In Polymer Materials for Electronic Applications, ed. by E.D.Feit, C.Wilkins Jr. (American Chemical Society 1982) p.1
11. R.P.Haelbich, J.P.Silverman, J.M.Warlaumont: Nuc. Inst. Phys. Rev. 222, 291 (1984)
12. R.W.Eason: In SERC Central Laser Facility Annual Report RAL-86-046 (Laser Divn. Rutherford Appleton Laboratory 1986) p.A4.1
13. K.S.Richards, A.D.Rush, D.T.Clarke, W.J.Myring: J. Microsc. 143, 321 (1986)
14. B.J.Panessa-Warren, G.T.Tortora, J.B.Warren: In Proc. 45th Ann. Meeting Microsc. Soc. America, ed. by G.W.Bailey (San Francisco Press, San Francisco 1987) p.882

Biological Applications
of Microtomography

J.C. Elliott[1], R. Boakes[1], S.D. Dover[2], and D.K. Bowen[3]

[1]Dept of Biochemistry, The London Hospital Medical College,
 Turner Street, London E12AD, UK
[2]Dept of Biophysics, King's College, 26–29 Drury Lane,
 London WC2B5RL, UK
[3]Dept of Engineering, University of Warwick,
 Coventry CV47AL, UK

1. Abstract

X-ray microtomography can be used to obtain, non-destructively, 3-dimensional information about biological structures at a resolution of a few microns. The information may be purely morphological, but if accurately monochromatic X-radiation is used, the results are in the form of the 3-dimensional distribution of the linear absorption coefficient.

The X-ray source used was a microfocus X-ray generator with a 100 μm source and molybdenum target. The transmitted X-ray intensity of the sample was measured with a high purity germanium detector which allows the measurements to be made with monochromatic radiation. This system was used to determine the quantitative distribution of X-ray absorption in a rat rib bone and elephant dentine (ivory).

2. Introduction

X-ray microtomography is a miniaturized version of the well-known medical radiographic technique of CT scanning and it enables "sections" of the X-ray absorption through an object to be determined non-destructively on a microscopic scale [1-8]. The principle of the method is illustrated in Fig.1. The specimen is mounted so it can be rotated about an axis 0 and stepped across a collimated X-ray beam. This enables the projected X-ray absorption in the plane of a thin slice normal to the axis of rotation to be determined. Projections are determined for a large number of orientations of the object, and these are used to reconstruct the distribution within the "section". A series of "sections" can be can be built up by translating the specimen parallel to the axis of rotation. Other experimental arrangements have been used, for example, those in which the X-ray source is moved by scanning the electron beam across the target [4], or by the use of a miniaturized 2-dimensional X-ray detector [6].

The choice of appropriate X-ray wavelength to be used for microtomography depends on the size of the specimen and its

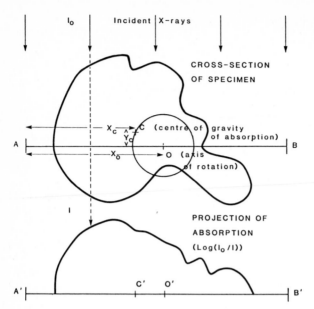

Fig.1 Formation of a
projection through a
"section"

I₀   Incident X-rays

CROSS-SECTION
OF SPECIMEN

X_c   C (centre of gravity
           of absorption)

A   X₀   O (axis   B
              of rotation)

PROJECTION OF
ABSORPTION

(Log(I₀/I))

A'   C' O'   B'

composition, and can be determined on the basis that the maximum
absorption that the specimen should produce is about 50 per
cent. Thus for a 1 mm thick specimen of liver approximately CuKα
(λ = 1.54 A) radiation is required, but for 1 mm of cortical
bone, the wavelength has to be reduced to that of AgK$_\alpha$ (λ = 0.56
A). Both these wavelengths are easily obtained from laboratory
or synchrotron sources, and both have been used for
microtomography work.

Biological applications of X-ray microtomography can be
rather arbitrarily divided into two types of investigations,
those in which purely morphological information is sought, and
more detailed quantitative studies in which the distribution of
the X-ray absorption coefficient is determined. The most
important aspect of the application of X-ray microtomography to
the determination of morphological structures is that it enables
3-dimensional information to be directly determined at a
resolution of a few microns in a non-destructive manner.
Biologists have normally tackled this type of problem by cutting
serial sections with a microtome and then making wax
reconstructions of the structure. This is a rather time
consuming process, but it has been very successfully used during
the last hundred years or so and is the basis for our
understanding of most biological structures at the level of the
light microscope. Therefore useful applications of X-ray
microtomography in this area would seem to be restricted to
special situations such as the study of valuable specimens where
a non-destructive technique is required, or to specimens where
it is difficult or impossible to make serial sections. Examples
of such specimens are valuable museum material, particularly
fossils, and hard and brittle specimens such as some types of
calcified tissues. Another potential application which would

only require a fairly modest improvement in resolution over that currently available would be to the study of 3-dimensional complex structures at a resolution from 0.2 - 1.0 μm. This is a region that can only be studied with great difficulty by electron microscopy serial sectioning because the sections can only be a few hundred angstroms thick. For example, in a study of synaptic interconnections, 1200 serial sections were required to cover a distance of 100 μm [9].

Experimental and theoretical aspects of the use of X-ray microtomography to obtain 3-dimensional morphological information have been discussed in an earlier study of the external morphology and internal flaws of a 3/4 X 1 mm crystal of copper sulphate [8]. The difficulties lie in the determination of a common origin for the projections so that the reconstruction algorithm will work properly, and the accurate registration of adjacent sections so that the 3-dimensional image will be correctly assembled. With macroscopic specimens, there are no problems because the position of the axis of rotation is well defined and can be accurately determined. It may therefore be used as the origin for the projections and the reference to line up adjacent sections. However with microscopic specimens, this is not so straightforward because of the requirement to locate the origin for the reconstructions to within a small fraction of a pixel (say 1/10 of 10 μm) and to overcome the possibility that the axis of rotation may move by a similar amount during rotation because of mechanical imperfections in the system. These difficulties can to a certain extent be overcome by using the projection of the centre of gravity of the absorption as the origin for the reconstruction for each section. The position of this can be determined accurately for each projection so that it does not matter if the axis of rotation moves between projections. However, this strategy means that there will be a different origin for the reconstruction for each "section". However if the assumption is made that the axis of rotation does not move, the locus of the centre of gravity will be the circle shown in Fig.1. The displacement between the axis of rotation and the projected centre of gravity (Xo - Xc) for a given orientation of the specimen can be calculated from a least squares analysis of the position of the projected centre of gravity as a function of the angular orientation of the specimen. The enables each section to be moved to a common origin (the axis of rotation) so that a correctly registered 3-dimensional image can be built up.

The second application of microtomography, namely the accurate determination of the distribution of the X-ray absorption coefficient, is the main subject of the experimental part of this paper, and will be illustrated by a study of the microscopic mineral distribution in bone. This is a subject of clinical importance in relation to a number of serious bone diseases, for example osteoporosis. The only way in which this is presently measured on an absolute scale is by density fractionation of ground bone by centrifugation [10]. The density range of interest is from 1.7 - 2.3 g cm-3 and typical studies require about 0.5 g of bone and take about a week for fractionation into 6 or 7 density values at 0.1 g cm-3 intervals. It should be possible to obtain the same information from microtomography studies, with the added advantage that positional information is retained. In this case, the X-ray absorption measurements will have to be made with accurately monochromatic radiation, a requirement that does not have to be so strictly met if only morphological detail or changes in absorption coefficient are to be observed.

3. Experimental

The X-ray source used was a Hilger & Watts Y33 microfocus generator with a molybdenum target run at 35 kV and 2.0 mA without filtration. The projected source size was 100 x 100 µm with a 10 µm electron microscope aperture placed 30 mm from the source with the bone sample mounted immediately behind it on a kinematically designed stage with an accurately defined axis of rotation described previously [7]. The transmitted X-ray intensity was measured with an EG & G Ortec high-purity germanium detector and preamplifier with feed-back resistor selected for high count rates, a 673 Spectroscopy Amplifier, and a pulse height analyzer set for MoK with a 10 per cent window. The count rates per second for the MoK and total radiation without window were approximately 20,000 and 110,000 respectively and no dead-time corrections were made. The linearity of the system was checked by measuring the absorption coefficient of aluminium for thicknesses in the range 0.25 - 2.0 mm at 0.25 mm intervals. During the X-ray microtomography measurements, the stability of the X-ray source and counting system was monitored every 15 minutes by counting for 5 times the normal counting time with the specimen withdrawn from the beam. Each projection was measured at 128 points and Fourier transform methods were used in the reconstructions. Other experimental details are given in the Figure captions.

4. Results and Discussion

The measurements of the linear absorption coefficient for the different thicknesses of aluminium foils gave values starting from 10.7 cm-1 for the thinnest, rising to a nearly constant value of 11.7 cm-1 when the thickness was over 1.0 mm. This compares with a theoretical value of 13.9 cm-1 for MoK radiation. The low value, particularly for thin specimens, is probably due to the fact that no dead-time corrections were applied, with the consequence that the measured intensity of the beam without a specimen will be too low. To correct for this, the measured absorption coefficients determined from the reconstructions have been multiplied by a factor 13.9/11.7 = 1.2.

A reconstructed "section" through a piece of dentine is shown in Fig.2, and through a rat rib in Fig.3. The range of absorption coefficients shown can be compared with the calculated value of 27.7 cm-1 for pure hydroxyapatite, the mineral component of bone. In the dentine, the dark and light bands seen parallel to the circumference are probably artifacts, but bands of differing mineral density running approximately in a north easterly to south westerly direction can also be seen.

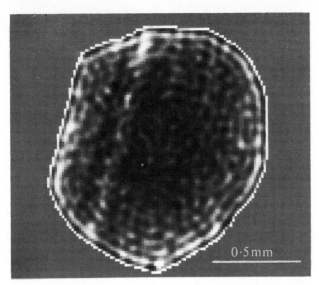

Fig.2 Reconstruction of a "section" through elephant dentine. The grey scale has been expanded to cover a range from 5.9 cm-1 (white) to 17.4 cm-1 (black). (1s counting time at each point, step size 13 μm, and 60 projections at 3 degree intervals)

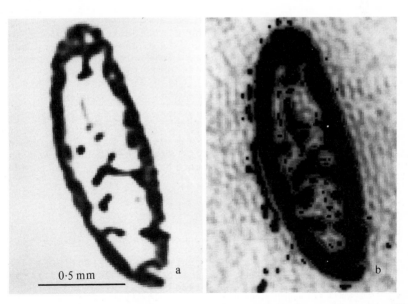

Fig.3 Reconstruction of a "section" through a rat rib. The grey scale has been expanded to cover the range from 9.0 cm-1 (white) to 23.4cm-1 (black) in (a),and from 5.3 cm-1(white) to 9.5cm-1 (black) in (b). (Counting time 5s at each point, step size 12 μm, and 120 projections at 1.5 degree intervals)

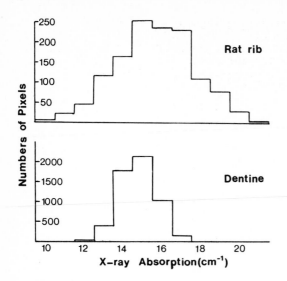

Fig.4 Mineral density distribution for rat rib bone and elephant dentine. The criterion for the selection of pixels to be included in the distribution is discussed in the text

The rat rib shows the more dense cortical bone surrounding trabecular bone. Mineral density distributions can be calculated from the values of the absorption coefficients for the individual pixels. Any pixel that lies across an edge will distort the distribution because it will have too low a value. To avoid this, a pixel was only counted if its 8 nearest neighbours did not differ from it by more than 7 per cent for dentine and 13 per cent for bone. The distributions for bone and dentine are given in Fig.4 and show that bone has a very much broader distribution than dentine. This is to be expected because bone is a much more heterogeneous tissue than dentine. However these results also show that the peak mineral density is significantly higher in bone than dentine.

5. Acknowledgements

This work was supported by the Medical Research Council and the Science and Engineering Research Council.

6. References

1. J.C. Elliott, S.D. Dover: Metab. Bone Dis. Rel. Res. _5_, 219 (1984)
2. J.C. Elliott, S.D. Dover: J. Microscopy _138_, 329 (1985)
3. D.K. Bowen, J.C. Elliott, S.R. Stock, S.D. Dover: In X-ray Imaging II, SPIE Vol.691 p.94 (1986)
4. A.Yu. Sasov: J. Microscopy _147_, 169 (1987)
5. A.Yu. Sasov: J. Microscopy _147_, 179 (1987)
6. B.P. Flannery, H.W. Deckman, W.G. Roberge, K.L.D'Amico: Science _237_, 1439 (1987)
7. J.C. Elliott, D.K. Bowen, S.D. Dover, S.T. Davies: Biol. Trace Elem. Res. _13_, (in press)

8. S.D. Dover, J.C. Elliott, R. Boakes, D.K. Bowen:
   J. Microscopy (in press)
9. J.E. Hamos, S.C. van Horn, D. Raczkowski, S.M. Sherman:
   J. Comp. Neurology $\underline{259}$, 165 (1987)
10. M.D. Grynpas, P. Patterson-Allen, D.J. Simmons: Calcif.
    Tissue Int. $\underline{39}$, 57 (1986)

X-Ray Microscopy – Its Application to Biological Sciences

P.C. Cheng[1], H.G. Kim[2], D.M. Shinozaki[3], K.H. Tan[4], and M.D. Wittman[2]

[1]Department of Electrical and Computer Engineering,
Department of Biology, State University of New York at Buffalo,
Buffalo, NY 14260, USA
[2]Laboratory for Laser Energetics, University of Rochester,
250 E. River Rd., Rochester, NY 14623, USA
[3]Faculty of Engineering, University of Western Ontario,
London, Ontario, Canada N6A 5B9, Canada
[4]Canadian Synchrotron Radiation Facility,
Physical Science Laboratory, University of Wisconsin,
Stoughton, WI 60598, USA

INTRODUCTION

It is not easy to examine bulky or hydrated biological material with an electron microscope (EM). In contrast, x-ray microscopy (XM) provides great penetration power that electron microscopy can not offer, and high resolution with which light microscopy can not compete. Therefore, x-ray microscopy could occupy a niche in biological research in the form of 3D imaging of thick (in comparison to the sample thickness of standard electron microscopy) and living specimens.

It has long been proposed and recently demonstrated that x-ray microscopy can be used to reveal the elemental composition of a specimen. But XM is incapable of distinguishing molecules which have similar elemental compositions but are very different in their physiological functions. For instance, proteins generally have very similar elemental compositions, but their structures and functions can be very different. Furthermore, many biologically important elements, such as calcium and magnesium, are present in a cell at very low concentrations. The detection of such low concentrations of ions in a living cell by x-ray absorption microscopy could be a sizable challenge, if even possible.

Recent advances in cellular biology have developed various fluorescent dyes and antibodies for tagging specific proteins, membrane receptors (1), and organelles (2) and to reveal intracellular pH values (3), calcium concentrations (4), and membrane potentials (5). The above-mentioned observations have all been carried out in various living cells by light microscopy. X-ray microscopy can be used for the study of living cells and thick specimens with a higher resolution than light microscopy. However, x-ray-dense probes which have sufficient specificity, such as the above mentioned molecular tags, are needed to give x-ray microscopy a competitive edge over both light and electron microscopy. Therefore, it is important not to restrict the application of x-ray microscopy to only unstained specimens.

356

The potential applications of x-ray microscopy are in the domain of wet/living cell studies; therefore, the time required to obtain an image is crucial. For instance, recent advances in video microscopy have provided a powerful tool for the study of dynamic changes in cells. The commonly used video microscope has a standard TV frame rate, but it is still too slow to study many biological phenomena. For instance, the standard TV scan rate is too slow to study the movement of flagella (6) and muscle contraction. Therefore, in order to fully explore the capability of x-ray microscopy, shortening the image gathering time becomes essential.

At present, five major types of x-ray microscopy are available. They are x-ray contact imaging (microradiography), scanning x-ray microscopy, x-ray imaging microscopy, x-ray photoelectron microscopy and shadow projection x-ray microscopy. In this article, we intend to review the present status of these various types of x-ray microscopy, and provide suggestions from a biological perspective. X-ray photoelectron microscopy is in its early developmental stage. Therefore, it is too premature to discuss the potential biological applications here.

X-RAY CONTACT IMAGING

This is the simplest form of x-ray imaging and was developed shortly after the discovery of x-rays. The technology, commonly referred to as x-ray microradiography, was extensively improved in the 40's and 50's in parallel with the development of electron microscopy. The research activities tapered off in the 60's and finally petered out. Not until the late 70's did the use of high molecular weight polymer resists as photochemical detectors make x-ray microradiography once again an attractive imaging technique. This modern form of x-ray microradiography has been commonly referred as x-ray contact microscopy. Figure 1 shows a typical example of an x-ray contact image formed

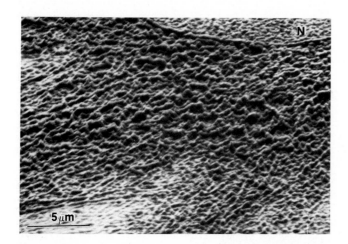

Figure 1. X-ray contact image of the cytoplasm region of a human fibroblast. The x-ray contact image was taken with monochromatic x-rays (3.5nm) on PMMA resist, magnified by using a scanning electron microscope operated in the low-loss electron (LLE) mode (E_{loss}= 200eV, E_o = 10keV). N:nucleus. (P. C. Cheng and O. C. Wells)

on PMMA resist and subsequently magnified using a scanning electron microscope operating in the low-loss electron mode. For a detailed description of the x-ray contact imaging technique, readers are suggested to consult some of the early publications (7,8) and articles in this proceedings.

X-ray microradiography is technically simple and inexpensive; however, the technology does have certain drawbacks. Because the imaging technique is a two-step process, namely a image formation step on the photoresist and the subsequent magnification step to magnify the miniature contact print formed on the resist, x-ray microradiography is, therefore, difficult to use in the study of dynamic biological processes. One may suggests that by using a high intensity pulsed x-ray source, a "frozen frame" of a dynamic process can be obtained. At present, it seems that only plasma x-ray sources, such as the z-pinch and laser-generated x-ray sources, can meet this requirement. However, modern biological research generally requires careful follow-ups and multiple recordings of a specimen in real time. For instance, the study of cell secretion requires multiple observations of a complex and dynamic process. Therefore, x-ray microradiography may have difficulties meeting the requirements in such research.

In the past few years, one has observed increased activity in x-ray microscopy, primarily in x-ray contact imaging, using both z-pinch and laser-generated x-ray sources. For biologists intending to use the pulsed x-ray sources for their research, it is important to note that both the z-pinch and laser-produced plasma sources generate not only x-ray radiation but also emit high intensity UV, visible and IR radiation. The photoresists, such as PMMA and PBS (poly butyl sulfone), used in x-ray contact imaging are also highly sensitive to UV radiation and subject to thermal damage caused by IR radiation. Therefore, it is essential to use suitable filters or monochromating devices to remove the UV and IR components emitted by the plasma source. Recent results indicate that, with a 25nm Al filter in place, PMMA resist does not provide the required sensitivity to form an image when an exposure is made using a single x-ray pulse (obtained by focusing a 100μm spot of 356nm, 40J, laser beam onto a Mo target). However, heavy exposure was observed when irradiation was made without the Al filter. Therefore, it is clear that an UV dominated contact image is obtained if the exposure is made without a suitable UV-blocking filter. Figure 2 shows the surface of

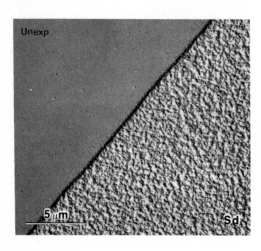

Figure 2. Surface roughness of a PMMA resist. The resist has been exposed to unfiltered radiation from a laser-generated plasma source. No chemical development was used. The surface of the unexposed resist (Unexp) is very smooth, but the exposed region (Sd, self-developed) becomes very rough.

the PMMA resist which has been exposed to unfiltered radiation from a laser-generated plasma source. The surface roughness is believed to be due to UV ablation of the resist. No ablation was observed when a 25nm Al filter was placed between the source and the resist (9).

The advantage of x-ray microradiography, beside being technically simple and inexpensive, is that the technique offers very high resolution images (if the contact image is obtained, viewed and magnified properly). Hence, x-ray microradiography could be used in a situation where real time imaging is not needed. In considering the positive and negative aspects of this technique, we believe that the primary applications of this technology are in the study of preserved materials or living specimens which do not require following dynamic processes.

SHADOW PROJECTION X-RAY MICROSCOPY

The shadow projection x-ray microscope (SPXM) was developed in the 50s, and few commercial units were built. The SPXM failed to develop into a common instrument primarily due to two factors. First, the attention of the biological communities was focused on the development of electron microscopes which had much higher resolution than the x-ray microscope. Second, the image intensity of the x-ray projection microscope was relatively low, and the lack of efficient detectors made observation and recording both difficult and time consuming.

Nearly four decades after the marketing of the first SPXM by General Electric Co., there is a resurgent interest in revitalizing the technology. During the past decades, electron microscopy has developed into a mature technology and has become an integral part of biological, medical and material sciences. However, due to the difficulties in handling hydrated specimens in electron microscopy, observations of hydrated biological samples primary remain in the hands of light microscopy. Bulky specimens are generally required to be serially sectioned to reveal their 3D structures. The availability of low intensity TV cameras, 2D solid state detectors and the capabilities of computer-based image processing systems should revitalize the SPXM. It is proposed that modern SPXM should achieve a real time or near real time imaging capability with resolution in the micrometer domain. The low intensity of the projected x-ray image formed on the fluorescent screen should be easily captured by using a low intensity TV camera (e.g. SIT camera) and processed by a computer to generate microtomographic images. Figure 3 shows a schematic of a proposed SPXM. For a detailed description of

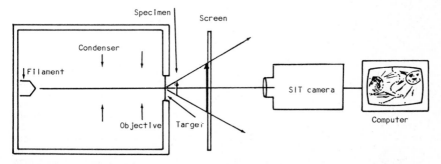

Figure 3. Schematic of a proposed shadow projection x-ray microscope.

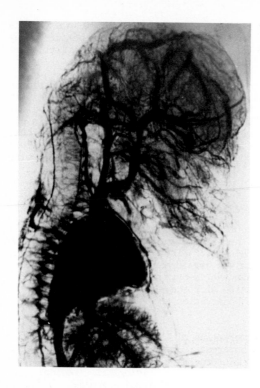

Figure 4. Rat embryo.
(Courtesy S. P. Newberry)

the SPXM, readers should refer to the articles by S. P. Newberry and others in this volume and elsewhere (10).

Although the resolution of the SPXM is lower than other forms of x-ray microscopy (typically 1-5μm), the shorter wavelength (fractions of an Angstrom to a few Angstroms) used in the SPXM provides much higher penetration power than those using soft x-rays in the range of 2nm - 5 nm. This high penetration power enables us the study of bulky specimens such as biopsy materials obtained for pathological investigations. A SPXM equipped with a real time imaging capability could assist, for instance, in the exploratory operation to ensure that a useable biopsy sample is obtained. Therefore, it can greatly reduce the unnecessary trauma to the patent if a second operation became necessary simply due to the first biopsy missing the affected tissue. Furthermore, SPXM should be very useful in the study of bulky botanical specimens to minimize time-consuming dissection and provide 3D views of the internal structure. Figure 4 shows a shadow projection x-ray image of a rat embryo.

SCANNING X-RAY MICROSCOPE AND X-RAY PROBE

Various scanning x-ray microscopes have been designed and constructed (11,12). For a detailed description of the instruments, the reader should turn to other articles in this proceedings. Figure 5 shows an x-ray image of a neuron outgrowth obtained by the Kirz's group at Stony Brook. In most of the SXM designs, a fine x-ray beam is focused on a specimen by a zone plate, and raster scanning of the specimen with respect to the stationary x-ray beam is performed by a precision mechanical stage. ·The transmitted intensity is detected and display on a CRT. In the SXM, not only the transmitted x-ray photons can be detected, but also the secondaries emitted from the specimen. For instance, it should be possible to collect the luminescence emitted from the specimen.

It has been shown that cathodoluminescence can be detected in a SEM from fluorochorms, such as fluorescein, fluorescein isothiocyanate (FITC) and acridine orange. It is possible that x-rays can also excite the above mentioned dyes. If it is so, one could use FITC tagged antibodies as fluorescent probes in scanning x-ray microscopy. Due to the high resolution, scanning x-ray microscopy should offer significant advantages over conventional fluorescence microscopy, and this could be a feature which other types of x-ray microscopy can not offer. The possibility of using fluorescent probes could allow SXM to study cellular structures which have similar elemental compositions but are different in their biological functions.

It is important to note that, in addition to imaging, the focused x-ray beam can also be used as a fine scalpel to selectively dissect or destroy fine cellular structures. The combination of an epi-fluorescence light microscope with the fine-focused x-ray probe could be an extremely useful instrument in cell biology for the study of cellular dynamics.

At present, the major drawback in the scanning x-ray microscope is the time required to form an image. It is important to note that the time scale can be greatly improved when a high brightness undulator beamline is used. Mechanical movement of the scanning stage could disturb the cell (e.g.

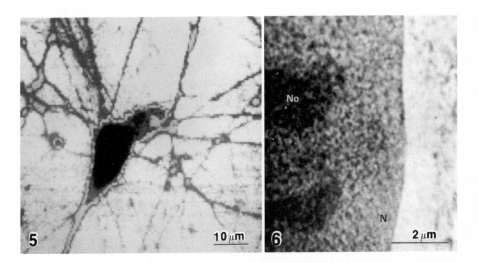

Figure 5. X-ray image of neuron outgrowth obtained by using the Stony Brook scanning x-ray microscope. (Courtesy J. Kirz)
Figure 6. X-ray image of a cultured human fibroblast. The cell was fixed in 1% glutaraldehyde, dehydrated in ethanol, critical-point dried in carbon dioxide. The image was taken by the Göttingen x-ray microscope at BESSY. (W. Meyer-Ilse, G. Nyakatura, P. Guttmann, B. Niemann, D. Rudolph, G. Schmahl and P. C. Cheng)

activate stretch channels on the cell membrane) and cause structure movements. A similar problem has been encountered with the scanning stage version of the confocal light microscope. Therefore, the latest version of the biological confocal light microscope employs a scanning beam method. However, the physical properties of x-rays may not allow an easy solution for a scanning beam x-ray microscope.

X-RAY IMAGING MICROSCOPY

The development of the x-ray imaging microscope has been primarily carried out by the Göttingen group (13). Figure 6 shows an absorption contrasted x-ray image of a human fibroblast (fixed and critical point dried) taken by the Gottingen microscope with 4.5nm x-rays. The possibility of constructing a phase contrast x-ray microscope has been proposed by G. Schmahl and his co-workers (14). In this conference, they reported the first phase contrast image of a human fibroblast. The result generated great excitement because x-ray images can now be obtained not only by absorption contrast but also by phase contrast.

The x-ray imaging microscope provides exciting possibilities for the biologist. The Göttingen group has shown biological images with superb resolution. Furthermore, the time required to obtain an image is quite short. They have installed an optical microscope for the purpose of both scanning and positioning of the sample with visible light. It would be wonderful if the optical microscope was equipped with an epi-fluorescence attachment so that the dynamic processes of living cells could be studied with the aid of various fluorescent probes. An x-ray image can then be obtained when higher resolution is required. The combination of an epi-fluorescence (light) microscope with the x-ray imaging microscope could become a very useful tool in biology.

WHAT MAKES A GOOD BIOLOGICAL MICROSCOPE?

If one asks what kind of features biological researchers would like to have in an x-ray microscope, the answer could be quite simple, but the solution might be a major challenge to the x-ray microscopy community. We all understand that it is impossible to build an x-ray microscope which will satisfy all of the biomedical researchers. Instead, a "standard" microscope with room for expansion should be the design goal. Based upon one of the evening discussions of this conference, we tried to summarized a "shopping" list which hopefully includes most of the general requirements for a "biological" x-ray microscope. Even though the "request" may sound radical to some people, nevertheless, it is the voice of the "shopper". Because both x-ray microradiography and shadow projection x-ray microscopy share different imaging technologies and applications, the following suggestions are primary for the scanning and imaging x-ray microscopes:

1. A horizontally placed stationary stage with room for specimen manipulation.
2. Time required for obtaining an image should be shortened. Real time imaging with 3D capability would be wonderful. Computer aided image processing will help a great deal, but it is not necessary, in most cases, to have false color images.
3. The combination of a light microscope, especially with an epi-fluorescence capability, with the x-ray microscope is highly desirable.
4. If it is necessary to use an environmental chamber, improvements should be made to the design of the specimen chamber so that both loading and

unloading can be done easily and quickly. Oxygen and carbon dioxide concentrations and temperature should be monitored and maintained.

5. For SXM, in addition to the flow proportional counter to detect transmitted x-ray intensities, detecting photoelectrons and photons from luminescence should be considered.

It has been frequently stressed that the design of an x-ray microscope should consider the need of the users (i.e. make the instrument user friendly). For instance, if a microscope is designed to be used in biological research, the specimen stage should be placed horizontally and remain stationary. This is the only reason why, for decades, biological light microscopes have always had a horizontal stage. This design is essential for the study of living specimens and performing micro-manipulation of specimens (e.g. microinjection). In addition to improvements in future x-ray microscopes, implementation of supporting facilities for biological research around synchrotron storage rings should be seriously addressed.

There is always a need in microscopy for the study of living cells at high resolution and with the capability of 3D imaging. A typical example of such a demand has been demonstrated by the recent advances in the confocal scanning reflected light microscope (CSRLM). The CSRLM has offered a powerful tool for nondestructive optical sectioning of living cells and provides near-real time 3D imaging by reconstructing the optically sectioned images (15). The confocal microscope, in conjunction with various well established specimen preparation techniques such as the use of molecular probes, begins to open a new era in the research of cellular functions. For x-ray microscopy to be competitive, it is clear that future development should target on real time and 3D capabilities. Furthermore, the biological communities should begin to develop techniques in specimen preparation for x-ray microscopy.

ACKNOWLEDGEMENTS

Parts of this work have been supported by the MRC, NRC and NSERC of Canada and DoE of USA. Special thanks to the Canadian Synchrotron Radiation Facility (CSRF) for providing the beamline facility. We thank the Laboratory for Laser Energetics (LLE) of the University of Rochester for providing a laser-produced plasma x-ray source.

REFERENCES

1. H. B. Peng, P. C. Cheng and P. Luther: Nature, 292, 831-834. (1982).
2. J. M. Collins and K. A. Foster: J. Cell Biol. 96, 94-99. (1983).
3. J. A. Thomas, R. N. Buschbaum, A. Zimniak and E. Racher: Biochemistry, 18, 2210. (1979).
4. C. H. Keith, B. Ratan, F. R. Maxfield, A. Bajer and M. L. Shelanski: Nature, 316, 848-850. (1985).
5. L. V. Johnson, M. L. Walsh, B. J. Bockus and L. B. Chen: J. Cell Biol. 88, 526-535. (1981).
6. S. Inoue: In Video microscopy, Plenum Press, NY. 219-221. (1986).
7. P. C. Cheng, D. M. Shinozaki, and K. H. Tan : In X-ray Microscopy - Instrumentation and Biological Applications, ed. by P. C. Cheng and G. J. Jan, Springer-Verlag, Berlin, 65-104, (1987).
8. P. C. Cheng, R. Feder, D. M. Shinozaki, K. H. Tan, R. W. Eason, A. Michette, and R. J. Rosser: Nucl. Inst. Methods Phys. Res., A246, 668-674. (1986).

9. P. C. Cheng, H. G. Kim and M. D. Wittman: <u>SPIE</u> <u>Proceedings</u> (1987).
10. S. P. Newberry: In <u>X-ray</u> <u>Microscopy</u> - <u>Instrumentation</u> and <u>Biological</u> <u>Applications</u>, ed. by P. C. Cheng and G. J. Jan, Springer-Verlag, Berlin, 126-141. (1987).
11. C. Jacobsen, J. M. Kenney, J. Kirz, R. Rosser and H. Rarback: <u>Proc.</u> <u>43rd</u> <u>Annual</u> <u>Meeting</u> of <u>Electron</u> <u>Microscopy</u> <u>Amer.</u>, 594-595.
12. B. Niemann: In <u>X-ray</u> <u>Microscopy</u> - <u>Instrumentation</u> and <u>Biological</u> <u>Applications</u>, ed. by P. C. Cheng and G. J. Jan, Springer-Verlag, Berlin, 39-52. (1987).
13. D. Rudolph and G. Schmahl: In <u>Ultrasoft</u> <u>x-ray</u> <u>microscopy:</u> <u>Its</u> <u>application</u> <u>to</u> <u>biological</u> and <u>physical</u> <u>sciences</u>, ed. by D. F. Parson. <u>Ann.</u> <u>N.</u> <u>Y.</u> <u>Acad.</u> <u>Sci.</u>, 342, 94-104. (1980).
14. G. Schmahl and D. Rudolph: In <u>X-ray</u> <u>Microscopy</u> - <u>Instrumentation</u> and <u>Biological</u> <u>Applications</u>, ed. by P. C. Cheng and G. J. Jan, Springer-Verlag, Berlin, 231-238, (1987).
15. J. G. White, W. B. Amos and M. Fordham: <u>J.</u> <u>Cell</u> <u>Biol.</u>, 105, 41-48. (1987)

Investigations of Biological Specimens with the X-Ray Microscope at BESSY

G. Nyakatura[1], W. Meyer-Ilse[1], P. Guttmann[1], B. Niemann[1],
D. Rudolph[1], G. Schmahl[1], V. Sarafis[2], N. Hertel[3], E. Uggerhøj[3],
E. Skriver[4], J.O.R. Nørgaard[4], and A.B. Maunsbach[4]

[1]Universität Göttingen, Forschungsgruppe Röntgenmikroskopie,
  Geiststraße 11, D-3400 Göttingen, Fed. Rep. of Germany
[2]Department of Applied and Environmental Science,
  School of Food Sciences, Hawkesbury Agricultural College,
  Richmond, 2753 N.S.W., Australia
[3]Institute of Physics, University of Aarhus,
  DK-8000 Aarhus C, Denmark
[4]Department of Cell Biology, Institute of Anatomy,
  University of Aarhus, DK-8000 Aarhus C, Denmark

1 Introduction

X-ray microscopy experiments on dry and wet specimens have been performed with the Göttingen x-ray microscope (XM) [1] using the x-ray wavelength of 4.5 nm. The microscope was operated in the phase contrast mode with a silver phase plate in the back focal plane of the objective as described in [2]. The resolution of the x-ray optical system (condenser KZP 3 and objective MZP 3 [3]) is between 50 nm and 60 nm. The microscope is installed at the BESSY electron storage ring in Berlin and utilizes the synchrotron radiation of a bending magnet. Micrographs were recorded on a Kodak 2462 film.

The aim of the present investigation was to explore the possibility of examining wet biological specimens with x-ray microscopy. For this purpose a special micro chamber was designed and different wet biological samples investigated. The results demonstrate that cellular structural details can be observed in wet specimens with the x-ray microscope.

2 Dry Specimens

2.1 Spores of *Dawsonia superba*

Images of dry spores of the Australian moss *Dawsonia superba* were made with an x-ray magnification of 350 and an exposure time of 15 s with a storage ring current of 350 mA. Figure 1 shows one example of such an image. The spore has a thickness of several micrometers. The spore wall ornamentation

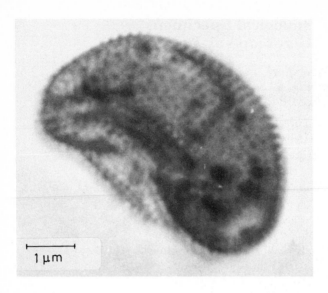

Figure 1: Dry spore of *Dawsonia superba*

can be seen. Internal structures are clearly visible, the round dark x-ray dense structures are probably lipid droplets.

This image demonstrates the capability of x-ray microscopy to look into the interior of relatively thick specimens with high resolution. The three-dimensional structure of this kind of object can be determined using a focus series as has been shown in [4].

2.2 Renal Proximal Tubules

X-ray images of kidney tubules were obtained after the same preparation procedure as for transmission electron microscopy. Rat kidneys were perfusion-fixed with glutaraldehyde, post-fixed in osmium tetroxide, dehydrated and embedded in Epon. Sections were cut to 1.0 µm. The embedding material was removed with KOH and the sections placed on polyimide foils.

Figure 2 shows part of the epithelium of a proximal convoluted tubule of the rat kidney. The epithelium extends between the tubule lumen (TL) and a peritubular capillary (C). On the luminal side of the epithelium the brush border (BB) is observed and in places individual microvilli, which in electron micrographs have a diameter of 0.1 µm, can be identified. In the cytoplasm several structural details can be identified, such as endocytic vacuoles (E), lysosomes (L) and mitochondria. The micrograph was recorded at an x-ray magnification of 390 using 15 s exposure time and a storage ring current of 145 mA.

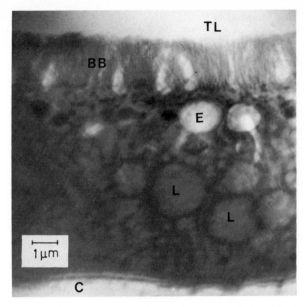

Figure 2: Section of renal proximal tubule of the rat

3 Wet Cells

During observation of wet cells with an x-ray microscope the cells have to be kept in a special environment. It is especially important to keep the cells in an aqueous layer of a defined thickness.

The wet cell experiments described in this paper have been performed with the environmental chamber shown in Fig.3. The regions outside the foils F1 and F4 are under vacuum. The small volume between the foils F1 and F2 and the space between F3 and F4 were pumped down to a pressure of about 300 mbar. The small volume between foils F2 and F3 contains the object surrounded by physiological saline solution. The distance between F2 and F3 is defined by a 5 μm thick aluminium foil. The foils F1, F2 and F4 are polyimide foils with a thickness of about 0.3 μm and an x-ray transmittance of about 85 %. F3 is a thin parlodion foil.

The contrast of cell organelles against water is highest in the "water window" between 2.4 nm and 4.4 nm. The experiments described in this paper were, however, carried out with 4.5 nm radiation because of lack of suitable condenser zone plates at the time. The comparatively low contrast at this wavelength does not allow the direct prefocussing of the cells using x-radiation and the built-in micro channel plate. It was therefore necessary to incorporate highly absorbing structures into the object plane. This was achieved by copying a silver grating, 120 nm thick and with a grating spacing constant of

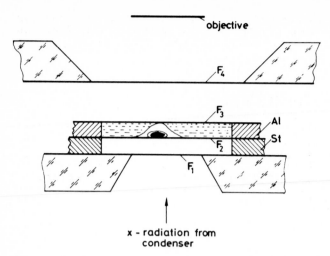

objective

F_4

F_3

Al

F_2

St

F_1

x - radiation from
condenser

Figure 3: Schematic arrangement of the environmental chamber
(not to scale)

50 μm, onto foil F2. This made it possible to focus on the
sharp edge of a highly absorbing silver bar in the object
plane. The experiments described in Sections 3.2 and 3.3 were
done using this technique.

3.1 Spores of *Dawsonia superba*

To obtain images of wet cells of *Dawsonia superba*, a few dry
spores were placed in the environmental chamber described
above and covered with a drop of water for several minutes
during which they swelled and became spherical. They were then
transferred to the microscope and imaged. Figure 4 shows an

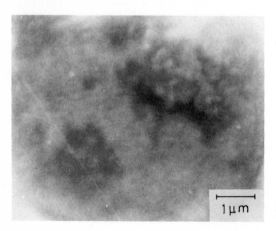

1 μm

Figure 4: Part of a wet
spore of *Dawsonia superba*

inner part of such a wet spore. Some internal structures can be made out. The image was taken with an x-ray magnification of 250, an exposure time of 30 seconds and a storage ring current of 220 mA.

3.2 *Nanochlorum eucaryotum*

Wet cells of the algae *Nanochlorum eucaryotum* are spherical and are about 2 μm in diameter. Under the experimental conditions at this stage of development in x-ray microscopy, the relatively small size of these cells makes them suitable for wet cell imaging. Figure 5 shows an x-ray micrograph of such an algae cell. The image was taken with an x-ray magnification of 250, an exposure time of 120 s and a storage ring current of 125 mA.

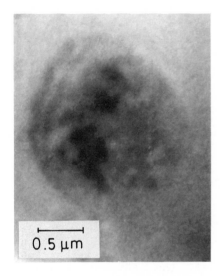

0.5 μm

Figure 5: Wet cell of *Nanochlorum eucaryotum*

3.3 Murine Epithelial Cells (FES)

Mouse cells of the Friend-Evelyn-suspension cell line (FES) infected with the retrovirus Friend murine leukemia virus (FMLV) were cultured in the environmental chamber described above and introduced into the microscope during the logarithmic growth phase to ensure maximum virus production. Figure 6 shows part of a FES cell. This image was taken with an x-ray magnification of 250, 70 s exposure time and a storage ring current of 300 mA. The micrograph has contrast and fine structures are discernible. Up to now, however, viruses could not be distinguished from other fine structures. These experiments will be continued with better contrast with 2.4 nm x-radiation.

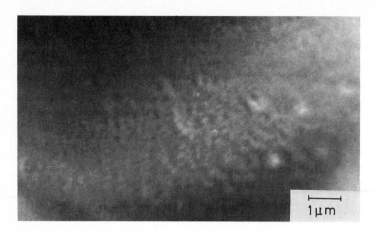

Figure 6: Part of a wet FES cell

4 Concluding Remarks

The experiments described above demonstrate that it is possible to make high resolution x-ray micrographs of biological samples. The advantage of this technique is that internal structures of comparatively thick objects can be studied. Progress in sample preparation of wet specimens has been made. Several high resolution images of wet objects are presented, however, further improvements of optical elements and of sample preparation methods are necessary.

Acknowledgements

We thank Dr. J. Schneider of the Deutsches Primatenzentrum, Göttingen, for providing us with the FES cell line, Dr. M. Reitz of the Institut für Physiologische Chemie, Universität Mainz, for the samples of *Nanochlorum eucaryotum* and the staff of the Berliner Elektronenspeicherring GmbH (BESSY) for providing good working conditions. This work has been funded by the German Federal Ministry for Research and Technology (BMFT) under contract number 05 320 DAB and by a grant of the Australia-Germany Bilateral Science and Technology Accord to Mr. Sarafis.

References

1. D. Rudolph, B. Niemann, G. Schmahl, O. Christ : "The Göttingen X-Ray Microscope and X-Ray Microscopy Experiments at the BESSY Storage Ring", in : <u>X-Ray Microscopy</u>, ed. by G. Schmahl and D. Rudolph, Springer

Series in Optical Sciences, Vol. <u>43</u>, (Springer, Berlin, Heidelberg 1984) p. 192-202

2. G. Schmahl, D. Rudolph, P. Guttmann, : "Phase Contrast X-Ray Microscopy Experiments at the BESSY Storage Ring", this volume

3. G. Schmahl, D. Rudolph, P. Guttmann, O. Christ : "Zone Plates for X-Ray Microscopy", in : <u>X-Ray Microscopy</u>, ed. by G. Schmahl and D. Rudolph, Springer Series in Optical Sciences, Vol <u>43</u>, (Springer, Berlin, Heidel- berg 1984) p. 63-74

4. L. Jochum : "Imaging X- Ray Microscopy with Extended Depth of Focus by use of a Digital Image Processing System", this volume

The Biology of the Cell and
the High Resolution X-Ray Microscope

S.S. Rothman[1], *N. Iskander, K. McQuaid*[1], *H. Ade*[4], *D.T. Attwood*[1],
T.H.P. Chang, J.H. Grendell, D.P. Kern[3], *J. Kirz*[4], *I. McNulty*[4],
H. Rarback[2], *D. Shu*[2], *and Y. Vladimirsky*[1]

[1]Center for X-Ray Optics, Lawrence Berkeley Laboratory,
  University of California, Berkeley, CA 94720, USA,
[2]National Synchrotron Light Source, Brookhaven National Laboratory,
  Upton, NY 11973, USA,
[3]IBM Research Center, Yorktown Heights,
  NY 10598, USA,
[4]State University of New York, Stony Brook, NY 11794, USA,
[5]University of California at San Francisco, San Francisco,
  CA 94143, USA

The biologist's dream is to observe cellular structure and content by natural contrast mechanisms and, if possible, to observe variations as a natural consequence of cell function. While the x-ray microscope will not likely allow full satisfaction of this dream, it does offer new and exciting possibilities in this direction. In what follows we will consider the basis for this view both in general and by means of a specific example, the important and ubiquitous cellular process of secretion.

The relationship between the physical and chemical dynamics of many cellular activities, and their geometric (anatomic or cytological) correlates, is often thought to have been in great part established by combining modern, electron-microscopy-based visual observation, with biochemical (and recently molecular biological) approaches, providing us with a more or less accurate accounting of the time-space relationships of cell processes. In fact, however, we often do not know whether this is the case or not, and the view that we have such knowledge often reflects as much wishful thinking as rigorous understanding. The core problem is that structures thought to be central to many cellular activities can only be resolved with the electron microscope. Although electron microscopy has unarguably provided a remarkable variety of new information about the cell since it came into general use in biology after World War II, and played a central role in the development of modern perceptions of cells and their activities, requisite preparation techniques raise difficult questions about the fidelity of those perceptions in both structural and dynamic terms.

As we are all aware, in order to examine samples in the electron microscope, radical alteration of the material is generally necessary. In an attempt to retain organic material in place (fixation), cross-linking agents such as glutaraldehyde must be added. To what extent such treatments are quantitatively effective and do not otherwise alter the object's structure, is unclear for most material. Furthermore, the sample must be stained, generally with heavy elements, to provide contrast; and as a result what is seen, or not seen, reflects the relative "staining" properties of material, and assumes that the stain does not alter structure. Water in the sample (about 75% of the content of the average cell) must be replaced by plastic in order to permit viewing in vacuo. The extent of structural alteration and geometric distortion introduced by dehydration and plastic embedding is another uncertainty. Finally, only thin sections of cells can be examined due to the short mean free path of the electrons on the order of 100 nm. Taken together these requirements, as well as the complexity of the material being viewed, leave the fidelity of the image to the natural object often, if not invariably, uncertain. That is, the image thought to represent a particular natural structure is of necessity an hypothesis for that structure, not an hypothesis-independent view of it. Unless we claim that the procedures used for preparation, in spite of the major chemical and physical alterations, reconstruct the natural situation exactly, the electron microscope provides us with a homunculus of a cell, a distortion related to the real object--we hope closely related--but a distortion nonetheless. More importantly, because we only know about the distortion, we cannot know unambiguously its relation to the natural, unknown object. Although substantial efforts have been made to overcome these uncertainties, for example, by using frozen hydrated specimens, there remains for commonly applied electron microscopic technique the problem of structural fidelity.

Even if faithful to its original appearance, the examined object, as a result of such treatments, can never "function." If our desire is to observe the dynamics of biological processes in the living cell, then electron microscopy does not give us access to them. One cannot observe the movement and metamorphosis of structures within cells in this fashion. In spite of attempts to overcome this limitation in a variety of ways, many cell-biological models remain merely plausible hypotheses for what might happen. Often, belief in a particular electron-microscopically-based functional hypothesis is bolstered by evidence from other types of investigation (such as biochemical studies), what we believe to be possible or not possible, and at times simply the hypothesis' plausibility in its own right. This is not to say that such hypotheses have no claim on the truth, but there is a question of dynamic fidelity, much as there is one of structural fidelity.

Perhaps the premiere example of such an hypothesis is the vesicle theory of secretion and intracellular transport. It is the central paradigm spawned by the electron-microscope-based discipline of "cell biology" as it emerged after WWII as part of the effort to obtain high resolution images of cells.

THE VESICLE PARADIGM

When the electron microscope permitted the first look into the ultrastructural world of the cell, the acinar cell of the pancreas -- a cell responsible for the secretion of large quantities of protein, the digestive enzymes (proteins) that break down the food we eat -- was found to be packed with fine structure (Fig. 1). Fully half of the cell's volume was occupied by diverse objects, such as the spherical 1-μm diameter zymogen granules (a type of secretion granule or vesicle) that contain some of the protein products secreted by the cell, centrally some 20 or so digestive enzymes. The granules, first seen in the light microscope almost 100 years earlier, occupied a variable fraction of the cell's volume (up to about 25%), and were primarily situated close to the pole of the cell where secretion in great part takes place (the apical end). In addition, some 20% of the cell volume was due to numerous parallel flattened membrane saccules, the endoplasmic reticulum (ER), situated primarily at the opposite end of the cell from the granules (the basal end) and separated from them by the nucleus. Attached to the membranes of the ER were small (about 15-nm diameter) roughly spherical structures, ribosomes, that are centrally involved in the synthesis of protein by the cell, including secreted proteins such as the digestive enzymes [1-3].

In addition to secretion granules and the ER, three other structures were observed that have been proposed to play a major role in secretion: the Golgi apparatus, stacks of membrane-enclosed cisterns lacking attached ribosomes, usually situated next to the nucleus; condensing vacuoles (another term for vesicle or granule), lucent, seemingly empty or partially empty granule-like structures, found amidst the darkly stained, seemingly full zymogen granules, that occupy about 3% of cell volume; and finally, small (about 40-nm diameter),

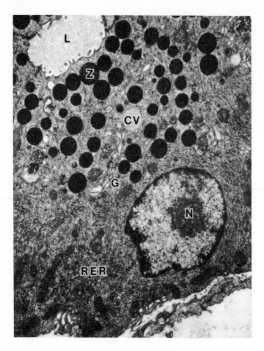

Fig. 1. Acinar cell from a rat pancreas. Uniformly dense zymogen granules (Z) cluster around the lumen (L) at the cell apex. Rough endoplasmic reticulum (RER) fills the base of the cell. CV, condensing vacuoles; G, Golgi complex; N, nucleus. Magnification 10,000 x. See text for a description of how these structures are thought to play a role in the process of secretion. [Reprinted, with permission, from [6].]

roughly spherical (vesicle-like) structures predominantly located laterally to the Golgi cisterns, called smooth-surfaced or microvesicles.

From this array of images, the vesicle theory of secretion evolved. If the proteins are manufactured by the ribosomes located at the basal end of the cell, and in great part end up in the medium outside the opposite, apical end, as well as within the zymogen granules situated towards that end, then the products must move from the basal site of synthesis to the apically-placed zymogen granules and out of the apical end of the cell. But how? It was proposed that this movement occurs by means of the passage of newly manufactured products through a series of discontinuous membrane-bounded compartments (namely, those just enumerated) due to the intermittent fusion and fission of their membranes with each other, and, as a result of this fusion-fission, the transfer of matter occurs directionally from one compartment to the next in an ordered sequence [1-3].

As each new protein is being formed, it is moved across the membrane enclosing the ER and "sequestered" within its confines. The products to be secreted then leave the ER in small (40 nm) vesicles that bud from the apical end of each ER stack, travel in these vesicles either to the Golgi apparatus or to condensing vacuoles, or both in sequence, at which sites they fuse with the recipient compartment and empty their contents therein. These small vesicles, emptied of their contents, shuttle back to the ER to receive a new load. From these roots the mature secretion granule is formed. The products stored within the zymogen granule are released from the cell en masse as the result of the fusion of the enclosing membrane of the granule and the cell membrane, producing a hole in both, and giving the granule's contents access to the extracellular world. This is called exocytosis.

This general model, with numerous addenda and modifications depending upon the particular circumstances, is widely held to account for the secretion of virtually all substances by all cells, proteins as well as small organic molecules, including such important secretion products in addition to the digestive enzymes, as hormones and neurotransmitters [1-6]. In spite of a wide range of evidence consistent with the proposed mechanism, a continuing accumulation of anomalous observations has led to the development of alternative models; either more complex vesicle models, or nonvesicular models that are based on the view that secreted molecules can move as a result of their own thermal motion, obeying mass action, crossing the relevant membranes individually, rather than being moved en masse by external processes [4-6]. In this latter case, vesicles are storage capacitors that discharge their contents, not transport-vesicles that carry them. Of course, these ideas are not mutually exclusive, and both events may occur.

The goal of our studies using the x-ray microscope is to directly test for the existence of both types of process in cells imaged at high resolution during active secretion; looking, in particular, for the fusion of granule and cell membrane and for changes in granule structure and content within the secreting cell.

PRELIMINARY RESULTS

As a preface to these experiments, we report here our preliminary observations on the interior of the zymogen granule. The granules were isolated from their cells of origin by differential centrifugation, after the tissue was homogenized in 0.3M sucrose (pH 5.5)[7]. The isolated granules were viewed in the soft x-ray (SXR) microscope using the X-17T undulator at NSLS, and Fresnel zone plate lenses of 50-70 nm outer zone widths fabricated at IBM. The characteristics of the apparatus are discussed in other articles in this volume. The granules were viewed whole, unaltered by chemicals, containing and suspended in water, at a resolution of approximately 100 nm with 3.2-nm photons.

Earlier work with electron micrographs has been interpreted to indicate that zymogen granule contents are of relatively uniform density, with a relatively fine grain (Fig. 1). Imaging with SXR photons provides a different appearance. A simple analysis of Fig. 2 indicates that the distribution of absorptive material within the granule does not conform to a model of uniformity or homogeneity. It deviates in that a uniform distribution of material within a sphere should show maximum absorption through the center, and these granules do not. In addition, substantial local variations in x-ray absorption are observed.

Inasmuch as carbon is a major constituent of granule protein, and is also the principal element absorbing 3.2-nm x-rays (and water absorption is minimal at this wavelength), we have been able to estimate protein concentration within the granule. This is done through knowledge of the photon flux and use of an average absorption coefficient for some of the major protein contents of the granule, assuming that all absorption is due to carbon in protein. We find the average concentration to be ± 25% of published chemical estimates [8]. This coincidence in values between C-based and chemical protein estimates, adds support to the widely held view that the major organic substance in the granule is protein, although no doubt there is at least a small amount of other organic material present.

The observed nonuniformity of absorption is consistent with other work that suggests that most granule protein is not in solution or suspension, but bound in site-specific arrays [9-11], and in addition that partially emptied granules viewed in the electron microscope appear to have a "reticulated" structure of interconnected dense strands bordering lucent regions (Fig. 3)[11].

It should be kept in mind that these conclusions are preliminary; only a few granules have been examined thus far, and the presence of background noise in

375

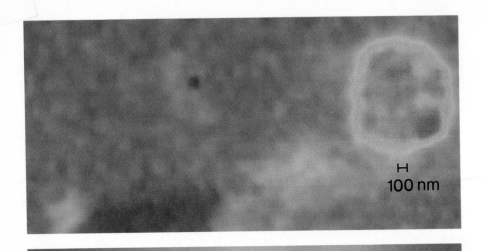

100 nm

06 FEB 20

Fig. 2. False color image of a one-micron diameter pancreatic secretion vesicle (a zymogen granule), imaged whole, unaltered by added chemicals, containing and suspended in water, resolved to approximately 0.1 micron with a zone plate microscope illuminated by undulator radiation at 3.2 nm wavelength. Red represents high carbon content and blue low content.

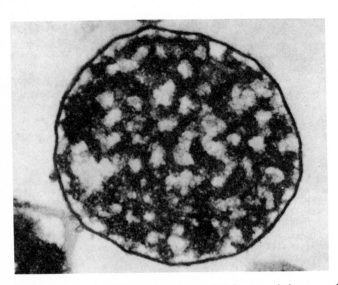

Fig. 3. Partially emptied zymogen granule containing a meshwork of electron-dense strands that make attachments to the membrane and outline lucent polymorphic spaces. Magnification 75,000 x. [Reprinted, with permission, from [11]].

Fig. 4. A single pancreatic acinar cell (fixed and air dried on an electron microscope grid) showing areas of high carbon concentration towards the cell's apex.

these early images give an indeterminacy to our best estimates. Future work should, we hope, increase the reliability of our estimates and allow for the development of increasingly specific structural models.

The granule-containing region within the cell was also observed under the x-ray microscope, appearing near the apical or secretory surface, as expected (Fig. 4). This suggests the possibility of measuring the contents of granules within the cell, as well as in vitro.

ACKNOWLEDGEMENTS: The authors would like to acknowledge valuable help from K. Conkling and D. Joel, as well as the generous assistance of the staff at the National Synchrotron Light Source. We also acknowledge the support of the National Science Foundation (BBS-8618066), the Department of Energy (DE-ACO2-76CH00016; DE-ACO3-76SF00098), and the Air Force Office of Scientific Research (F49620-87-K-0001).

LITERATURE

1. G. Palade: Science 189, 347 (1975).
2. J.D. Jamieson: In Cell Membranes, ed. by G. Weissmann and R. Claiborne (H.P. Publishing Co. Inc., New York 1975), p.143.
3. J.D. Jamieson and G.E. Palade: In International Cell Biology, ed. by B.R. Brinkley and K.R. Porter (Rockefeller Univ. Press, New York 1977), p.308.
4. S.S. Rothman and J.J.L. Ho, Eds., Nonvesicular Transport (John Wiley & Sons, New York 1985).
5. S.S. Rothman: Science 190, 747 (1975).
6. S.S. Rothman, Protein Secretion -- A Critical Analysis of the Vesicle Model (John Wiley & Sons, New York 1985).
7. C. Niederau, J.H. Grendell, and S.S. Rothman: Amer. J. Physiol. 251, G421 (1986).
8. J.J.L. Ho and S.S. Rothman: Biochim. Biophys. Acta 755, 457 (1983).
9. S.S. Rothman: Biochim. Biophys. Acta, 241, 567 (1971).
10. S.S. Rothman: Am. J. Physiol. 222, 1299 (1972).
11. T.H. Ermak and S.S. Rothman: J. Ultrastruct. Res. 64, 98 (1978).

X-Ray Microscopy: A Comparative Assessment with Other Microscopies

R.E. Burge and S. Tajbakhsh

Department of Physics, King's College, University of London,
The Strand, London WC2R 2LS, UK

1. Introduction

Some analogies will be considered here between imaging in the scanning transmission X-ray microscope (STXM) and imaging in both the scanning transmission electron microscope (STEM) and the scanning optical microscope (SOM). Such analogies are valuable especially when the influence of source coherence and of detector configurations is being considered. The theory of scanned image formation first evaluated for the STEM, and elaborated for the SOM can, in general, be carried over to a discussion of imaging in the STXM. Similarly, discussions of convergent beam electron diffraction (CBED) assist in the consideration of convergent beam X-ray diffraction (CBXD) relevant to imaging with a convergent X-ray probe.

Sheppard and Wilson and coworkers (see [1]) in respect of SOM imaging have introduced the nomenclature of Type I and Type II scanning microscopes (Fig 1). In Type I, a point source is focused on a specimen and the transmitted light is collected by a large area detector. There may, or may not, be a lens following the specimen. In these terms, the STXM is a Type I microscope, where the scanning probe, in the form of a hollow cone, is formed by diffraction-limited imaging of a small source aperture by a zone plate, and the specimen is followed directly by a large-area detector. If the specimen is illuminated by a broad monochromatic source, the image may be considered to be made up of an incoherent superposition of point images in parallel, exactly similar to the image formed if the specimen is scanned sequentially across a single small probe. Thus the Type 1 scanning microscope is equivalent, in imaging performance, to the conventional full field microscope.

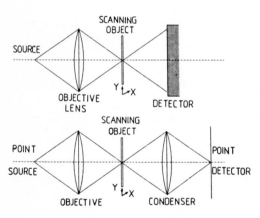

Figure 1 Upper diagram shows simplest (Type I) microscope arrangement, lower diagram shows Type II (confocal) microscope

In the Type II SOM, also known as the confocal scanning microscope, there is a point source and a point axial detector. Two lenses are used, one before and one after the specimen which are independently focused on the specimen. Imaging is coherent, and as the point spread function is the product of the point spread functions of the two lenses, the Rayleigh-limited point to point spatial resolution may be somewhat improved over the Type I configuration. Type II microscopes, in the STXM context, would add a further focusing element to the Type I configuration, downstream of the specimen. If the additional element were a zone plate with 5% focusing efficiency, the large increase in exposure time resulting would vitiate all the present activity using undulators and high brightness lattices directed towards the increase in the spatially coherent flux of X-rays in the scanning probe. There would, also, be a corresponding increase in radiation damage to the specimen.

Thus the X-ray microscope, whether in the form of a scanning or full field imaging microscope, is constrained currently by practical factors to be a Type I microscope. Assuming a diffraction-limited X-ray probe at the specimen, we consider the possible ways for observing different images from the same object region, with the hope that the simultaneous recording of a number of images may provide more information about the object. Of the instrumental variables available which may give different images, the angle of incidence of the probe at the specimen can be varied, bulk specimens may be examined at grazing incidence, and photoelectron emission and fluorescent X-ray signals may be detected as well as the usual X-ray absorption in transmission through thin specimens. However, the instrumental variable of principal concern here is the detector. More particularly, on the basis of published work for STEM and SOM we wish to draw attention to the value, in principle, of configured (pixellated) detector surfaces, without, at this stage, considering experimental aspects. An immediate advantage of detectors inside and outside the incident cone of X-rays is to record independent bright field and dark field images.

We consider below first convergent beam X-ray diffraction, which deals with the geometry of the interaction of the probe with an object element, followed by a discussion of the effects of detector configurations on imaging; we note the spatial differentiation accomplished by a split detector. In order to make an analogy with published work on image analysis in STXM, we shall adopt a specimen model with weak interaction with the transmitted X-ray beam ie a weak phase, weak amplitude model. In practice, considering for example a real biological specimen, this is more likely to be a weak phase, strong amplitude object, but the assumption of weak interaction provides a useful understanding of the mechanism of image formation.

2. Convergent Beam X-ray Diffraction (CBXD)

There is a close analogy between the diffraction geometry of CBXD and convergent beam electron diffraction. A description of the diffraction geometry is given by Steeds [2]. For the short (compared with soft X-rays) electron wavelength the radius of the Ewald sphere is much larger than the reciprocal spacing of lattice planes and diffraction from both zero and higher-order Laue zones (HOLZ) are readily observed in STEM. Because of the convergent electron beam, the diffraction pattern is a series of discs, each centred at a Bragg spot, of diameter proportional to the cone angle of the beam.

For the X-ray case, with a wavelength in the 1 nm to 3 nm range, and both stationary probe and stationary specimen, the effect of the hollow cone X-ray probe may be considered by its spatial decomposition into sets of plane waves. For a typical zone plate with 50 μm diameter and 1 mm focal length, the probe semi-angle $\alpha = 25$ mrad; the focused hollow-cone radiation from an apodised zone plate is shown in Figure 2. Diffraction will be observed both inside and outside the angular range of the X-ray probe. For an X-ray wavelength of 2 nm, diffraction angles of 5.7° and 0.57° arise for object spacings of 10 nm (dark field) and 100 nm (bright field) respectively.

It is of interest to comment on the (soft) X-ray pattern of a single diatom observed by Yun, Kirz and Sayre [3], which has a first order spacing of 200 nm and was observed to the 5th

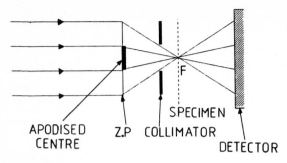

Figure 2 Schematic diagram of hollow cone X-ray probe produced by apodised zone plate. First order focus only is shown, zero order and higher order radiation is blocked from the specimen by the collimating aperture

order. This work was carried out with a 20 μm diameter X-ray beam, but is, nevertheless approximately the performance expected in the zero order Laue zone.

The point behind this consideration of CBXD is to show that there is actually structure due to diffraction in the diverging X-ray beam transmitted by the specimen to the detector. This structure is normally lost in X-ray microscopy by the use of a large integrating detector. Further, an important feature of CBED is the three-dimensional information carried by the HOLZ structure,which may be detected by slight changes in electron energy (wavelength); analogies in soft X-ray diffraction may be worth pursuing.

The use of the far-field diffraction pattern for each position of the X-ray probe introduces the question of configured detectors in X-ray microscopy which is discussed in the following section. In respect of a defocused probe, again STEM provides information which may carry over to the X-ray case in the form of the comprehensive experiments on defocused and shadow images carried out by Cowley [4] with a wide-angle probe and for both amorphous and crystalline specimens. Perhaps the most important point from Cowley's work is the presence in the defocused shadow images of interference fringes at a scale of resolution improved beyond the instrumental resolution of STEM; a method of image reconstruction has been given [5].

3. Configured Detectors

We consider the bright-field region, that is the far-field diffraction pattern within the probe angle of divergence as seen by the detector, and assume that an annular collector for independent dark field imaging surrounds the bright field detector.

It is useful to list the contributions from electron microscopy, and from optical microscopy in respect of configured detectors:

(i) Split detectors First proposed by Dekkers and de Lang [6,7] for both optical and electron microscopy, in the form of two signals each corresponding to an integration over a semi-circle. This was elaborated theoretically by Hawkes [8] and implemented experimentally [9] into four quadrant detectors to vary the selection of signals; a further extension into a first moment detector was introduced by Waddell and Chapman [10].

The split detector configuration was introduced experimentally into SOM applications by Hamilton and Sheppard [11].

(ii) <u>Circular-division detectors</u> This was a suggestion first made by Rose [12] from a consideration of the phase contrast transfer function (PCTF) within the confines of a weak phase model for the electron-specimen interaction for electron imaging. The PCTF and the point spread function for phase imaging form a Fourier transform pair. In the absence of astigmatism both functions are circularly symmetrical. Rose showed for electron imaging that, at a given focal value, the PCTF had positive and negative lobes and therefore, that individual circular (at the centre) and annular detectors, which could provide separate signals recorded individually, avoided the partial cancellations of the signal when integrated across the entire bright field cone. The Rose detector was implemented experimentally by Burge et al [13] using a system of scintillators and bundles of optical fibres; a difficulty with this detector is the dependence of the ring radii on defocus.

(iii) <u>Generalised-detector</u> The question whether it is possible to define an optimum set of weighting factors for the separate divisions of a pixellated detector surface was considered theoretically by Huiser and van Toorn [14], and, in a sense, this is the system adopted by Cowley [15] who used an image intensifier and a series of optical lenses to image the entire diffraction plane. Detector instrumentation directed towards the same end, with up to eight independent integrating detectors sampling the STEM CBED pattern in bright and dark field was implemented by Burge et al.[16].

The discussion of the operation of configured detectors in STEM is assisted by assuming weak electron-specimen interaction. The transmission function at position $\underline{x} = (x,y)$ may be represented

$$t(\underline{x}) = (1-s(x)) \exp i\phi(\underline{x}) \tag{1}$$

for an incident electron beam of unit amplitude. $s(\underline{x})$ represents specimen absorption (or, for electrons, scattering outside the detector), $\phi(\underline{x})$ gives the phase distribution. For a weak phase, weak amplitude model (1) becomes

$$t(\underline{x}) \simeq 1 - s(x) + i\phi(x) \tag{2}$$

Correspondingly, for X-ray imaging, with object of complex refractive index

$n(\underline{x}) = 1 - \delta(\underline{x}) - i\beta(\underline{x})$, then

$$t(\underline{x}) = \exp\left[\frac{-2\pi i}{\lambda} n(\underline{x}) \, p(\underline{x})\right] \tag{3}$$

for object thickness $p(\underline{x})$.

To produce a similar form for weak X-ray imaging, it is convenient to call $p(\underline{x}) = p$ a constant over the image and to deal with the ratio of the transmission of the actual X-ray specimen to that which would be transmitted by a specimen of the same thickness and unit (real) refractive index, ie.

$$T(\underline{x}) = \exp\left(\frac{+2\pi ip}{\lambda}\right) t(\underline{x}) \simeq 1 - kp\beta(\underline{x}) + ikp\delta(\underline{x}) , \tag{4}$$

where $k = 2\pi/\lambda$ the wave number.

If the source intensity function projected to the specimen (image of monochromator output slit) is $S(\underline{x})$, and the point spread function of the zone plate is $h(\underline{x})$ then the X-ray probe intensity is $P(\underline{x}) = S(\underline{x}) * h(\underline{x})$, where $*$ represents the convolution operation.

381

The wave emerging from the specimen is

$$\psi(\underline{x},\underline{\xi}) = t(\underline{x}). \quad P(\underline{\xi} - \underline{x}) ,$$

where the specimen transmission is t, and the probe is centred on $\underline{\xi}$ at a given instant.

This wave is propagated to the detector, where it becomes $\psi_d(\underline{x}_d,\underline{\xi})$ and is integrated over the probe area on recording to give current

$$j_d(\underline{\xi}) = \int |\psi_d(\underline{x}_d, \underline{\xi})|^2 D(\underline{x}_d) \, d\underline{x}_d , \tag{5}$$

where $D(\underline{x}_d)$ represents the two-dimensional sensitivity function of the detector surface.

This approach was introduced by Zeitler and Thomson [17] for a coherent probe in STEM, and developed further by Rose [12] among others. The theory of image formation in STEM was extended to the partially coherent case by Burge and Dainty [18]. A discussion on the effect of coherence is also given by Cowley [19], and the strongly related developments in imaging in optical scanning microscopy are summarised by Wilson and Sheppard [1].

The evaluation of (5) depends critically on the detector sensitivity function. If a point detector is used, then $D(\underline{x}_d) = \delta(\underline{x}_d)$, a delta function, then

$$j_d(\underline{\xi}) = \left| \int t(\underline{x}) P(\underline{\xi} - \underline{x}) d\underline{x} \right|^2 , \tag{6}$$

and [18,19] the resultant expression has the same form in STEM as in the scanning X-ray microscope. For a weak phase weak amplitude object we have, to first order

$$j_d(\underline{\xi}) = 1 - s(\underline{\xi}) * q_s(\underline{\xi}) + \phi(\underline{\xi}) * q_\phi(\underline{\xi}) , \tag{7}$$

where $q_s(\underline{\xi})$ and $q_\phi(\underline{\xi})$ may be regarded as point spread functions for amplitude and for phase imaging.

The important point about (7) is that the measurement intensity is linear in the amplitude and the phase terms for a weak scatterer. In principle, changing the relationship between the components δ and β of the refractive index, for example by making use of absorption edge structure, opens up the possibility of the determination of independent two-dimensional maps of the phase and amplitude components.

Image formation for increasing detector aperture diameter has been examined [19] and, for weak specimens, the form of (7) is retained, but the effective point spread functions become increasingly complex. Obviously the various configured detectors correspond to giving $D(\underline{x}_d)$ particular values eg the Rose ring detector has a central disc with value +1, adjacent ring -1, next ring +1 and so on.

4. Conclusion

This has been a brief consideration of the large body of published literature on imaging by scanning probes, which provides a framework for the future discussion of imaging in scanning X-ray microscopy. The scanning X-ray probe, in the form of a hollow cone as produced by an apodised zone plate, is unique and the precise form of the probe should be considered in a discussion of phase contrast and amplitude contrast transfer on imaging. It will be important to establish early the nature of the object-X-ray interaction in respect of using object models to evaluate image contrast. Currently rather few measurements on optical constants have been made in the soft X-ray region, and an evaluation of image contrast from known specimens will provide experimental values for the optical constants.

The weak phase, weak amplitude model which is used here for convenience is probably not directly relevant to X-ray microscope images, but it does provide a simple framework by which to enhance image understanding.

There is an essential equivalence between the scanning and the full field X-ray microscopes, but, just as in STEM relative to the conventional electron microscope, important diffraction information is contained in the diffraction pattern both within and outside the probe for each and every object point. It is an open question as to whether the information in a series of independent detectors (eg for quadrants in bright field plus a dark field detector) can be made use of satisfactorily in the STXM. Success will depend on the source brightness, the necessary irradiation time and the associated radiation damage, and whether or not suitable configured detectors for STXM can be made. We note again the improved resolution beyond the Rayleigh limit found in images taken with a defocused probe [4].

Another facet of interpretation of electron microscope images concerns solutions to the phase problem, expressed within the weak phase and weak amplitude model (related remarks could be made for the general case) by the need to determine these two independent two-dimensional object distributions. Sufficient to say that a wide range of approaches to the solution of the phase problem has been developed for the electron case, which may carry over to the X-ray case.

5. Acknowledgements

The work on X-ray microscopy is supported by the UK Science and Engineering Research Council. Thanks are due to Dr G R Morrison for helpful discussion.

6. References

1. T. Wilson and C.J.R. Sheppard: Theory and Practice of Scanning Optical Microscopy, Academic Press, London, 1984.
2. J.W. Steeds: In Introduction to Analytical Electron Microscopy, Edited by J.J. Hren, J.I. Goldstein and D.C. Joy, Chapter 15, Plenum Press, London, 1979.
3. W-B Yun, J. Kirz and D. Sayre, Acta Cryst **A43**, 131, 1987.
4. J.M. Cowley, Ultramicroscopy, **7**, 19 (1981)
5. J.M. Cowley and D.J. Walker, Ultramicroscopy **6**, 71 (1981)
6. N.H. Dekkers and H. de Lang, Optik, **41**, 452 (1974)
7. N.H. Dekkers and H. de Lang, Philips Tech. Rev. **37**, 1 (1977)
8. H. Rose, Optik **39**, 416 (1974)
9. J.F.L. Ward, M.T. Browne, R.E. Burge: In Proceedings of EMAG '79, Ed T. Mulvey, IOP Conf. Ser. No **52** (IOP, Bristol 1980), 85
   and
   G.R. Morrison, J.N. Chapman, A.J. Craven: In Proceedings of EMAG '78, Ed T. Mulvey, IOP Conf. Ser. No **52** (IOP Bristol 1980), 257
10. E.M. Waddell and J.N. Chapman, Optik **54**, 83 (1979)
11. D.K. Hamilton and C.J.R. Sheppard, J. Microsc. **133**, 27 (1984)
12. H. Rose, Optik **39**, 416 (1974)
13. R.E. Burge, M.T. Browne, S. Lackovic and J.F.L. Ward, SEM 1979/I, 127, SEM Inc. AMF O'Hare, 1979.
14 A.M.J. Huiser and P. van Toorn, J Phys D **15**, 747-755, 1982

15 J.M. Cowley, SEM/1980/I, P53 SEM Inc AMF O'Hare (1980)
16. R.E. Burge, M.T. Browne, P. Charalambous, A. Clarke and J.K. Wu, J. Microsc. **127**, 47 (1982)
17. E. Zeitler and M.G.R. Thomson, Optik **31**, 258 and 359 (1970)
18. R.E. Burge and J.C. Dainty, Optik, **46**, 229 (1976)
19. J.M. Cowley, Ultramicroscopy **2**, 3 (1976)

X-Ray Microscopy in the Study of Biological Structure: A Prospective View

E. Lattman

Department of Biophysics, Johns Hopkins Medical School,
Baltimore, MD 21205, USA

1. INTRODUCTION

As summarized in this volume, x-ray microscopy is making very rapid progress
on many fronts. Thus, it would be foolhardy and perhaps counterproductive to
attempt to predict the full impact of x-ray microscopy on the study of
biological structure. Instead, this paper focuses upon a particular problem
of great interest to biologists (the structure of the chromosome), for which
present technologies such as electron microscopy and light microscopy are
inadequate. The characteristics of this problem suggest that it may just be
accessible to x-ray microscopy in the near future. The problem also serves
to illustrate both the strengths of biological x-ray microscopy - for
example, its ability to distinguish substances by elemental composition - and
the weaknesses - for example, marginal resolution and lack of ability to
acquire full three-dimensional information.

2. STRUCTURE OF THE EUKARYOTIC CHROMOSOME

2.1 Hierarchical Organization

Eukaryotes are organisms which have nuclei within their cells. The DNA
resides in these nuclei and is organized into units called chromosomes which
contain a single long strand of DNA complexed with many protein molecules.
The DNA in the chromosome is highly compacted. In humans, for example, the
length of the fully extended DNA in a single cell is about 1.2 m. This
corresponds to an average of approximately 2 cm of DNA per chromosome. The
DNA as packed in the chromosome is shortened by a factor of about 10^4 from
this extended length. This compaction is accomplished in a highly organized
and controlled way, since the DNA remains accessible to the machinery of the
cell. The compaction is also hierarchical, taking place at several different
levels. As is explained below, the arrangement of the DNA in the lower
levels of the chromosomal hierarchy is well understood. A key element in
this hierarchy is the 30 nm fiber. In this unit, the DNA, which is complexed
with protein, is already shortened by a factor of approximately 40. The 30
nm fiber can be regarded as a relatively featureless cable that takes a
complex and tortuous path in space in the final structure of the chromosome.
Because 30 nm is comparable to the resolution to be expected of the best
microscopes in the near future, the elucidation of this path in space by
x-ray microscopy becomes a possibility. These topics are discussed in an
excellent monograph by Saenger [1].

2.2 Structure of the Nucleosome

Figure 1a shows a model of DNA, the fundamental component in the hierarchical
structure of the chromosome. The first level of compaction of DNA involves
the use of protein cores or spools around which the DNA is wrapped. The core
contains four different proteins called histones, designated H2a, H2b, H3 and
H4. Two copies of each of these four species form an octameric unit called
the core octamer. The core octamer binds DNA in a relatively specific
fashion. As shown in Figure 1b, the DNA molecule wraps about 1-3/4 times
around the core octamer and continues on roughly along the direction from
which it entered. There are about 146 base pairs of DNA bound to the core
octamer. This unit – core octamer and 146 base pairs – is termed a
nucleosome. At low ionic strength, when viewed in the electron microscope,
the nucleosomes in DNA appear like a series of beads on a string, with
lengths of spacer DNA running between them. As the ionic strength is raised,
the nucleosomes stack one on top of another like a tilted stack of coins to
form a more compact, stable structure known as the 10 nm fiber. The details
of the structure of the nucleosome and of the 10 nm fiber have been
elucidated by a combination of x-ray crystallography, electron microscopy and
sophisticated biochemical studies [2,3].

In the next hierarchical level the 10 nm fiber is wrapped into a flat
solenoid, in which there are approximately six nucleosomes per turn. This
solenoid has come to be known as the 30 nm fiber. The transformation from
the 10 to the 30 nm fiber is accomplished by an additional increase in ionic
strength. As mentioned above, in the 30 nm fiber DNA is compacted by about a
factor of 40. Figure 2 shows the 10 nm and 30 nm fibers.

2.3 The Role of the 30 nm Fiber in the Chromosome

Figure 3 shows a photograph of human chromosome 12. In the next stage of the
cell cycle the two lobes of this chromosome will separate into independent
chromosomes and segregate into daughter cells. The filamentous character of
the chromosomal structure is clear, and measurements show that this filament
is indeed the 30 nm fiber. The mature chromosome contains a large number of

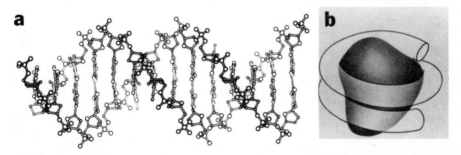

Figure 1. (a) shows about $1\frac{1}{2}$ turns of a double stranded DNA molecule; the
base pairs, separated by .34 nm, run vertically and the helix formed by the 2
sugar-phosphate backbones runs horizontally: (b) shows a schematic DNA
molecule wrapped around the histone core octamer (see text) to form a
nucleosome, about 10 nm in diameter: the magnification in (a) is ~6X that in
(b): (Reprinted from [1] with the kind permission of the author)

385

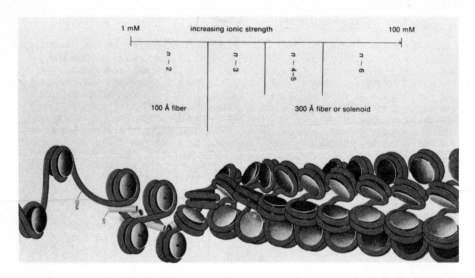

Figure 2. Displays 3 states of compaction of histone-DNA complexes; on the left are separated nucleosomes connected by linker DNA; in the middle is the partially condensed 10 nm fiber and its transition to the 30 nm fiber on the far right; transitions mediated by increasing salt concentration (reprinted from [1] with the kind permission of the author)

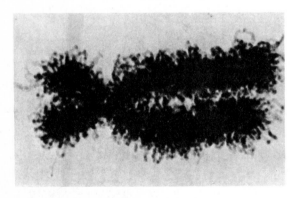

Figure 3. Human chromosome 12 in a condensed form (metaphase). The apparent filamentous structure arises from the 30 nm fiber (reproduced from [1] with the kind permission of the author)

proteins in addition to the histones mentioned earlier. These make the structure compact and opaque to the electron microscope. Thus, it is likely that some proteins have been removed from this chromosome prior to photography to improve the visibility of the 30 nm fibers. Unhappily this has the consequence of relaxing the structure so that, even if the path could be determined in a case like this, it would not necessarily represent the path of the native state. One clue to the structure of the chromosome has been obtained from experiments in which almost all of the protein, including histone, was removed. Electron micrographs taken of these disrupted structures reveal a spine or backbone of insoluble protein running longitudinally through the middle of the chromosome. DNA is linked to this scaffold matrix at many points. This suggests an overall model in which the 30 nm filament loops in and out from a central protein scaffold, producing a structure very much like a bottle brush used in biological and chemical laboratories. This is a very featureless view of the structure, however, and one that is totally inadequate to understand structure-function relationships at the level of the whole chromosome. In the remainder of this paper we discuss the possibility that x-ray microscopy may be of use in determining the in vivo structure of the 30 nm fiber as it exists in the chromosome.

3. MICROSCOPY OF THE CHROMOSOME

3.1 Electron Microscopy

The model of the 30 nm fiber is critically dependent on electron microscopic observations. The primitive notions that we now have about the path of the 30 nm fiber in the whole chromosome also come from electron microscopy. However, there are hosts of technical reasons why the electron microscope is poorly suited to the determination of the in vivo structure of the 30 nm fiber. First, native chromosomes are thick specimens by the standards of electron microscopy. Thus, beam penetration is poor and specimens tend simply to look black. Also, the chromosome has intrinsically low contrast between the nucleoprotein 30 nm fiber and densely packed, surrounding protein. Even using special high voltage instruments does not improve the situation a great deal. In order to make the chromosome open enough to be visible in the electron microscope adherent protein has to be removed and the ionic strength has to be lowered, procedures which disrupt the native structure.

3.2 X-ray Microscopy

It is immediately clear why the characteristics of soft x-ray microscopes make them particularly suitable for examining this structure. A whole chromosome is perhaps a few tenths of a micrometer thick, a comfortable distance for soft x-ray microscopy. Use of radiation at wavelengths at which water is transparent means that fully hydrated specimens can be examined under near native-conditions. Nucleic acids and proteins differ significantly in the fraction of carbon and nitrogen in their structures. The difference in absorption of these elements at appropriately chosen wavelengths provides a basis for distinguishing the 30 nm fibers, which contain nucleic acid and protein, from the pure protein components, which obscure the view in the electron microscope. While the resolution of 30 nm necessary to resolve the fibers is not yet routinely available on current instruments, the rapid progress presented in this volume strongly suggests that it will be available within the next few years. Thus, almost all of the instrumental characteristics necessary to provide useful images of the eukaryotic chromosome are either in hand or within reach. The single exception is the ability to provide genuine three-dimensional information.

4. THREE-DIMENSIONAL RECONSTRUCTION

4.1 Three-Dimensional Reconstruction in X-ray Microscopy

Methods for reconstructing images in two and three dimensions are in use in a wide variety of fields. There is therefore a large well of experience for x-ray microscopy to draw upon. In the succeeding sections we review the principal methods of three-dimensional reconstruction and discuss how they are being or may be applied to the different modalities of x-ray microscopy.

4.2 Reconstruction Techniques Based on Direct Images

Three-dimensional reconstruction techniques can be divided into two broad categories, those for which the primary data are a set of images of the object or specimen and those for which the primary data are diffraction patterns. Image-space techniques are further divided into two subclasses, those that involve images of sections of the specimen, which are later assembled to make a three-dimensional reconstruction, and those that involve images of projections. We ignore here techniques in which the object is physically divided into sections which are individually imaged.

Sections through a specimen are commonly obtained by making a throughfocal series of pictures. An elegant example is the work of Agard [4] using light microscopy in the three-dimensional reconstruction of the polytene chromosome of the fruit fly. The technique has generally been applied with imaging microscopes. A series of pictures is made in which the focal plane of the objective lens is moved through the object. Because of the non-zero depth of focus of the lens, each picture contains not only a focused view of the desired section but out-of-focus contributions from adjacent sections. If the point-spread function of the optical system is known, it can be deconvoluted from these views to give a much better approximation to the desired image of a particular section. These in turn can be stacked to produce a three-dimensional image. In confocal microscopy, an elegant version of this procedure, a beam of light is focused by the objective lens and scanned rapidly across the focal plane. It is viewed by a second objective lens, or in special cases the same one, focused onto the same plane. Because of the introduction of the second lens, the depth of focus of the optical system is determined by the square of the point-spread function. Out-of-focus contamination from adjacent sections is greatly reduced and the problem of deconvoluting their influence is minimized. This technique is obviously applicable in principle to both scanning and imaging x-ray microscopes using zone plate optics. Existing zone plate objectives, however, have small numerical apertures and consequently a large depth of focus. This will increase the difficulty of obtaining adequate resolution along the optical axis. It makes the introduction of confocal microscopy particularly attractive. However, the small numerical aperture of the second objective lens causes difficulty since it will significantly decrease the photon counting rate.

4.3 Image-Space Projection Methods

In this set of methods a parallel or divergent beam is used to make images of a number of different projections of the specimen. These are later reassembled to form an image. Pertinent to the study of biological structure is the pioneering work of DeRosier and Klug [5] on three-dimensional reconstructions in the electron microscope. Most electron micrographs are actually projections of the specimen, since the depth of focus of the electron microscope is much greater than any practical specimen thickness. Modern electron microscopes are available with tilting stages that allow many projections of the specimen to be recorded. Fourier transformation of these projections yields a family of central sections through the three-dimensional transform of the specimen. Interpolation leads to an approximation to the full three-dimensional transform which can be inverted to provide the desired reconstruction.

It seems plausible that the requisite series of projections could be obtained in either scanning or imaging zone plate microscopes through the introduction of a tilting stage. In contact microscopy one might be able to rotate the specimen and the resist holder as a unit about the axis in order to obtain the desired views.

4.4 Diffraction-Based Methods

The most familiar example of diffraction-based three-dimensional reconstruction is probably x-ray crystallography. Here the discrete three-dimensional diffraction pattern or Fourier transform of a crystal is measured and is inverted to provide the desired image. Direct measurement of the phase of the Fourier transform is not possible in most x-ray diffraction experiments. The missing phase is supplied by one of a number of ingenious methods. A basically similar technique is involved in the soft x-ray, far-field diffraction camera of Sayre and Yun [6]. Here extremely careful construction of the optical components has allowed the diffraction pattern from a single diatom to be observed on film. In this camera a series of diffraction patterns acquired at different orientations of the specimen around a spindle axis is required to achieve a full three-dimensional reconstruction.

A number of efforts are also underway to record x-ray holograms from specimens. In this context holograms may be considered as diffraction patterns in which the phase has been interferometrically encoded. Although holography is popularly believed to be a three-dimensional method, it must be emphasized that any individual hologram samples the three-dimensional diffraction pattern only on the surface of Ewald's sphere. Because this sphere is not flat, the Fourier transform values that it samples correspond to spatial density variations that provide some resolution along the direction of illumination. However, as in any other far field diffraction method, a complete set of views of the object must be obtained in order to have adequate information for a full three-dimensional reconstruction.

In addition to this work in far-field diffraction, Sayre has made an original and innovative proposal of the use of near field diffraction for three-dimensional reconstruction [7]. Unlike the far-field diffraction pattern, the near-field pattern varies rapidly with distance from the specimen. Sayre has shown that the measurement of the diffraction field in a three-dimensional region of space contains sufficient information for the three-dimensional reconstruction of the object. In this method, as in the through focal sectioning procedure, the specimen need not be moved.

4.5 Short-Term Prospects for X-Ray Microscopic Three-Dimensional Reconstruction

The author is relatively pessimistic about the short-term prospects for three-dimensional reconstruction in x-ray microscopy. In the case of imaging methods, the introduction of tilting stages or comparable devices into microscopes seems far off. The use of through focal sections is hampered by the small numerical aperture of zone plate lenses. In addition, the immediate priorities of the instrument makers appear to be elsewhere. In the case of diffraction-based methods, developments as a whole are in a more primitive state. Sayre's far field diffraction camera is a wonderful achievement. However, the high resolution diffraction which is in principle possible has yet to be observed, and a reliable method of providing the missing phases for the diffraction pattern has yet to be demonstrated. The holographic methods face many difficulties in providing adequate magnification, and in using reconstructions from single views. Finally, the near-field diffraction microscope proposed by Sayre requires considerable developments in detector technology before it comes of age.

5. CONCLUSION

5.1 Summary

The higher order structure of the eukaryotic chromosome is based on a long filamentous nucleoprotein unit called the 30 nm fiber. How this fiber is arranged in the native chromosome is of great biological interest but represents a structural problem that has proved very difficult for existing technologies to solve. Because of its ability to see through water and to distinguish components based on elemental composition, x-ray microscopy represents a significant new avenue of attack on this problem. Although the required 30 nm resolution appears to be within reach, the capability for three-dimensional reconstructions necessary to visualize the complex path of this filament in the chromosome is more problematic in the short term.

5.2 Acknowledgements

The author thanks Drs. David Sayre and Janos Kirz for helpful discussions and for hospitality. This work was supported by NIH Grant GM 36358.

6. LITERATURE

1. W. Saenger: Principles of Nucleic Acid Structure, (Springer, Berlin, Heidelberg 1984)
2. J.D. McGhee, G. Felsenfeld: Ann. Rev. Biochem. 49, 1115, (1980)
3. B. Lewin: Genes, 3rd edition (John Wiley, New York, 1987), pp.519-545
4. D. Agard: Ann. Rev. Biophys. Biophysical Chem. 13, 191, (1984)
5. R.A. Crowther, D.J. DeRosier, A. Klug: Proc. R. Soc. Lond. A 317, 319-340 (1970)
6. W.B. Yun, J. Kirz, D. Sayre: Acta Cryst. A43, 131, (1987)
7. D. Sayre: American Crystallographic Association Abstracts Series 2, 14, 42, (1986)

X-Ray Microscopy Studies on the Pharmaco-Dynamics of Therapeutic Gallium in Rat Bones

R. Bockman[1], M. Repo[1], R. Warrell[1], J.G. Pounds[2], W.M. Kwiatek[2], G.J. Long[2], G. Schidlovksy[2], and K.W. Jones[2]

[1]Memorial Sloan-Kettering Cancer Center, New York, NY 10021, USA
[2]Brookhaven National Laboratory, Upton, NY 11973, USA

1. INTRODUCTION

Since the original observation of exogenous gallium accumulation in bones ANGHILERI et al. [1], several studies have demonstrated that gallium nitrate is extremely effective in preserving bone mineral content both in vivo and in vitro (WARRELL et al., [2]; BOCKMAN et al. [3]). Gallium nitrate therapy normalized serum calcium levels in a study of patients with cancer-related hypercalcemia, resistant to standard hydration and diuretic therapy. Recently, gallium nitrate treatment has been shown to halt the accelerated bone resorption that is frequently associated with cancers metastatic to bone (BOCKMAN et al. [4]). Several lines of evidence from in vitro studies (CHUN et al. [5]) recently led to the demonstration of increased bone calcium and improvement in hydroxyapatite crystallinity in adult gallium-treated rats (BOCKMAN et al. [6]). Evidence is rapidly accumulating that gallium nitrate is an effective, new therapeutic agent for inhibition of accelerated bone resorption associated with cancer-related hypercalcemia. It has also been suggested that gallium could have wide clinical applications in disorders characterized by accelerated calcium loss from bone (BOCKMAN et al. [6]). However, while gallium is known to accumulate in bone, the effects of gallium on the kinetics of bone calcium uptake and loss are unknown. Nor is it known in which compartments (epiphysis, metaphysis, diaphysis, endosteal or periosteal surfaces) the gallium metal is preferentially localized. Gross analytical measurements of gallium content have previously been performed on excised bone segments, but these lack precise spatial resolution. We describe here our preliminary results on gallium and calcium quantitation and localization using x-ray microscopy techniques at the X-26 beam line of the National Synchrotron Light Source (NSLS) at 50-100 μm resolution and 10^{-6} g/g detection levels.

2. MATERIALS AND METHODS

Young female Sprague-Dawley rats (100-120 grams) (Charles River, Kingston, NY) were given 7 injections, with either 0.5 or 5 mg of gallium nitrate/kg in saline solution. The gallium nitrate solutions were injected i.p. every other day to achieve a total dose of 3.5 and 35 mg of gallium nitrate/kg received per animal. Littermates were injected with saline solutions only, following the same protocol. At the end of the injection cycle, the rats were anesthetized and sacrificed. The long bones (femurs, tibias, and humerii) were removed, cleaned of all muscles and adventitious tissues, frozen in liquid nitrogen, and embedded in tissue Tek®, sectioned longitudinally at 15 μm on a freezing cryostat, transferred to polyimide supporting film (8.5 μm thick) and freeze-dried. The supported sections were framed into 2 × 2 in. cardboard slide mounts and studied by x-ray fluorescence at the NSLS and photographed in a scanning electron micro-scope (Amray 1400, Bedford, MA) after deposition of a conductive layer of spectroscopic-grade carbon by evaporation. A scanning electron micrograph of a typical thin section of adult rat tibia is shown in fig. 1. The por-tions of the bone investigated are indicated. The elemental concentra-tions in the sections were measured using the collimated x-ray microscope (CXRM) located at the X-26C beam line at the NSLS. The instrument is used to give elemental concentrations at single points, along line scans, or to fully map the area of interest (GORDON et al. [7]).

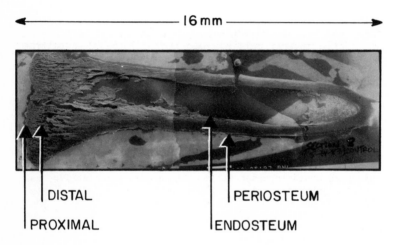

Fig. 1. Scanning electron micrograph of 15-μm thick section of adult rat tibia. XRF scans that were made of the growth plate from proximal to distal regions and of the diaphysis from endosteum to periosteum   are shown in fig. 3. The gallium and calcium distributions shown in fig. 3 were obtained around the growth plate.

3. RESULTS AND DISCUSSION

A typical x-ray fluorescence spectrum and map of trace elements observed in a foetal rat bone treated with gallium nitrate in organ culture are also given in the work of GORDON et al. [7]. A contour plot of Ca and Ga in the growth plate region of a tibia from a rat treated with 5 mg of Ga is shown in fig. 2. The figure shows that the gallium and calcium dis-tributions in the bone are similar. The ratio of gallium to calcium atomic concentrations obtained in a line scan across the growth plate and diaphysis are shown in fig. 3. The Ga/Ca atomic ratios are higher in the growth plate than in the diaphysis while the values for the gradient of the ratio are higher in the diaphysis. A summary of the results that were obtained is given in table 1. At 5-mg treatment level there is evidence

392

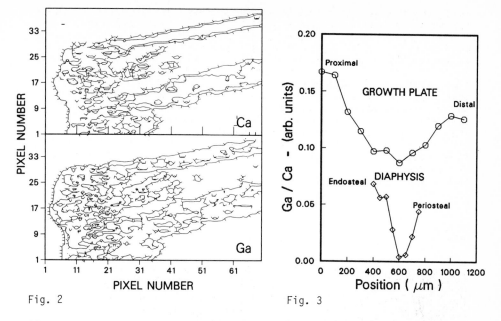

Fig. 2   Fig. 3

Fig. 2. Contour plots for Ca and Ga count distributions for 70 x40 pixel scan of adult rat bone section after treatment with 5 mg Ga. The contour intervals are 400 and 300 counts for Ca and Ga, respectively. The tick marks point in the direction of fewer counts. The pixel size is 100 x 100 µm².

Fig. 3. Ratio of Ga/Ca atomic concentrations obtained in scans across the growth plate and diaphysis of an adult rat tibia. The locations of the scans are indicated in fig. 1.

Table 1. Ratio of gallium concentration to calcium concentration for several different regions on the tibia of adult rats. Values are given for a control animal and for ones administered gallium doses of 0.5 and 5.0 mg. The concentration values are in arbitrary units.

Region of Bone	Arbitrary Units		
	Control	0.5 mg	5.0 mg
Growth Plate			
Proximal	0.002	0.167	1.002
Mid	0.001	0.106	0.944
Distal	0.0025	0.127	0.719
Diaphysis (shaft)			
Endosteal	0.001	0.062	0.498
Mid	0.0015	0.024	0.105
Periosteal	0.047	0.033	0.202

that the rate of gallium accumulation is much less than at the .5 mg
treatment level.

4. CONCLUSIONS

The CXRM has been demonstrated to be a uniquely useful instrument for
studies of the accumulation of gallium in bone. A more detailed investigation of the pharmaco-dynamics involved in the use of gallium will be
carried out along the lines of the work discussed here.

5. ACKNOWLEDGMENTS

Work supported in part by Processes and Techniques Branch, Division of
Chemical Sciences, Office of Basic Energy Sciences, US Department of
Energy, under Contract No. DE-AC02-76CH00016; National Institutes of
Health as a Biotechnology Research Resource under Grant No. P41RR01838;
National Institutes of Health Grant No. ES04040; and Public Health Service
Grant No. CA38645.

6. LITERATURE REFERENCES

1. I. Anghileri: Strahlentherapie 142, 456 (1971)
2. R.P. Warrell, Jr., R.S. Bockman, C.J. Coonley, M. Isaacs,
   H.J. Staszewski: J. Clin. Invest. 73, 1487 (1984)
3. R.S. Bockman, A. Boskey, N. Alcock, R.P. Warrell, Jr.: ASBMR Abstract
   A30, 1985
4. R.S. Bockman, R.P. Warrell, N.W. Alcock, A.L. Bockey, A. Skelos:
   Clin. Res. 33, 576A (1985)
5. H. Chun, R.S. Bockman, A.L. Boskey, N.W. Alcock, R.P. Warrell, Jr.:
   Calcif. Tissue Int. (abstract) 36, 456 (1984)
6. R.S. Bockman, A.L. Boskey, M.C. Blumenthal, N.W. Alcock, R.P. Warrell,
   Jr.: Calcif. Tissue Int. 39, 376 (1986)
7. B.M. Gordon, A.L. Hanson, K.W. Jones, W.M. Kwiatek, G.J. Long,
   J.G. Pounds, G. Schidlovsky, P. Spanne, M.L. Rivers: these
   proceedings.

Location and Mapping of Gold Sites in Thin Sections of Unoxidized Carlin-Type Ores Using Complementary Micro-analytical Techniques

J.R. Chen[1], J.M. Back[2], J.A. Minkin[2], E.C.T. Chao[2], A.L. Hanson[3], K.W. Jones[3], M.L. Rivers[4], and S.R. Sutton[4]

[1]State University of New York, Geneseo, NY 14454, USA
[2]U.S. Geological Survey, Reston, VA 22092, USA
[3]Brookhaven National Laboratory, Upton, NY 11973, USA
[4]University of Chicago, Chicago, IL 60637, USA

1. INTRODUCTION

Determining how precious metals occur in mineral deposits is of importance in resolving how they formed, where to explore for additional, similar ore bodies, and how to most efficiently extract the metal from the host rock. Determining what minerals are associated with previous metals is particularly difficult when the metal is finely disseminated and submicroscopic, as it is in unoxidized Carlin-type gold ores. The geologic setting of Carlin-type ore deposits was described by BAGBY and BERGER [1]. We combined data from four tools to find submicroscopic gold in thin sections of unoxidized Carlin-type gold ore and to describe the mineral associated with each particle. The tools are a synchrotron-induced x-ray fluorescence (SXRF) microprobe, an optical microscope, a scanning electron microscope with energy dispersive x-ray analysis capabilities (SEM-EDX), and an electron probe used for microanalysis (EPMA). The procedure derived is a relatively quick way of finding the location of discrete gold particles $\geq$ 0.2 μm diameter that occur at the surface of a sample.

2. EXPERIMENTAL METHOD

Doubly polished thin sections of the ore samples were examined with a petrographic microscope to determine phases, textural relationships, and lithologies; SEM-EDX was used to verify identification, as well as textural relationships, of the phases. SXRF analyses of the different components in the samples were then used to determine gold-bearing areas. To retain low minimum detection limits at relatively short counting times, a beam diameter of 50 μm (table 1) was commonly used. Thus, most SXRF analyses included more than one grain of a mineral, and often also more than one phase. To determine site-specific occurrences of gold in the gold-rich areas, both SEM-EDX and EPMA were tested. Comparative features of these methods are given in table 1.

X-ray maps were generated with both SXRF and SEM-EDX to determine their effectiveness in locating gold in the samples. Operating conditions for generating these maps are listed in table 1. A gold-rich area 1.0 x 1.4 mm^2 was chosen for the SXRF-generated x-ray map (fig. 1). In order to decrease acquisition time to about 1 hour, the energy window included some ZnK$_\beta$ and

Table 1. Comparison of analytical parameters used.

Analysis	SXRF	SEM-EDX	EPMA
Minimum detection limit (ppm)	1-3	~5,000	300-500
Au peaks examined	L	L and M	M_α
Counting times	15 min.	$\geq$100 sec.	120 sec.
Volume analyzed:			
Diameter (μm)	20-70	~2	~3
Depth analyzed (μm)	30*	1-2	1-2

X-ray Maps		
Map areas	1.0 x 1.4 mm²	35 x 35 μm²
Area/pixel (μm²)	50	0.13
Acquisition rate	10 sec/pixel	0.6 sec/pixel
Time to complete map	~1 hour	~13 hours
Au peak mapped	L_α	L_α

*SXRF analyzes the entire thickness of the thin section.

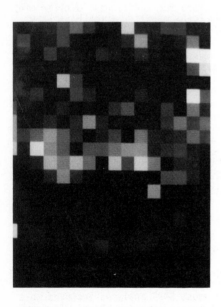

Fig. 1. SXRF map of AuL_α x-rays, including some ZnK_β and HgL_α x-rays.
Lighter areas indicate larger concentrations of Zn, Au, or Hg.

HgL_α x-rays in addition to AuL_α x-rays. To determine the limitations and
applications of SEM-EDX x-ray maps, areas where discrete gold particles had
already been located were mapped. SEM-EDX x-ray maps of ZnK_α and HgL_α were
generated simultaneously with those of AuL_α, in some instances, to check for
possible overlaps of data for these elements with data for gold on the Au
map.

3. ANALYTICAL RESULTS

The principal results of this study are (1) only a single occurrence of gold with pyrite was observed, and none was observed associated with carbonaceous material, and (2) gold was determined to be concentrated in areas of pre-existing illite and secondary quartz. The applicability of SXRF to these studies is described in CHAO et al. [2], CHEN et al. [3], and MINKIN et al. [4].

Site-specific occurrences of gold in gold-rich areas were successfully determined with SEM-EDX, but not with EPMA. In addition, two limitations of the x-ray maps as carried out were found: (1) some of the light areas on the SXRF x-ray map were attributed to the presence of mercury, and (2) no gold particles < 0.5 μm were detected on the SEM-EDX x-ray maps.

4. DISCUSSION

Without the use of SXRF, the task of characterizing which lithologies are gold-rich would have been exceedingly tedious and very inefficient. SXRF-generated x-ray maps may be a useful method of surveying to determine areas of gold concentration in a sample.

The use of backscattered-electron images (BEIs), at a large range of magnifications, with SEM-EDX proved to be the most effective method of determining precisely the location of gold particles in areas determined by SXRF to be gold-rich. Even small gold particles ($\gtrsim 0.2$ μm diameter) in the ore are strongly contrasted in BEI because gold has an atomic number higher than those of most of the other elements present in the ore. SEM-EDX spectra enable one to examine many M and L lines for gold, instead of just the one wavelength examined with EPMA. Partial overlaps of Au x-ray lines with those for other elements (e.g., AuL_α may overlap ZnK_β) are more likely to be noticed, and, thus, the misidentification of, for example, Zn for Au is less likely with SEM-EDX than with EPMA.

Although EPMA has lower minimum detection level for gold than does SEM-EDX, the small beam diameter used for EPMA makes searching for gold particles with this technique alone extremely time consuming. EPMA with SEM-EDX attachments might provide a useful combination of techniques.

Some problems exist in using SEM-EDX to characterize the trace-metal concentrations indicated by SXRF. First, SEM-EDX is limited to analyzing only a few micrometers below the surface of the sample, whereas SXRF analyzes the entire thickness of the thin section. Therefore good correlation of SXRF results with SEM-EDX in samples where gold is inhomogeneously distributed is dependent on thinning the samples in steps to examine the occurrence of gold with depth. Another weakness of SEM-EDX with respect to Carlin-type samples is the inability to analyze individual gold particles <0.2 μm using EDX. HOCHELLA et al. [5] and HOCHELLA (oral communication, 1987) have found gold particles 0.01-0.04 μm in diameter associated with clay in Carlin-type samples.

In summary, SXRF was found to be an essential tool for locating gold-rich areas in thin sections of ore samples. With further development of the SXRF microprobe, such as improved spatial resolution (to 10 μm), use of wavelength dispersive spectrometers, and installation of a monochromator to improve signal to background ratios and thus lower minimum detection limits, the SXRF microprobe will offer even more attractive features as a microanalytical tool. The combined use of optical, SXRF, and SEM-EDX techniques should be valuable in the study of the occurrence of precious metals in Carlin-type and other deposits.

5. ACKNOWLEDGMENTS

This research was supported in part by the U.S. Department of Energy contract no. DE-AC02-76CH00016, by the National Science Foundation grant no. EAR-8618346, and by the U.S. National Aeronautics and Space Administration grant no. NAG 9-106.

6. LITERATURE REFERENCES

1. W. C. Bagby, B. R. Berger: In Geology and Geochemistry of Epithermal Systems, ed. by B. R. Berger, P. M. Bethke, Reviews in Econ. Geol., Vol. 2, p. 169-199 (1986).
2. E. C. T. Chao, J. R. Chen, J. A. Minkin, J. M. Back: In Process Mineralogy VII (TMS/AIME), in press.
3. J. R. Chen, E. C. T. Chao, J. A. Minkin, J. M. Back, W. C. Bagby, M. L. Rivers, S. R. Sutton, B. M. Gordon, A. L. Hanson, K. W. Jones: Nucl. Instrum. and Methods B22, p. 394-400 (1987).
4. J. A. Minkin, E. C. T. Chao, J. M. Back, J. R. Chen: In Microbeam Analysis - 1987, p. 329-331 (1987).
5. M. F. Hochella, B. M. Bakken, A. F. Marshall, D. W. Harrison, A. M. Turner: 116th AIME Annual Meeting, Abstracts, p. 37 (1987).

Exploration of the Demyelinated Axon of the Medullated Shrimp Giant Nerve by Soft X-Ray Scanning Microscopy

Shih-Fang Fan[1], *H. Rarback*[2], *H. Ade*[3], *and J. Kirz*[3]

[1]Dept. of Anatomical Sciences, Health Sciences Center,
  State University of New York, Stony Brook, NY 11794, USA
[2]National Synchrotron Light Source,
  Brookhaven National Laboratory, Upton, NY 11973, USA
[3] Physics Department, State University of New York,
  Stony Brook, NY 11794, USA

The giant nerve fiber of the abdominal nerve cord of the shrimp (Penaeus) has several unique features both in its structure and in its electrophysiological characteristics:

1. It has the fastest reported conduction velocity for a nerve. In the case of the giant median nerve fiber, it may reach more than 200 m/sec at about $20°C$[1-3].

2. The giant fiber is not segmented as in the case of the crayfish and earthworm but has cell bodies in each abdominal ganglion (Fan et al. unpublished result).

3. The fiber is surrounded by a thick and compact myelin sheath layer. For a fiber with external diameter of about 170 μm, the thickness of the myelin layer may reach 50 μm[1,2,4]. There is no typical node of Ranvier, yet the myelin layer thins or almost disappears where the giant fiber makes synaptic contact with the giant motor axon and where the neurite joins the axon in the ganglion. These regions are thought to serve functionally as the nodes[3,4].

4. There is a wide gap between the axon proper and the myelin sheath. For a fiber with an external diameter of about 170 μm, the inner diameter of the myelin layer is 70-80 μm, yet the diameter of the axon is only about 12 μm, except at the region where it makes synaptic contact with the giant motor fiber, where it broadens to more than half of the fiber diameter[3].

399

5. As the tip of the microelectrode is situated within the wide gap between the myelin sheath and the axon, an intracellular-like action potential with amplitude around 60-80 mV can be recorded, though no resting membrane potential can be detected. However, the resting potential can be detected as the electrode tip is penetrated into the axon, yet the action potential thus recorded shows little difference from that recorded from the submyelin gap[2,3,6,7].

6. A steady potential is established by charged protein(s) in the submyelin space[6,7] and there is indication from the experiments done with potential sensitive dye that the resting membrane potential of the internodal axon membrane has a smaller value than that of the functional node regions (Fan and Brink, unpublished result).

7. The axon proper is surrounded by a thick layer of longitudinally disposed microtubules[8]. For an axon with a diameter of about 12 μm, the thickness of the microtubular layer may reach 3-4 μm. As far as we know, in nerve fibers of other animals, microtubules could only be found in the axoplasma. They are supposed to provide the mechanical strength of the fiber and play a role in the axonal transport. Such a peculiar arrangement of microtubules in the shrimp nerve is probably associated with unknown function(s) of the microtubule in this nerve fiber.

Soft x-ray scanning microscopy using tunable wavelengths enable us to study the distribution of element(s) of particular interest in biological specimens by comparing the images taken with wavelengths above and below the absorption edge of that particular element(s). In the future it will be possible to study the specimen in solution. A single axon of the shrimp giant nerve with the thick layer of microtubules but without the myelin sheath can be isolated by microdissection. With such a preparation, the following aspects can be studied: 1. By using specific element(s) attached to channel-binding molecules one may study the distribution of channels in the internodal segment of the myelinated nerve fiber.
2. Using specific element(s) attached to molecules which are usually transported by the microtubular system one may study what relationship, if any, the microtubule layer has with the transport of material inside this nerve fiber. As a first step toward such studies, in this report we show that it is feasible to take soft x-ray scanning micrographs of the shrimp giant nerve fiber.

The work was done with the scanning microscope at the NSLS X17T beamline at Brookhaven National Laboratory. The instrument forms the subject of another paper in these Proceedings (Rarback et al). A wavelength of 3.2 nm was used. The axon was isolated from the giant nerve fiber in the abdominal nerve cord of shrimp (Penaeus duorarum). After isolation the axon was dried on an electron microscope grid without coating. Figure 1 shows one example of the x-ray micrographs obtained. It shows both longitudinally oriented and elliptical dense structures. From the shape and the dimension of those structures we tentatively suggest that the former might be the image of clusters of microtubules and the latter might be that of the nucleus. Quantitative data on the composition and the density of the constitutents of the relevant structures are required before definite conclusions can be drawn.

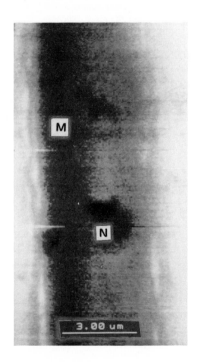

Fig. 1. The scanning x-ray micrograph of the demyelinated axon of medullated shrimp nerve. The x-ray wavelength is 3.2 nm. M and N are the structures we tentatively identify as the cluster of microtubules and the nucleus respectively.

We thank Peter Brink, Dieter Kern, DeMing Shu, Yuli Vladimirsky and the NSLS staff for their help and contributions. The work at Stony Brook is supported by the NSF under grant #BBS 8618066. The NSLS is supported by the DOE under Contract # DE-AC02-76-CH00016.

REFERENCES

1. S.F. Fan, K. Hsu, F.S. Chen and B. Hao: On the High Conduction Velocity of the Giant Nerve Fiber of Shrimp Penaeus Orientalis. Kexue Tongbao (Sci. Bull., Peking) April 51-52 (1961).
2. S.F. Fan and K. Hsu: The Structure and the Physiological Characteristics of the High Conduction Velocity Shrimp (Penaeus Orientalis) Giant Nerve Fiber. Work of the Institute of Physiology, Chinese Academy of Sciences, 1-12 (1961).
3. K. Kusano: Electrical Activity and Structural Correlates of Giant Nerve Fibers in Kuruma Shrimp (Penaeus Japonica). J. Cell. Physiol., 68:361-384 (1966).
4. S.K. Huang, Y. Yeh and K. Hsu: A Microscopic and Electron Microscopic Investigation of the Myelin Sheath of the Nerve Fiber of Penaeus Orientalis. Acta Physiol. Sinica, 26:39-42 (1963).
5. K. Kusano and M.M. LaVail: Impulse Conduction in the Shrimp Medullated Giant Fiber with Special Reference to the Structure of Functionally Excitable Areas. J. Comp. Neurol., 142:481-494 (1971).
6. S.F. Fan, P. Brink and M.M. Dewey: Some Structural Features and the Paradox of Near Zero Resting Membrane Potential of the High Conduction Velocity Shrimp Giant Nerve Fiber. Abs. Int. Conf. Structure and Function in Excitable Cells, Woods Hole M.A., USA (1981).
7. S.F. Fan, P. Brink and M.M. Dewey: The Paradox of Near Zero Resting Potential of the High Conduction Velocity Shrimp Giant Nerve Fiber. Biophys. J., 47:268a (1985).
8. K. Hama: The Fine Structure of the Schwann Cell Sheath of the Nerve Fiber in the Shrimp (Penaeus Japonica). J. Cell Biol., 31:624-632 (1966).

Comparison of Soft X-Ray Contact Microscopy with Other Microscopical Techniques for the Study of the Fine Structure of Plant Cells

T.W. Ford[1], A.D. Stead[1], W. Myring[2], C.P.B. Hills[3], and R. Rosser[4]

[1]Dept. of Biology, RHBNC, Egham, Surrey TW20 0EX, UK
[2]SRS, Daresbury, Warrington, Cheshire, WA4 4AD, UK
[3]Dept. of Physics, Kings College, The Strand,
 London WC2R 2LS, UK
[4]Rutherford Appleton Laboratory, Chilton, Oxon., OX11 0QX, UK

1. INTRODUCTION

The maximum resolution of light microscopy is about 250 nm, so for most biological ultrastructural work the electron microscope with its higher resolution
of approximately 5 nm is used. However, since electrons are readily absorbed
by biological material, thin sections (usually 50-100 nm) are essential, with
subsequent heavy metal deposition to enhance contrast. The preparative procedures necessary for electron microscopy (fixation, embedding, sectioning and
post-staining all have the potential for introducing artefacts into the final
image obtained. During the development of electron microscopy such problems
have been reduced and this technique is now widespread and accepted as the only
practical method for examining the fine structure of cells. The technique of
soft x-ray contact microscopy offers the potential for examining wet (living)
cells with a higher resolution (<100 nm) than light microscopy though its resolution is unlikely to approach that of transmission electron microscopy (TEM).
Initial work at Daresbury using the lower energy synchrotron-generated soft
x-rays, required relatively long exposure times (10-20 secs.). The possibility
of cell damage before imaging necessitated the use of critical point dried
(CPD) rather than wet cells. The higher energy of laser-produced plasmas at
the Rutherford-Appleton Laboratory (RAL) permitted very short exposure times
(1-50ns) so enabling wet cell imaging to be attempted. We report here preliminary results on the imaging of whole algal cells which were selected because
their dimensions (1-10μm) were within the theoretical penetration limit of soft
x-rays. The images obtained are compared with those produced by light microscopy (Nomarski interference) and TEM.

2. METHODS

The test organisms used were the prokaryotic, filamentous cyanobacterium *Anabaena cylindrica* (CCAP 1403/2a) and two unicellular, eukaryotic algae *Cyanidium
caldarium* (CCAP 1355/1) and *Trebouxia* sp. (RHBNC B3a/1). All three organisms
were maintained in axenic culture and samples taken as required. Living cells
were examined by Nomarski interference light microscopy using an Olympus BH-2.
For TEM, cells were fixed in 3% glutaraldehyde in 0.05M phosphate buffer pH 7.4
followed by 1% osmium tetroxide, dehydrated in an ethanol series then embedded
in low viscosity resin. Thin sections were post-stained with uranyl acetate

403

followed by lead citrate before viewing under a Zeiss EM109 or Hitachi H600. The methods for imaging CPD or wet cells using soft x-rays, and for the subsequent development and examination of resists, are described elsewhere /1/.

3. RESULTS

Biological structures such as cell walls, membranes etc. are composed of closely packed macromolecules of high carbon content (e.g. protein, lipid, carbohydrate) which will strongly absorb soft x-rays. The less dense cytoplasm and the watery vacuolar sap also contain such molecules but less ordered and concentrated. The vacuoles of plant cells, and to a lesser extent their cytoplasm, will therefore be relatively transparent to soft x-rays whilst membranes around and within organelles such as chloroplasts, mitochondria and nuclei will be highly absorbent. When cells are critical point dried, the position and structure of membranes and organelles should be retained as in the living cell presumably with similar soft x-ray absorbing properties. However the removal of water will convert the vacuole essentially into an air sac, whilst the cytoplasm will be reduced to a thin meshwork or local deposition of dried cytoplasmic contents resulting in areas of minimal absorbance. Contact images of these cells should therefore emphasize carbon-dense structural features against a low absorbing background of residual cytoplasm.

The two main cell types of an *Anabaena* filament, vegetative cells and heterocysts, can be easily distinguished by interference light microscopy (Fig. 1). Examination of filaments by TEM reveals intracellular detail of these two cell types. The thin-walled vegetative cell contains a random arrangement of internal membranes which permeate most of the cytoplasm (Fig. 2) whilst these membranes are less prolific, but of a similar distribution, in the thick-walled heterocyst. At the junction of vegetative cells and heterocysts, the cell wall of the latter is breached by a narrow cytoplasmic plug. Resist images of heterocysts, produced by an exposure factor of 6000mA.sec. and developed to a depth of 750 nm, show clear delimitation of this thick wall from the bulk of the cytoplasm and of the cytoplasmic plug interconnecting the heterocyst and adjacent vegetative cell when examined by scanning electron microscopy (SEM) (Fig. 3). At this low exposure little internal discrimination of the vegetative cells is visible. However, at higher exposures (30000mA.sec.) and deeper development (1μm) resist images of vegetative cells have a spongy appearance representing the random proliferation of cytoplasmic membranes and resulting from the differential soft x-ray absorption of membranes and inter-membrane spaces (Fig. 4). Such discrimination is not possible in heterocysts due to the high x-ray absorption by the surrounding cell wall. End walls of these vegetative cells are also visible in resist images, but not the side walls, due to enhanced soft x-ray absorption by the former.

Nomarski light microscopy of the alga *Trebouxia* reveals the large chloroplast occupying half the cell but little other intracellular detail (Fig. 5). The cytoplasmic membrane-bound components of this eukaryote, chloroplast, nucleus and mitochondria, are clearly seen by TEM as are the internal membranes and dense pyrenoid of the chloroplast (Fig. 6). The cell is bounded by a thick cell wall. Resist images produced by relatively low exposure (6000mA.sec. and 875 nm development), and viewed by SEM, show this cell wall separated from the main cytoplasmic inclusions (Fig. 7). After much higher exposure (15000mA.sec.) and deeper development (1μm), the cell wall image becomes completely separated from images of the cytoplasm which itself is segregated into regions which may represent nucleus and chloroplast. Smaller cytoplasmic inclusions are visible between this central mass and the image of the cell wall (Fig. 8).

404

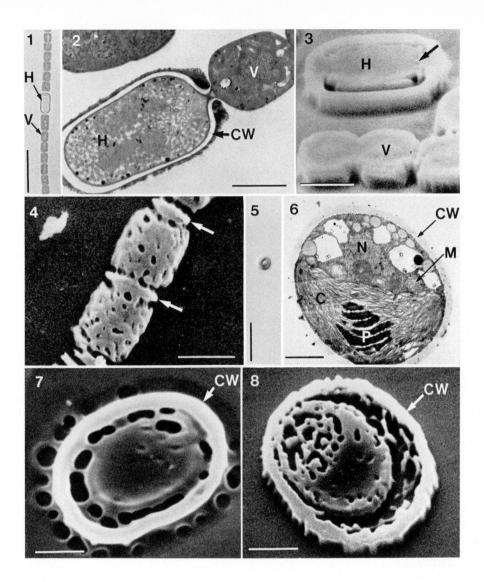

Figures 1-4 *Anabaena cylindrica* 1) light micrograph of filament bar = 15 μm; 2) thin section of vegetative cell and heterocyst viewed by TEM, bar = 2 μm; 3) SEM of resist images of CPD vegetative cells and heterocyst after 6000mA sec exposure showing cytoplasmic plug in end wall of the latter (arrow), bar = 2μm; 4) SEM of resist image of CPD vegetative cell after 30000mA sec exposure showing images of internal membranes and of end walls of cell (arrow), bar = 2μm

Figures 5-8 *Trebouxia* 5) light micrograph of single cell, bar = 15 μm; 6) thin section of cell viewed under TEM, bar = 1 μm; 7) & 8) resist images of CPD cells after exposure at 6000mA sec and 15000mA sec respectively viewed by SEM, bar = 2 μm. V-vegetative cell, H-heterocyst, CW-cell wall, C-chloroplast, N-nucleus M-mitochondrion, P-pyrenoid

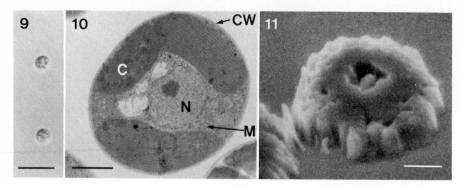

Figures 9-11 *Cyanidium caldarium* 9) light micrograph of single cell, bar = 15μm; 10) thin section of cell viewed under TEM, bar = 1 μm; 11) resist images of CPD cell after 30000mA sec viewed by SEM, bar = 1 μm. C-chloroplast, N-nucleus, M-mitochondrion, CW-cell wall

Under interference light microscopy the eukaryotic alga *Cyanidium caldarium* is seen as a spherical cell with highly granular contents (Fig. 9). Examination by TEM reveals a dense cytoplasm containing peripheral chloroplasts, nucleus and mitochondria with a relatively thin cell wall (Fig. 10). Such a cell is relatively homogeneous with respect to soft x-ray penetration and resist images obtained, when examined by SEM, show little internal discrimination (Fig. 11) though the central cavity may be due to less x-ray absorbance by the centre of the cell compared to the peripheral chloroplasts. These results were all obtained using CPD cells and synchrotron-generated x-rays. Imaging of wet cells was also attempted though this required the higher energy of laser-produced plasmas in order to reduce the exposure time to nanoseconds. Images of wet cells obtained so far show little internal detail though the cell wall of *Anabaena* and *Trebouxia* could again be resolved from the cytoplasm.

4. CONCLUSIONS

Three microorganisms (as CPD material) have been successfully imaged using synchrotron-generated soft x-rays. Examination of developed resists by SEM reveals features which correlate with those observed in thin sections of cells viewed by conventional TEM e.g. cell wall in *Anabaena* and *Trebouxia*, internal membranes in *Anabaena*, chloroplast/nucleus complex in *Trebouxia*. However these high soft x-ray absorbing structures must be located in a soft x-ray transparent matrix since in the relatively homogeneous *Cyanidium* cell, resolution of cellular contents was poor. Images of wet cells using laser-produced plasmas revealed little cellular detail though this may improve as the methodology is refined.

5. ACKNOWLEDGEMENTS

The assistance of Grahame Lawes, Anton Page, Lynne Etherington and David Ward (RHBNC), Dave Clark (SRS Daresbury) and Alan Ridgeley (RAL) is gratefully acknowledged, also the Central Research Fund of University of London for purchase of the interferometer.

6. REFERENCES

1. A.D. Stead, T.W. Ford, R. Eason, A.G. Michette, W. Myring, R. Rosser: This volume

Biological Applications at LBL's Soft X-Ray Contact Microscopy Station

G.D. Guttmann and M.R. Howells

Center for X-Ray Optics, Lawrence Berkeley Laboratory,
Berkeley, CA 94720, USA

1 Introduction

The soft x-ray contact microscopy station at LBL has been in operation for about a year and has just recently yielded some images from a couple of biological samples. The station uses a commercial electron-beam evaporation source with a well-cooled target to generate monochromatic soft x-rays in a relatively high vacuum (10^{-7} torr)[1]. The targets are interchangeable and presently we are using vanadium, which yields L_α x-rays with energy 0.511 keV. Other targets are aluminum, carbon and the bare copper target. At present, we are imaging biological samples with the intent of establishing a sample preparation protocol and standardizing our resist development procedures.

Currently, normal red blood cells have been x-ray imaged and the resists examined by scanning electron microscopy (SEM). Our current knowledge indicates that mammalian red blood cells have no internal cytoskeletal architecture but retain their shape with a submembranous meshwork of actin and spectrin [2]. The soft x-ray contact micrographs of the normal red blood cells did not show an internal cytoskeleton.

2 Materials and Methods

The x-ray images have been recorded onto copolymer resists, 90% polymethylmethacrylate (PMMA) and 10% methacrylic acid (MAA). The etching solution was methyl isobutyl ketone (MIBK) and isopropyl alcohol (IPA) with a ratio of 1:5 respectively.

The normal and sickled red blood cells were fixed in glutaraldehyde and then washed in deionized water. The exposure period was 6 hours, which corresponds to a dose of 1.1 megarads, and the development time was 8 minutes.

3 Images

The soft x-ray contact micrographs of normal red blood cells correlate very well with known images made from both transmission and scanning electron microscopy. Figures 1 and 2 illustrate the light micrograph and SEM, respectively, of the x-ray resist image from normal red blood cells.

4 Summary

The normal red blood cell images definitely correlate very well with the TEMs and SEMs and our current knowledge of normal mammalian erythrocytes.

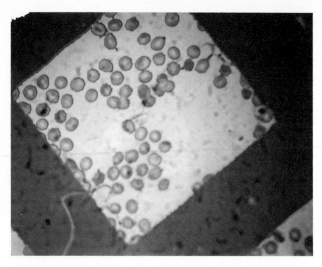

Figure 1. Light micrograph of the x-ray resist image from normal red blood cells (Magnification 1000X)

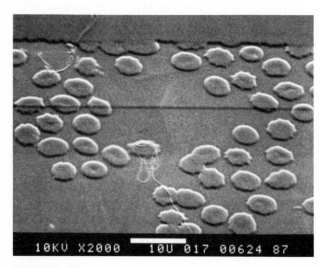

Figure 2. SEM of the preceding x-ray resist (Magnification 2000X)

We have found 'readout', SEM examination of the resist, to be a difficult problem and are now experimenting with various methods, i.e. shadowing and sharper SEM examination angles [3,4], to improve our images.

5 Acknowledgements

We wish to thank J. Bastacky (LBL) for his assistance with the normal red blood cells and also for his scanning electron microscopy expertise. This

408

research was supported by the U.S. Department of Energy under Contract No. DE-AC03-76-SF00098.

6 References

1. E. Spiller and R. Feder: In X-ray Optics. Applications to solids, ed. by H.-J. Queisser, Springer Series, Topics in Appl. Phys., Vol. 22, p. 35 (Springer, Berlin, Heidelberg 1977)

2. A. B. Fulton: The Cytoskeleton: Cellular architecture and choreography, pp.24-26 (Chapman and Hall, London, England 1984)

3. Private communication with D. M. Shinozaki

4. R. Feder, E. Spiller, J. Topalian, A. N. Broers, W. Gudat, B. J. Panessa, Z. A. Zadunaisky, and J. Sedat: Science, $\underline{197}$, 259 (1977)

X-Ray Microscopy of Single Neurons in the Central Nervous System

J.E. Hamos

Department of Neurology,
University of Massachusetts Medical Center,
Worcester, MA 01655, USA

Since the original application of light microscopic techniques to analyses of the nervous system over 100 years ago, neurobiologists have believed that a detailed examination of the structure of neurons will provide a foundation for understanding the elaboration of neuronal function. More recently, with the advent of high resolution electron microscopic methods, this logic has been carried to an ultrastructural level where the goal has been to define the organization of synaptic connections onto the postsynaptic dendrites of neurons in order to determine the circuitry underlying various physiological responses. To date, studies utilizing the electron microscope have provided few insights into the origins of these responses and have largely confirmed the synaptic nature of connections that cannot be resolved with the light microscope.

My research focuses on three-dimensional aspects of circuitry in the belief that the responses of a neuron are largely determined by the spatial distribution of many inputs onto a neuron's dendritic surface. The neuron then integrates the activity from these inputs according to the specific electrical properties of its membranes and the timing of the inputs. The goal of this research is to precisely describe the patterns of input to single neurons and then model the effectiveness of these inputs in activating the cells (HAMOS et al. [1,2]).

Present electron microscopic methods to define these patterns are time consuming and labor intensive since the resolution that this technique affords requires the analysis of sections of tissue that are roughly $0.08\mu m$ thick while a typical neuron may fill a region that spans $250\mu m$ X $250\mu m$ X $250\mu m$. A technology utilizing an X-ray microscope may save time by permitting the analysis of more tissue at one time since this microscope can penetrate thick sections with a high level of resolution. In this way, studies of synaptic circuitry which require resolution in the range of tenths of microns can be performed in thicker sections that generally have much lower resolution.

1. Light Microscopy

The first insights into the intricacies of the nervous system were gained by light microscopists who examined the constituents of different neuronal regions and compared the morphological variety of neurons residing there. Towards this end, many methods were developed to reveal the structural features of neurons. Among the first techniques applied to studies of the nervous system was the Golgi technique. This method utilizes a protocol in which chromate and dichromate salts are deposited on individual neurons and their processes so as to make them visible. A more recently developed approach to identify single neurons is the intracellular injection of cells with a marker that allows the neuron to be identified after processing the tissue through various histochemical reactions (Fig. 1).

Utilizing these methods, a wealth of structural data on single neurons has been generated but it has been difficult to incorporate this information into a functional understanding of neuronal networks for several reasons. Firstly, the nervous system is organized in an extremely complex manner as indicated by the vast differences in size and shape of neurons within any region. Secondly, further complicating the organizational scheme is that the processes of these neurons, to varying degrees, have minute appendages (Fig. 1) whose structural contributions must be appreciated since they are important receptive structures for inputs to the cells. Finally, to understand how these elements work in specified circuits that must perform different functional tasks, it is necessary to examine the substrate for neuronal connections, i.e. the synapse. Since pre- and postsynaptic elements of synapses are evident only with the resolution of components that are roughly 0.1µm or less in size, the resolving

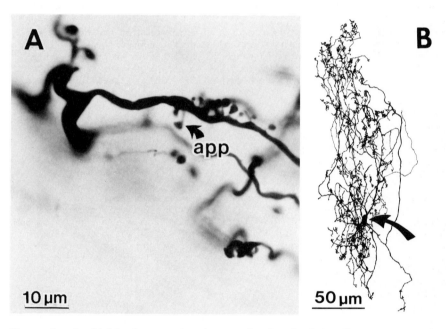

Figure 1. A: Light microscopic micrograph of a dendrite from a neuron that was injected with a dye and then recovered after using a histochemical procedure. Fine appendages (app) emanate from dendritic processes. These appendages and the adjacent membrane surface are important in that they are major postsynaptic targets for inputs to the cell. B: Drawing of a single neuron that was intracellularly injected with a dye and reconstructed from four 50µm thick sections. Neurons such as this are structurally complex, as illustrated by the extensive branching of the dendritic arborization from the origin of these processes at the cell body (arrow). This neuron has previously been described in HAMOS et al. [2].

411

power of the light microscope is not a sufficient tool to examine neuronal circuitry.

2. Electron Microscopy

It is only at the electron microscopic level that synaptic connections between neurons are identifiable. To prepare tissue for the electron microscope, several methods have been developed that alleviate the limited ability of an electron beam to penetrate tissue sections. Typical preparations used for this approach are exceedingly thin (roughly 0.08μm thick). Production of such sections requires blocks of tissue to first be embedded in plastic resins. Thereafter, the tissue may be thin sectioned on an ultramicrotome and the sections analyzed with the electron microscope.

Ultrastructural features of synapses have been well documented (PETERS et al. [3]). A synapse is identified by the presence of synaptic vesicles in the presynaptic terminal, a density in the postsynaptic process, and an enlarged synaptic cleft in the extracellular space between the two elements (Fig. 2A). However, even when such a synaptic site is localized, one rarely knows the identity of the pre- or postsynaptic elements as the electron microscopic view is of processes separated from their cell body of origin.

Ultimately, the synaptic organization of any region of the nervous system includes neurons of varying form and dendritic complexity as well as the many synapses formed onto the neurons. Single thin sections grossly oversimplify this organization as they expose only a glimpse of the relationships between the constituent elements (Fig. 2B). Plausible interpre-

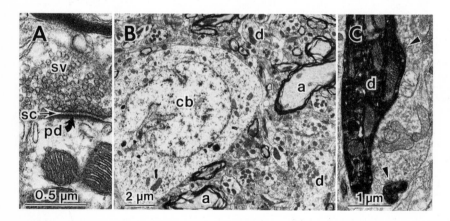

Figure 2. Electron micrographs illustrating ultrastructural features of the nervous system. A: Synapses are indicated by the presence of synaptic vesicles (sv), a postsynaptic density (pd) and a synaptic cleft (sc). B: Single thin sections reveal various neuronal elements including a cell body (cb), dendrites (d) and axons (a). However, only a portion of the inputs to specific neurons can be identified in this way. C: By combining labeling of individual neurons (e.g. Fig. 1B) with electron microscopy, it is possible to localize synaptic inputs (arrows) onto identified dendritic segments (d) of neurons whose three-dimensional structure is known.

412

tations of neuronal circuitry are made even more difficult by the fact that processes of neurons with different form and, one suspects, different function intermingle within the tightly compacted space of the nervous system.

Recent technical advances permit analyses of both the three-dimensional form of single neurons and the pattern of inputs onto dendrites of the same neurons. A powerful approach has been to inject an electron opaque dye into a single neuron, identify and draw the neuron with a light microscope (Fig. 1B), and then examine the neuron through the electron microscope. As the dye is electron dense, dendrites are easily identified (Fig. 2C) and synapses formed onto specific segments can be related to the original drawing of the cell. While this method provides data on circuitry of functional import to the neuron, a study of the full distribution of synapses formed onto a single injected neuron requires that thousands of electron micrographs are taken and related to the original drawing of the cell. The technical dilemma is that sections available for electron microscopic analysis are exceedingly thin when compared to the full dimensions of the individual neurons. Studies utilizing this approach (e.g. HAMOS et al. [1,2]) require years to complete with a resultant analysis of a few neurons.

3. X-ray Microscopy

The inherent weakness in electron microscopic studies is the amount of time required to reconstruct and analyze the full complement of inputs to a few cells. An aim for the future, therefore, is to resolve synaptic circuitry in thicker sections. Development of X-ray microscopes to probe the nervous system may provide a valuable approach to synaptic analyses.

Recently, I joined with J. Kirz and colleagues to examine a potential use of the X-ray microscope (at the National Synchrotron Light Source) in studies of synaptic circuitry by directing the microscope at $1\mu m$ thick sections of a neuron that had been injected in the manner described above. We identified dendritic segments of the injected neuron, receiving digitized images of the segments with contrast related to the opacity of the injected dye. Significantly, we gained greater than an order of magnitude in the tissue thickness that may be analyzed (from $0.08\mu m$ thickness required for the electron microscope to roughly $1um$ for the X-ray microscope) without theoretically sacrificing the resolution requirements to identify synapses.

This exploratory study revealed that the X-ray microscope identifies portions of single neurons. Additional improvements will ensure that neurons are retrieved in sections of tissue that are thicker than $1\mu m$. Moreover, the future development of X-ray holographic techniques will allow for complete three-dimensional views of the processes. In combination with the higher resolution that may be achieved by the X-ray microscope, we will receive digitized views of neurons that are significantly enhanced compared to images currently available through the light microscope.

More importantly, development of new staining methods will ensure that the full capabilities of the X-ray microscope are used for synaptic analyses that can currently be achieved only through exhaustive electron microscopic studies. These methods should take into account an ability to alter the energy of the X-ray beam such that chemical elements can be detected on the basis of their absorption properties. These elements may be localized at synapses through various immunohistochemical techniques that are available to identify neurotransmitters in presynaptic terminals (e.g. SOMOGYI

413

and SOLTESZ [4]) or receptors for these transmitters that are in the post-synaptic membrane (e.g. TRILLER et al. [5]).

In summary, procedures are presently available to reveal the constitu-ents of synaptic sites in the nervous system including methods to localize neurotransmitters in presynaptic terminals, the synaptic complex itself, and receptors in the postsynaptic membrane. When these synaptic components are linked to chemical elements with distinct absorption properties, the X-ray microscope may resolve the patterns of synapses on individual neurons.

4. Literature

1. J.E. Hamos, S.C. Van Horn, D. Raczkowski, S.M. Sherman: J. Comp. Neurol. 259, 165 (1987).
2. J.E. Hamos, S.C. Van Horn, D. Raczkowski, D.J. Uhlrich, S.M. Sherman: Nature 317, 618 (1985).
3. A. Peters, S. Palay, H.F. Webster: The Fine Structure of the Nervous System (W.B. Saunders Company, Philadelphia, London, Toronto 1976).
4. P. Somogyi, I. Soltesz: Neuroscience 19, 1051 (1986).
5. A. Triller, F. Cluzeaud, F. Pfeiffer, H. Betz, H. Korn: J. Cell Biol. 101, 683 (1985).

X-Ray Introscopy of Mediastinal Lymph Nodes Using Synchrotron Radiation

G.N. Kulipanov[1], *N.A. Mezentsev*[1], *V.F. Pindyurin*[1], *A.S. Sokolov*[1], *M.A. Sheremov*[1], *Y. I. Borodin*[2], *V.A. Golovnev*[2], *G.N. Dragun*[2], and *E.L. Zelentsov*[2]

[1]Institute of Nuclear Physics, Novosibirsk, 630090, USSR
[2]Institute of Physiology, Novosibirsk, 630090, USSR

The results are presented of the examination of normal and pathological human lymph nodes. The results have been derived by X-ray scanning difference microscopy and microtomography using synchrotron radiation from the storage ring VEPP-4 (Novosibirsk) and have been supplemented by the information gained from direct radiography using an X-ray tube as well as from the examination of the morphological structure of lymph nodes. The results of different methods are compared. The advantages of X-ray difference microscopy for research of the mechanism of functioning of the lymph nodes are indicated.

1. Introduction

A need for improving the diagnostics of various diseases, particularly at the early stages, requires the developing of instrumental and laboratory research methods. In recent decades some new introscopy methods - the visualization of inner organs with the acquiring of matrix images using a computer, have been brought into medical practice and are in progress. These methods are computer X-ray tomography, sonography, radionuclide diagnostics and some others based on an analysis of the anatomic structure and the topographic interrelationships of inner organs, or on a quantitative analysis of the agent and its distribution in different tissues for the examination of the functional conditions of the human organs and systems.

Digital subtraction angiography using synchrotron radiation is one of the developing introscopy methods /1-4/. In comparison with the other X-ray methods, subtraction angiography has **significant** advantages enabling the contrast studies of the vascular system to be carried out, with the subtraction of a "background" which is due to an absorption of X-ray by tissues not containing the contrast agent. These advantages open broad possibilities for contrast studies of the arterial and venous systems of an organism.

Lymphography - the contrast examination of the lymphatic system - is based on the same principle as contrast arterio - and phlebography but has the features of its own. These are

mainly due to the structure of the lymphatic system (LS) including both the net of vessels as well as the set of lymph
nodes distributed over the organism according to the regional
principle. As for the functional aspect, the moving of X-ray
contrast agents along the LS occurs much more slowly as against
the circulatory system. This characteristic property turns the
LS into a system of "slowly variable structures" which is more
accessible for an X-ray and radiological examination in comparison with the circulatory system. The contrast agents conserve
in lymph nodes for long periods (hours, days). This permits
one to observe the dynamics of their moving and, in many cases,
to avoid a repeat introducing.

The necessity to study the LS is most dictated by insufficient knowledge of morphofunctional transformations occurring
in normal conditions of lymphodynamics and in pathological
states, as well as by the important role of the LS in the body
to maintain its medium constant. It is the LS that immediately
reacts to an external agent. The possibility of fixing these
early changes in the morphostructure and LS functioning with
the use of modern research methods determines to a great extent the early diagnostics of a great deal of pathologies and
allows their course to be observed.

A slow moving of X-ray contrast agents enables one to examine carefully the structure of lymph nodes using radiography
for the purpose of   revealing the roentgenological indications characterising the different conditions of the organism.
In addition it is possible to utilize and adapt the data obtained on static objects (in vitro) with the help of the modern microstructure analyses which reveal the mechanism of
lymph node contrasting.

2. Experimental results

As the static objects, the lymph nodes of human mediastinum
have been studied by means of X-ray scanning difference microscopy and microtomography using the synchrotron radiation (SR)
from the storage ring VEPP-4 /5,6/. For these experiments, a
contrast agent, the colloidal solution of thorium dioxide, has
previously been introduced into the lymph node structure.

In the experiments, the differences have been revealed in
the images of lymph nodes on the conventional and difference
radiographs taken during the scanning with a 0.1 mm step and
with a spatial resolution of 0.065 mm . The differences in the
images varied depending on the sizes and density of lymph nodes
and on the concentration of the contrast agent contained in
them. The SR studies have clearly exhibited the advantages of
the use of synchrotron radiation. In particular, it offers the
possibility of "selecting" an X-ray contrast agent in the lymph
node structure thereby allowing the spatial distribution of
an agent to be studied irrespective of the sizes and density
of lymph nodes and the concentration of the agent being introduced to be found. To check the data, electroradiography of
the lymph nodes has been performed with a 4-5 X magnification
of an X-ray image; in addition, their morphological structure

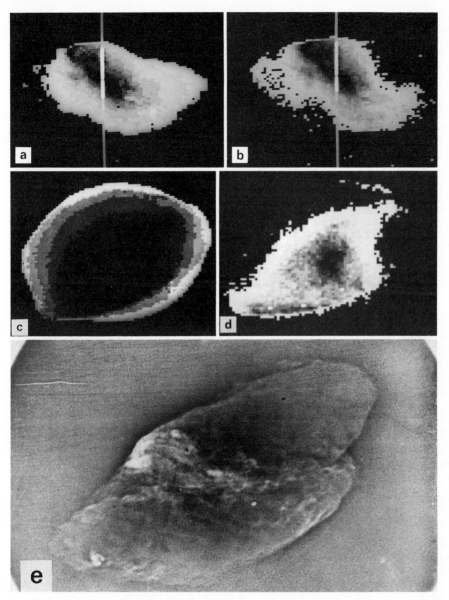

Fig.1. The conventional (a) and difference (b) radiographs of the mediastinal lymph node; the conventional (c) and difference (d) tomographs taken over the section shown as the line on Figs. (a) and (b); the electroradiograph of this node is shown in Fig.1-e

has been studied as well. The data which have been obtained
using difference microscopy and electroradiography substantially supplement each another and confirm the results earlier
derived when the lymph nodes of animals had been examined by
means of the method of X-ray fluorescent analysis using SR /7/.

Figures 1-a and -b illustrate the conventional and difference
radiographs of a mediastinal lymph node partially filled with
the contrast agent. The difference radiograph gives deep insight into the real distribution of the contrast agent and its
concentration in different sections of the lymph node. The line
in the centre of the lymph node picture indicates the section
at which the computer tomographs have been made. The conventional and difference tomographs are shown in Figs.1-c and -d,
respectively. The computer difference microtomograph of the
lymph node substantially contributes to the data derived by
means of microscopy through the visualization of the spatial
distribution of the contrast agent in the node structure. Figure 1-e illustrates the electroradiograph of this node which
yields much less information.

Fig.2-a,-b and -c demonstrate the conventional, difference
and electroradiographs of the normal mediastinal lymph node

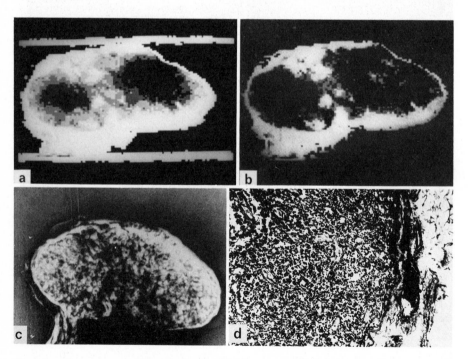

Fig.2. The conventional (a), difference (b) and electroradiograph (c) of a normal mediastinal lymph node; a part
of the histological section of the node is illustrated
in (d)

completely filled with the contrast agent. The comparison of
the images on the conventional and difference radiographs shows
that the former gives no real distribution of the agent in the
lymph node structure. On the difference radiographs, the spatial distribution of the agent is shown with high reliability
excepting the low-concentration regions such as the lymph node
hilus, vascular crus and the marginal sinus. In histological
examinations of the normal lymph nodes the sinus is seen dis-
tinctly in all the lymph node sections with its architectonics
and the sizes of vessels conserved (a part of the histological
section is shown in Fig.2-d). The uniform distribution of the
contrast agent in the lymph node allows a conclusion to be
drawn on the conservation of its transport function.

Fig.3-a and -b show the conventional and difference radiographs of the lymph nodes of the bronchopulmonary section of

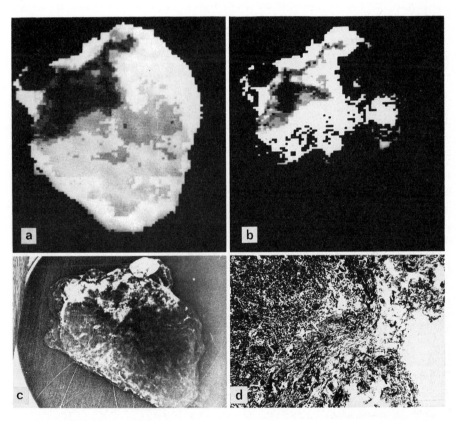

Fig.3. The conventional (a), difference (b) and electroradiograph (c) of the lymph node in the bronchopulmonic
group of mediastinum of a patient with chronic inflammation; a part of the histological section of the
node is shown in (d)

mediastinum of a patient with chronic inflammation. The non-uniformity of contrasting the lymph node is clearly seen. The real distribution of the contrast agent is given by the difference radiograph. The histological examination (see Fig.3-d) indicates a sharp change in the architectonics of the sinus system followed by its deformation, sclerosis, obliteration and by the development of collateral lymphatic vessels. The contrast agent is distributed non-uniformly over the lymph node.

In the case of chronic inflammation the capability of the lymphatic system of mediastinum to transport contrast agents drastically worsens /5/. Lymphostases, lymphothrombosis, sclerosis and obliteration of the sinus system, which eliminate in this case some parts of the lymph nodes tissue from the normal lymph circulation or decrease sharply its capability to transport, determine the "mosaic" character and the irregularity in the X-ray images of the lymph nodes.

The electroradiograph of Fig.3-c confirms the data of the histological analysis and the data obtained using the conventional and difference microscopy of the lymph node under examination which undergoes severe or less severe pathological changes in the lymphodynamics. The newly formed lymph node at the top of the image, which probably implements the compensatory transport function, is the most contrast on radiographs.

The presented results enable the conclusion to be drawn that the character of X-ray imaging of lymph nodes is determined by their morphological structure thereby reflecting, to a considerable extent, the functional state of their conducting system and its capability to lymph transport in the period of examination. The X-ray images of lymph nodes exhibit the specific nature of the contrasting.

3. Conclusion

Investigation of the lymph node microstructure by means of scanning difference computer X-ray microscopy and microtomography is a direct technique to reveal the spatial distribution of agents in the lymph node structure. The technique is highly resolved in space and offers the possibility of a quantitative analysis of the agent at low concentrations. At the same time, this technique has no disadvantages of the other methods of X-ray introscopy not allowing the structure studies to be performed with "background" subtraction.

References

1. E.N.Dementyev et al.: Nucl. Instr. Meth. A246, 726 (1986)
2. W.-R.Dix et al.: Nucl. Instr. Meth. A246, 702 (1986)
3. A.Akisada et al.: Nucl. Instr. Meth. A246, 713 (1986)
4. E.B.Hughes et al.: Nucl. Instr. Meth. A246, 719 (1986)
5. E.A.Bir et al.: Proceedings of the 6th National Meeting on SR Usage SR-84, Novosibirsk, 1984, p.212
6. Yu.I.Borodin et al.: Nucl. Instr. Meth. A246, 649 (1986)
7. V.B.Baryshev et al.: Proceedings of the 6th National Meeting on SR Usage SR-84, Novosibirsk, 1984, p.341

Absorption Edge Imaging of Sporocide-Treated and Non-treated Bacterial Spores

B.J. Panessa-Warren[1], G.T. Tortora[2], and J.B. Warren[3]

[1]Allied Health Resources Dept., SUNY, Stony Brook,
NY 11794, USA
[2]Medical Technology/Clinical Microbiology Dept., Stony Brook,
NY 11794, USA
[3]Instrumentation Division, Brookhaven National Laboratory,
Upton, NY 11973, USA

When deprived of nutrients, spore-forming bacilli produce endospores which are remarkably resistant to chemical sterilization.[1,2] Little is known about the morphology and response of these spores following exposure to sporocidal agents. Light microscopy does not provide sufficient resolution for studying the rupture of the spore coat and fate of intracellular material. Transmission and scanning electron microscopy offer superior resolution but require specimen preparation methods that induce physiologic as well as morphologic changes in the spores, thereby making accurate interpretation of micrographs difficult. To eliminate the possible artifacts induced by chemical fixation, dehydration, embeddment, staining and sectioning, treated and non-sporocide-treated endospores of B. thuringiensis and B. subtilis were imaged by x-ray contact microscopy using monochromatic x-rays.

Soft x-rays (1-10 nm wavelength) offer the biologist distinct advantages for imaging samples that are particularly sensitive to electron or proton beams, or to conventional specimen preparation procedures.[3] These x-rays produce less specimen damage than electrons and protons, and they penetrate the biological specimen more easily than electrons without necessitation of a vacuum environment.[4] By using monochromatic x-rays for exposure, it is possible to augment selective absorptivity and produce x-ray contact replicas that reveal the location of specific elements within the tissue.[5] With the advances in synchrotron technology, it is now possible to obtain monochromatic x-rays with sufficient coherence, brightness and dose for the brief exposure times required with biological hydrated samples. Although quantitative x-ray microanalysis of frozen biological tissue can also identify specific elements within the sample, soft x-ray absorption edge microscopy offers a less damaging method that requires little specimen preparation or skill; has an inherent contrast mechanism (selective x-ray absorptivity) which eliminates the need for staining or chemically fixing a specimen;[3] and the ability to image elements (such as oxygen, nitrogen, carbon, etc.) that cannot readily be viewed or identified by conventional x-ray microanalysis methods. This paper represents some of our first attempts at absorption edge imaging and in no way presents this methodology as a quantitative trace analysis technique.

1. METHODS

1.1 Endospore Preparation

Bacillus subtilis var niger ATCC 7932 was obtained from American Type Culture collection and Bacillus thuringiensis was obtained from Dipel, Abbot Laboratories. Spores were grown on slants of AK Sporulating Agar No. 2 (BBL, Cockeysville, MD) in 16 oz prescription bottles at 35°C aerobically. After maximal sporulation, the spores were harvested in deionized water and heated to 80°C for 20 min to kill any remaining vegetative cells. The spores were washed three times with sterile deionized water and stock suspensions were stored at 2-8°C until needed. For comparison, spore samples were either vacuum dried onto aluminum scanning mounts or fixed in 1.5% glutaraldehyde in 0.1 M cacodylate buffer, followed by acetone dehydration and critical point drying in liquid CO_2. These samples were rotary coated with 7-10 nm of gold-palladium in an Edwards vacuum evaporator prior to examination in an AMRAY 1000A SEM at 20 or 30 keV accelerating voltage. Spores treated with a vanadium-containing sporocide for 30 sec, 10 min and 50 min were either vacuum dried or prepared in the aforementioned way and viewed by scanning electron microscopy.

1.2 X-ray Detector and Exposures

Polymethyl methacrylate (PMMA) dissolved in chlorobenzene was spun at 1 μm thickness onto silicon wafers and baked for 1 hr at 156°C. B. thuringiensis and B. subtilis spores were smeared as a monolayer onto the photoresist and excess fluid wicked-off with filter paper. These spores were exposed to monochromatic x-rays near the nitrogen absorption edge (30.99 nmλ) and above and below the V L_{III} absorption edge. B. thuringiensis and B. subtilis spore preparations treated with sporocide for 30 sec, 10 min and 50-80 min were exposed to monochromatic x-rays near the nitrogen edge and above and below the V L_{III} edge. For this study the exposures were made at a dedicated soft x-ray beamline (U15) at the high brightness synchrotron radiation facility at Brookhaven National Laboratory, National Synchrotron Light Source. A toroidal grating monochromator provided a monochromatic beam at the 750 MeV VUV storage ring. All of the x-ray exposures ranged from 20.7 to 28.3 μA minutes.

Following x-ray exposure, the spores were washed off the photoresist and the photoresist developed in 1:1 isopropanol/methyl-isobutyl ketone (MIBK). The x-ray contact replicas were rotary coated with 5-8 nm gold-palladium in an Edwards vacuum evaporator and viewed at 60-75° tilt, 20 or 30 keV accelerating voltage.

2. RESULTS AND DISCUSSION

X-ray contact replicas of B. subtilis and B. thuringiensis endospores made following exposure to x-rays near the nitrogen edge revealed details of the spore coat and exosporium. Nitrogen absorption maps of the spores were used to monitor spore morphology and general protein distribution on the silicon wafer. At 30 sec after sporocide treatment, nitrogen x-ray maps of B. thuringiensis revealed alterations in the spore coat and at 5 min post-treatment complete disruption of the spore coat. B. subtilis treated with sporocide for 1-5 min produced nitrogen x-ray maps with wrinkling and blebs on the spore coat but complete disruption of the spore coat was not found even after 80 min of sporocide treatment.

X-ray exposures of sporocide-treated B. subtilis spores made below the absorption edge of vanadium (V) produced x-ray contact replicas with no clearly intact spores. The outline of the disrupted spore coat could be seen surrounded by a puddle of material with numerous globular inclusions (Fig. 1). Areas of the photoresist lacking the hazy images of the spores also lacked the puddles of globular material, suggesting that the sporocide treated B. subtilis spores may exude or lose their intracellular contents without the total destruction of the spore coat. Above the absorption edge of V the same sporocide-treated B. subtilis spores revealed the hazy image of the spore coat and a surrounding puddle of material, however, in this case the formerly globular inclusions in the puddle now appear dendritic with six or more arms radiating from a central core (Fig. 2). These rosette-like arrays surrounding the spore coats were only seen in exposures made just above the V L $_{III}$ absorption edge. The nitrogen x-ray replicas and replicas made below the V L$_{III}$ edge using the same specimen showed no rosettes but rather indistinct globular material in the puddles surrounding the spore coats. One would expect areas of the speci-

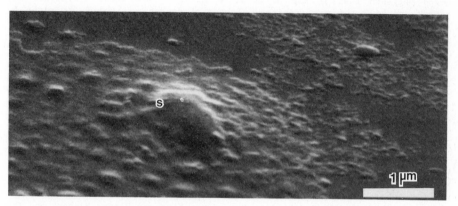

Fig. 1. Sporocide treated B. subtilis x-ray contact replica taken below the absorption edge of vanadium shows hazy image of spore coat (s) surrounded by a region of material with elliptical inclusions.

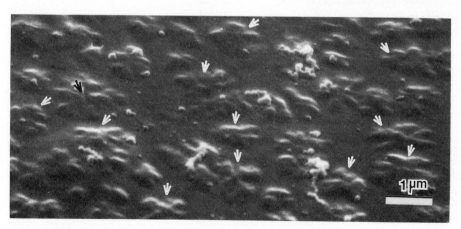

Fig. 2. Same sample as figure 1 when imaged with x-rays above the absorption edge of vanadium shows same puddle surrounding the spore coat, but the inclusions appear crystalline or as rosettes (arrows).

men originally containing V to be strongly imaged when exposed to monochromatic x-rays just above the absorption edge of the V L_{III} line and these same structures to be poorly imaged at x-ray wavelengths below the absorption edge of the element in question.[6]

X-ray contact replicas of sporocide treated B. thuringiensis spores revealed such extensive fragmentation of the spore coat and cellular debris at 30 sec or more post-sporocide treatment that it was not possible to identify individual structures in the micrographs. This was also the case with the B. thuringiensis glutaraldehyde fixed and vacuum dried control preparations.

The capability to reveal compositional information from the same specimen by imaging it with different monochromatic x-rays poses many exciting possibilities. The fact that the rosette-like images seen above the V absorption edge may represent the V in the sporocide (bound to endogenous dipicolinic acid) suggests that it may be possible to learn how these sporocides may attack and kill the more resistant bacterial spores. Although it might be considered that this type of analysis could be done by electron or x-ray induced x-ray microanalysis, it is not possible to produce images of, or analyze these puddles around the disrupted spores following most specimen preparation procedures. By x-ray fluorescence spectrometry it would be impossible to obtain this spatial resolution, and by electron microprobe analysis the problems of mass loss, and orientation of an unstained sample, would make this type of analysis extremely difficult. The ability to image partially hydrated, unstained unfixed biological material and gain immediate information about the possible localization of a specific element without the beam damage and specimen trauma experienced with other analytical techniques, is an exciting step forward in the realm of biological analysis.

3. ACKNOWLEDGMENT

The authors are deeply indebted to Drs. Janos Kirz and Harvey Rarback and the entire staff of the U15 beamline at the National Synchrotron Light Source who so generously gave their time and expert advice during the course of this work. We would also like to thank Mrs. Barbara Kponou for her skill in preparing the manuscript.

4. REFERENCES

1. P. Fitz-James and E. Young: In The Bacterial Spore, ed. by G. Gould and A. Hurst, (Academic Press, NY, 1969) p. 39.

2. C. R. Phillips: Bact. Rev. 16, 135 (1952).

3. B. J. Panessa-Warren and J. B. Warren: Ann. NY Acad. Sci. 342, 350 (1980).

4. J. Kirz and D. Sayre: In Synchrotron Radiation Research, ed. by S. Doniach and H. Winick, (Plenum Press, NY, 1980) p. 277.

5. J. Kirz: Ann. NY Acad. Sci. 342, 273 (1980).

6. L. Beese et al., Biophys. J. 49, 259 (1986).

The Role of High-Energy Synchrotron Radiation in Biomedical Trace Element Research

J.G. Pounds, G.J. Long, W. M. Kwiatek, K.W. Jones, B.M. Gordon, and A.L. Hanson

Division of Atomic and Applied Physics,
Department of Applied Science, Brookhaven National Laboratory,
Upton, NY 11973, USA

1. INTRODUCTION

Trace elements are intimately involved in biological function and dysfunction at all levels of biological organization. At the molecular level, trace elements perform innumerable catalytic and structural roles in macromolecules and other cell constituents. At the cellular level, trace elements are necessary for maintenance and regulation of compartmentation of cell function, stimulus-response coupling, gene regulation, etc. Perturbations in trace element homeostasis and utilization at the molecular and cellular level is manifested in many disease states. Biological systems have evolved elaborate and diverse control mechanisms to provide for trace element homeostasis at the subcellular, cellular, and organismal levels. These critical functions of trace elements have profound influence on human health and disease states. Trace elements are well recognized as contribution factors in modifying development, aging, oncogenesis, and many chronic diseases including cardiovascular disorders.

X-ray fluorescence analysis with high-energy synchrotron radiation has several characteristics, including high sensitivity, multi-elemental capabilities, good spatial resolution, and minimal matrix effects and sample preparation which make it a powerful tool for understanding the role of trace elements in biological systems. The application of the high-energy x-ray microscope is directed at three general areas: first, the role of trace element interactions in normal biochemical and physiological processes; second, the microscopic distribution and metabolism of toxic elements, and third, the microscopic distribution and metabolism of the metal moiety of metal containing drugs. A description of the high-energy x-ray microscope is presented by GORDON *et al.* /1/ and an investigation of the microscopic distribution of gallium, following administration of gallium nitrate, an experimental drug for treating tumor related bone loss is given by BOCKMAN *et al.* /2/.

This paper will present the results of an investigation of the distribution of essential elements in the normal hepatic lobule. The liver is the organ responsible for metabolism and storage of most trace elements. Although parenchymal hepatocytes are rather uniform histologically, morphometry, histochemistry, immunohistochemistry, and microdissection with microchemical investigations have revealed marked heterogeneity on a functional and biochemical level. Hepatocytes from the periportal and perivenous zones of the liver parenchyma differ in oxidative energy metabolism, glucose uptake and output, ureagenesis, biotransformation, bile acid secretion, and plasma protein

425

synthesis and secretion /3/. Although trace elements are intimately involved in the regulation and maintenance of these functions, little is known regarding the heterogeneity of trace element localization of the liver parenchyma. Histochemical techniques for trace elements generally give high spatial resolution, but lack specificity and stoichiometry. Microdissection has been of marginal usefulness for trace element analyses due to the very small size of the dissected parenchyma. The characteristics of the high-energy x-ray microscope provide an effective approach for elucidating the trace element content of these small biological structures or regions. The preliminary results of this investigation are presented below.

2. METHODS AND MATERIALS

2.1 Specimen preparation

Male, Sprague-Dawley rats of 300 - 350 grams were used for this study. The animals were fasted for 12 hours to deplete liver glycogen stores prior to sacrifice. The animals were deeply anesthetized with ether, and the liver perfused *in situ* for one minute with 300 mM Sucrose / 5 mM HEPES buffer, at pH 7.0 and 4° C, to flush the liver sinusoids of both the cellular and acellular components of blood. This buffer was previously shown to remove unbound extracellular elements while minimizing the loss and exchange of intracellular elements /4/. Five mm thick sections of the caudal lobe were removed, placed in cryomolds with Tissue Tek and frozen in liquid nitrogen. The livers from five rats were sectioned at 20 μm (approximately 1 cell thick) with a freezing microtome, mounted on 7.3 μm Kapton support film, and lyophilized prior to trace element analysis.

2.2 Microprobe analysis

Filtered white synchrotron radiation was used for trace element analyses. The beam was filtered by 100 μm aluminum before irradiating the target. The fluorescence signal was filtered by 50 μm of Kapton placed in front of the Si(Li) detector. Tantalum slits were used to collimate the incident x-ray beam to 50 x 50 μm (approximately 2-3 cells wide). The fluorescence spectra were collected from at least 10 locations within 100 μm of the periportal and perivenous regions of the parenchyma from each rat.

2.3 Data analysis

The peak areas for each measurement were determined using the program HEX as slightly modified for synchrotron radiation. The efficiency for x-ray fluorescence for each element studied was determined from cryosections of gelatin with selected elements added to 10 μg per ml gelatin. The gram atoms of Ca, Cr, Mn, Fe, Cu, Zn, Se, and Br, were normalized to the gram atoms of K for each measurement. The rationale for this normalization is that there may be regional and section-to-section variation in section thickness making normalization to the volume tenuous. Moreover, the liver is highly vascular and each measurement would include a variable volume of blood space which had to be flushed of elements. However, K is distributed uniformly, for the purposes of this study, in the cell water, and is proportional to cell water. Thus the normalization of elements to potassium provided an appropriate and rational approach to normalize without concern for variations in section thickness or the relative amounts of parenchyma and sinusoidal space irradiated during measurement. The differences in the trace element content of the periportal and perivenous regions of the parenchyma was determined by a paired t-test.

The elemental content of liver parenchyma was determined in tissue volume of approximately 7×10^{-4} μl , that is to say, a tissue volume approximately equivalent of 4 to 8 parenchymal cells. The concentrations of K, Ca, Cr, Mn, Fe, Cu, Zn, Se, and Br, averaged 1700, 88, 1.8, 1.4, 35, 1.0, 32, 0.8, and 1.2 μg / cm^3 fresh tissue respectively. These values are consistent with values for fresh liver determined by conventional chemical analysis /5/. A representative spectrum is shown in Figure 1. The relative distribution of these elements, normalized to potassium was not significantly different between the periportal and perivenous regions (Table 1).

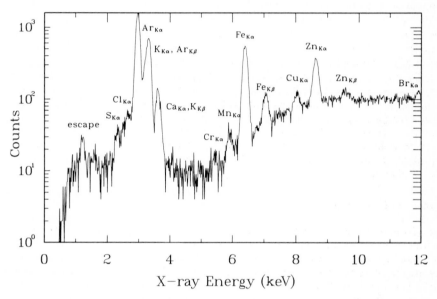

Figure 1. Representative x-ray fluorescence spectrum from periportal liver parenchyma

CONCLUSIONS

The current study provides a basis to evaluate the distribution, kinetics, and function of these elements in periportal and perivenous liver parenchyma in situation where the whole liver metabolism is known to be altered. These situations include acute phase reactions, acute and chronic alcohol intoxication, iron and copper over load, and intoxication with non- essential elements. This example illustrates how a high-energy x-ray microscope may be used to gain information about the trace element content of biological structures which are too small or complex in shape to dissect for chemical analyses.

This lack of a difference between regions is somewhat surprising considering the important and multiple roles of Fe in biotransformation, Zn in the dehydrogenases of

Table 1. The ratio of selected essential elements in periportal to perivenous region of normal liver parenchyma.

Animal	Ca	Cr	Mn	Fe	Cu	Zn	Br
#1	0.83	0.83	1.17	0.94	0.94	0.87	1.15
#2	1.14	1.05	1.04	0.94	1.07	0.97	0.73
#3	1.17	0.86	0.99	0.71	1.43	1.26	1.53
#4	0.95	1.32	0.87	1.11	0.96	1.03	0.54
#5	0.58	0.85	0.91	1.05	0.88	1.07	0.57

The gram atoms of each element in the beam spot were normalized to potassium. The data represent the ratio of the element/K in the periportal region to the element/K in the perivenous region. No statistical differences were detected with a paired t-test.

intermediary metabolism, and Cu and Se in enzymes of biotransformation. Certainly the biochemical form of the trace elements. i.e. the functional trace element complex is the important form. These measurements provide the distribution of these elements in the normal, fasted liver parenchyma and in the current study do not provide information on the dynamic state of the elements.

4. ACKNOWLEDGEMENTS

Research supported by US Department of Energy, Processes and Techniques Branch, Division of Chemical Sciences, Office of Basic Energy Sciences, Contract No. DE-ACO2-76CH00016 and the National Institutes of Health as a Biotechnology Research Resource under Grant No. P41RR01838.

5. REFERENCES

1. B.M. Gordon, A.L. Hanson, K.W. Jones, W.M. Kwiatek, G.J. Long, J.G. Pounds, G. Schidlovsky, P.Spanne, M.L. Rivers, S.R. Sutton, and J.V. Smith. This volume.

2. R. Bockman, M. Repo, R. Warrell, J.G. Pounds, W.M. Kwiatek, G.J. Long, G. Schidlovsky, and K.W. Jones. This volume.

3. K. Jungerman and N. Katz. Hepatology 2, 385 (1982).

4. J.G. Pounds, R.W. Wright, and R.L. Kodell. Toxicol. Appl. Toxicol. 66, 88 (1982).

5. R. Zeisler, S.H. Harrison, and S.A. Wise. Biol. Trace Elem. Res. 6, 31 (1984).

X-Ray Contact Microscopy
of Human Chromosomes and Human Fibroblasts

K. Shinohara[1], H. Nakano[1], M. Watanabe[2], Y. Kinjo[2], S. Kikuchi[3], Y. Kagoshima[4], K. Kobayashi[4], and H. Maezawa[4]

[1]Department of Radiation Research,
 Tokyo Metropolitan Institute of Medical Science,
 Bunkyo-ku, Tokyo 113, Japan
[2]Tokyo Metropolitan Isotope Research Center,
 Setagaya-ku, Tokyo 158, Japan
[3]Toshiba VLSI Research Center, Saiwai-ku, Kawasaki-shi,
 Kanagawa 210, Japan
[4]Photon Factory, National Laboratory for High Energy Physics,
 Oho-machi, Ibaraki 305, Japan

1.1 Introduction

It has been discussed that x-ray microscopy has potential advantages for observing intact biological materials at high resolution /1/. At present, x-ray contact microscopy is the best developed method in this field /2-4/. One of the expected advantages of x-ray microscopy for observing biological specimen is that biological components themselves have high absorption co-efficient to the x-rays and the contrast can be obtained with no staining process which may induce artifact. We applied x-ray contact microscopy to observe unstained human chromosomes dried with no fixative as a step to ob-serve intact biological specimen in a wet condition. Three-dimensional ob-servation of biological structure is another attractive application of x-ray microscopy. X-Ray holography is quite promising for this purpose /5,6/. We also observed fixed human fibroblasts grown on silicon nitrite film (0.6 -0.7 μm thickness) for a step to make a holographic recording of the cells and to observe the structure of chromosomes in interphase nucleus in com-parison with those in metaphase.

1.2 Materials and Methods

Chromosomes from human lymphocytes (RPMI 1788) were spread on a surface of distilled water and whole-mounted directly on polymethylmethacrylate (PMMA) (Fig. 1) /7/. They were dried immediately without fixation. The PMMA with chromosomes was exposed to the 2.98 nm monochromatic undulator radiation /8/ from BL-2B at the Photon Factory, National Laboratory for High Energy Phys-ics, in a vacuum. A zone plate monochromator (stopping material, tantalum; thickness, 1 μm; innermost diameter, 62.9 μm; outermost diameter, 2 mm) was installed. The exposed specimens were treated as described elsewhere /9/. Briefly, chromosomes were removed with sodium hypochlorite from PMMA. The PMMA was developed with a mixture of methylisobutylketone and isopropanol. The developed images were observed either with differential interference microscope or with transmission electron microscope by the use of replica method with the plasma polymerization-film in a glow discharge. Cells (N14

429

SURFACE-SPREADING WHOLE-MOUNT TECHNIQUE
FOR
X-RAY CONTACT MICROSCOPY

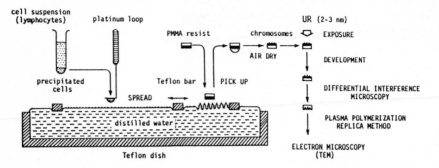

Fig. 1 Surface-spreading whole-mount method for the preparation of chromosomes of human lymphocytes

OS, fibroblasts derived from normal human skin) were cultured directly either on silicon nitrite (0.6-0.7 μm thickness) window or on PMMA, fixed
with glutaraldehyde, and dehydrated. The PMMA with cells was exposed to the
2.98 nm undulator radiation and treated as shown above. The silicon nitrite
with cells was placed in front of a PMMA with a 0.6 mm distance and exposed
to the 2.18 nm undulator radiation without zone plate monochromator. The
exposed PMMA was developed and observed under differential interference
microscope.

1.3 Results and Discussion

Figure 2 shows the x-ray images of human chromosomes observed under differential interference microscope. The image was produced by the chromosomes
themselves since they are not stained and no fixatives are added. Figure 3
shows a human chromosome observed under transmission electron microscope
with the use of replica method with plasma polymerization-film in a glow
discharge. Stretched portion of a chromosome is shown in Fig. 4. Chromatin
fibrils of 10-20 nm thickness are seen in this figure. These results indicated that the observation of chromosomes dried with no fixative was possible at high resolution without further staining process which may induce
artifact.

To observe inside structure of thick specimen such as nucleus of interphase cells, x-ray holography may be worth to try. We tried holographic recording of human cells. Figure 5 shows the x-ray image of human fibroblast
(N140S) directly attached to PMMA and Fig. 6 shows the image of N140S distant from PMMA by 0.6 mm. There are some interference fringes inside the
nucleus besides the images of nucleoli when the cells were distant from the
recording PMMA (Fig. 6), while no such pattern was observed inside the nucleus when the cells were directly attached to PMMA (Fig. 5). These results
suggested that holographic observation may reveal the organization of chromosomes in the interphase nucleus.

430

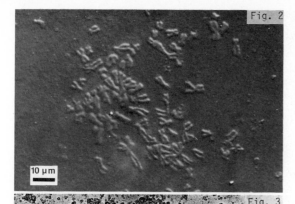

Fig. 2. X-ray image of chromosomes of human lymphocytes as seen in differential interference microscope

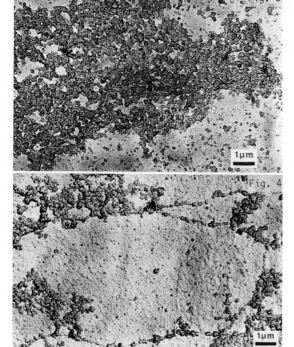

Fig. 3. X-ray image of a human chromosome as seen in transmission electron microscope

Fig. 4. Stretched portion of a chromosome

In summary, we showed that the observation of human chromosomes dried with no fixative was possible by x-ray contact microscopy at high resolution without further staining process. The present results suggested that x-ray microscopy may be applicable to observe not only intact metaphase chromosomes in the cells at wet condition, but also three-dimensional structure of chromosomes of interphase nucleus.

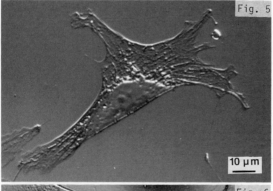

Fig. 5. X-ray image of
human fibroblast (N140S)
on PMMA as seen in dif-
ferential interference
microscope

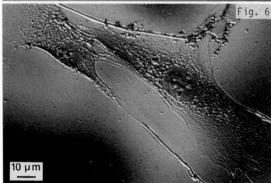

Fig. 6. X-ray image of
N140S distant from PMMA
by 0.6 mm

Acknowledgements

The authors are grateful for the kind cooperation of Dr. Y. Horiike, Dr. A.
Tanaka and the staff at the Photon Factory, National Laboratory for High
Energy Physics. This work was supported in part by Grants-in-Aid from the
Ministry of Education, Science and Culture, Japan.

References

1. J. Kirz, D. Sayre: In Synchrotron Radiation Research ed. by H. Winick
   and S. Doniach (Plenum Press, New York and London 1980) pp. 277-322
2. R. Feder, D. Sayre, E. Spiller, J. Topalian, J. Kirz: J. Appl. Phys. 47,
   1192-1193 (1976)
3. W. Gudat: In Uses of Synchrotron Radiation in Biology ed. by H. B.
   Stuhrmann (Academic Press, London 1982) pp. 23-50
4. B. J. Panessa-Warren: In X-Ray Microscopy ed. by G. Schmahl and D.
   Rudolph, Springer Ser. Optical Sci., Vol. 43 (Springer-Verlag, Berlin,
   Heidelberg, New York, Tokyo 1984) pp. 268-278
5. S. Aoki, S. Kikuta: Jpn. J. Appl. Phys. 13, 1385-1392 (1974)
6. C. Jacobsen: this volume
7. M. Watanabe, N. Tanaka: Jpn. J. Genetics 47, 1-18 (1972)
8. H. Maezawa, Y. Suzuki, H. Kitamura, T. Sasaki: Nucl. Instrum. Meth. Phys.
   Res. A246, 82-85 (1986)
9. K. Shinohara, S. Aoki, M. Yanagihara, A. Yagishita, Y. Iguchi, A.
   Tanaka: Photochem. Photobiol. 44, 401-403 (1986)

Soft X-Ray Contact Microscopy of Botanical Material Using Laser-Produced Plasmas or Synchrotron Radiation

A.D. Stead[1], T.W. Ford[1], R. Eason[2], A.G. Michette[3], W. Myring[4], and R. Rosser[5]

[1]Dept. of Biology, RHBNC, Egham, Surrey TW20 0EX, UK
[2]Dept. of Physics, University of Essex, Wivenhoe Park, Colchester CO4 3SQ, UK
[3]Dept. of Physics, Kings College, The Strand, London WC2R 2LS, UK
[4]SRS, Daresbury, Warrington, Cheshire, WA4 4AD, UK
[5]Rutherford Appleton Laboratory, Chilton, Oxon OX11 0QX, UK

1 INTRODUCTION

The use of soft x-ray contact microscopy (SXCM) allows biological specimens to be imaged with a resolution better than that possible with the light microscope (LM), but not as good as that obtained with the transmission electron microscope (TEM). Unlike TEM however, it is possible to image wet biological specimens that require no pretreatment before imaging. For TEM specimens need to be chemically fixed, dehydrated, embedded and sectioned before examination, any one of these preparative steps can, and does, introduce artefacts, thus the development of any technique which eliminates or reduces the possibility of artefacts will be of great advantage.

2 METHODS

The imaging process is remarkably simple, but the interpretation of the final image requires a thorough understanding of the way in which that image has been produced /1/. The specimen is placed in direct contact with the photosensitive recording medium and, after exposure, the specimen is removed and the resist developed, in a mixture of isobutyl methyl ketone (MIBK) diluted with isopropyl alcohol, to produce a relief pattern which corresponds to the integrated mass absorption coefficient of the specimen. This can then be viewed by either interference light microscopy or, after sputter coating, by scanning electron microscopy (SEM).

To image wet biological specimens it is necessary to use the minimal possible exposure time to reduce the risk of radiation damage, the specimen must also be physically isolated from the vacuum. Using laser-produced plasmas (Nd: glass laser at RAL) the latter requirement was met by mounting the specimen behind a thin (120nm) Si_3N_4 window - itself about 70-80% transparent to soft x-rays. The selection of the x-ray wavelengths used was accomplished by careful choice of the target materials - in this case mylar.

Using synchrotron radiation the longer exposure times required (10 to 20s) dictated the use of aldehyde-fixed, critical point dried material. The cells to be imaged were placed on formvar coated EM grids and held adjacent to Balzers 6x6cm plates coated with Phillips PM15. Previous work /2/ suggested that the optimum exposure was 3000mA.s. We have used this exposure, and others, to optimise the exposure for whole cells. Additionally, since the depth of development is related to the length of development time, the images obtained after various development times are also compared. A Michelson interferometer was used to measure development depth and fringe displacement measured to the nearest 0.25 of a fringe; using white light the minimum depth difference was therefore approx. 50nm on coated resists and 100nm on uncoated ones /3/.

3 RESULTS and DISCUSSION

3.1 Using laser-produced plasmas

In the multicellular floral epidermal hairs of the foxglove flower, some internal structure can be seen when examined by transmitted white light but when examined by TEM little internal detail is seen, presumably because it is destroyed during the lengthy chemical fixation and dehydration that is required (Fig. 2). However, imaging these cells by SXCM reveals considerable internal detail when the resists are viewed by LM (Fig. 1) and even more detail when viewed by SEM (Fig. 3). The cytoplasmic strands are more abundant than other techniques indicate; within, or associated with, these strands are numerous organelles, the larger of which are thought to be plastids and the smaller, more frequent ones mitochondria.

3.2 Using synchrotron radiation

Following exposure for 3000mA.s and the use of 25% MIBK, complete development occurred after 60s, and 10s was sufficient to give a development depth of

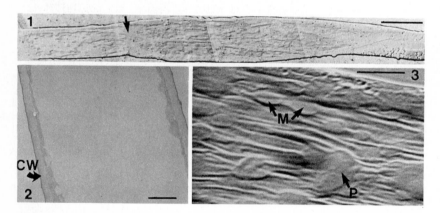

Figure 1   LM of the developed resist showing numerous cytoplasmic strands and transverse x-ray opaque region representing the junction between two adjacent cells (arrowed).   Bar = 40μm
Figure 2   TEM of conventionally prepared material. Median LS through a basal cell showing peripheral cytoplasm lining the cell wall (CW). Bar = 2μm
Figure 3   SEM of the developed resist showing details of the cytoplasmic strands and associated mitochondria (M) and plastids (P). Bar = 5μm

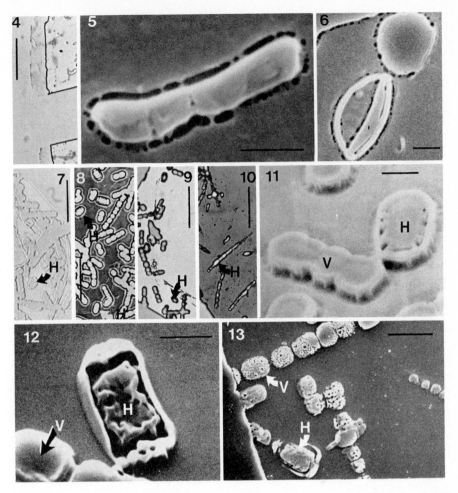

Figure 4 LM image of *Digitalis* floral epidermal hair (3000mA.s; 60s in 25% MIBK). Bar = 20μm

Figure 5 SEM image of *Anacystis* (3000mA.s; 21s in 25% MIBK). Bar = 2μm

Figure 6 SEM image of *Cyanidium* cells (3000mA.s; 17s in 25% MIBK) apparently shows internal detail (lower cell) in some cells but not others (upper cell). Bar = 2μm

Figures 7-10 LM showing the effect of increased exposure on images of *Anabaena*, in all cases the development was in 15% MIBK for 60s. The heterocyst cells are labelled H. Bar = 20μm 7) 750mA.s 8) 6000mA.s 9) 15000mA.s and 10) 30000mA.s
Figures 11-13 SEM of similar material, after 11) 6000mA.s; bar = 2μm 12) 15000mA.s; bar = 2μm and 13) 30000mA.s; bar = 5μm

250nm. Under LM the resist images suggested that little internal cell detail had been revealed when the development depth was less than maximal. With fox-glove hairs, the largest specimens imaged, some detail was visible after 60s development (Fig. 4) and the structures seen are similar to those revealed using laser-produced plasmas (Fig. 3). Under SEM the resists showed only cell outlines after 10-15s development without any indication of the underlying cell structure. After longer development the images of *Anacystis* showed that the cells were elongated, often curved and aggregated into short filaments. The differential x-ray absorption by the cells (Fig. 5) is probably due to concentric rings of photosynthetic lamellae that occur in *Anacystis*. Greater differential absorption was occasionally seen in the cells, but these images (Fig. 6) are thought to be of senescent cells with little or no cell content /4/. The images of the foxglove hair showed no detail when viewed by SEM, presumably because the features were too shallow.

The greater thickness of these whole cells, compared to the thin sections previously imaged /2/, and the featureless images obtained in the experiments above suggests that the exposure was insufficient for these larger cells. A range of exposures between 750 and 30000mA.s were compared; each resist was developed in 15% MIBK for 60s and the depth of development measured. For ease of comparison only images of the filamentous cyanobacterium, *Anabaena*, are in-cluded in this discussion. Using the highest exposure, complete development of the resist occurred when developed for 60s, lesser exposures producing shallow-er development. The clarity of the LM images obtained varied considerably de-pending on the exposure used (Figs. 7-10). Although the images are clearly present in Fig. 7 (750mA.s) these were found to be too shallow for viewing by SEM, even those produced with 3000mA.s were difficult to see by SEM even though good LM images were obtained. When the exposure was 6000mA.s good LM and SEM images were obtained (Fig. 8 and 11). Greater exposures appear to have been excessive when seen under LM, in Fig. 9 (15000mA.s) the images of the folds in the formvar coating are almost lost, suggesting that similar small details may be lost from the cell images. In Fig. 10 (3000mA.s) many of the images of the filaments seem to disappear - again indicative of excessive development. How-ever when examined by SEM both 15000 and 30000mA.s exposures showed good inter-nal cell detail (Figs. 12, 13). In Fig. 12 (15000mA.s) the cell wall of the heterocyst is distinct from the cell contents, after increased exposure how-ever, the heterocyst wall almost developed away but internal detail is visible in the images of the vegetative cells (Fig. 13). The biological significance of these results is discussed by FORD et al. /4/.

4 CONCLUSIONS

The wet cell images of foxglove hairs imaged using laser-produced plasmas show abundant internal cell detail - detail which is not apparent when examined by conventional TEM. The longer exposures using the Daresbury synchrotron have precluded the use of wet cells to date, the use of more sensitive resists and the installation of HBL should overcome this in the near future.

5 ACKNOWLEDGEMENTS

We are grateful to SERC for facilities at RAL and SRS and for the collaboration of Dr. O'Neil, Dr. Ridgeley (RAL) and Mr. Clarke (SRS); to the CRF of London University for provision of the interferometer. Thanks are also due to Mr. Feder (IBM) for resist materials and to Mr. Catcheside and Mr. Lawes for assis-tance in examining the exposed resists and to Ms. Etherington for producing the photographs.

6 REFERENCES

1. P.C. Cheng, R. Feder, D.M. Shinozaki, K.H. Tan, R.W. Eason, A. Michette, R. Rosser: Nucl. Instr. and Methods in Phys. Res. A. 246, 668, (1986)
2. K.S. Richards, A.T. Rush, D. Clarke, W.J. Myring: J. Microscop. 142, 1 (1986)
3. C. Buckley: The fabrication of gold zone plates and their use in scanning x-ray microscopy. Ph.D. Thesis, Univ. of London (1987)
4. T.W. Ford, A.D. Stead, W.J. Myring, C.P. Hills, R. Rosser: This volume

Advances in Geochemistry and Cosmochemistry: Trace Element Microdistributions with the Synchrotron X-Ray Fluorescence Microprobe

S.R. Sutton[1], *M.L. Rivers*[1], *J.V. Smith*[1], *and K.W. Jones*[2]

[1]Department of the Geophysical Sciences,
5734 S. Ellis Avenue, The University of Chicago,
Chicago, IL 60637, USA

[2]Brookhaven National Laboratory, Upton, NY 11973, USA

1. INTRODUCTION

Geochemistry studies rely heavily on knowledge of elemental partitioning behavior under given physio-chemical conditions such as temperature, pressure and bulk system composition. An important aspect of this research is the ability to determine microdistributions for elements of differing geochemical character among minute, coexisting phases in natural rocks and synthetic, laboratory run-products. The high energy x-ray fluorescence microprobe (XRM) at the National Synchrotron Light Source (beamline X26-C), Brookhaven National Laboratory, is giving geoscientists a powerful new tool in these endeavors offering ppm sensitivity with 20 micrometer spatial resolution.

Synchrotron radiation induced x-ray fluorescence analysis offers several advantages for chemical analyses of geological specimens over conventional geochemical techniques, such as electron microprobe and neutron activation analysis [1,2]. Most important, high spatial resolution coupled with high elemental sensitivity allows analyses of trace elements on individual crystals, as opposed to bulk rocks or mineral separates, and the essential ability to avoid commonly occurring inclusions within these grains. The capability for detailed trace element mapping of geological specimens opens up new areas of studies [3]. For example, laboratory studies of geochemical behavior can be performed under pressure, temperature and geochemical conditions which approximate geological environments more closely than ever before. The non-destructive nature of the analysis leaves specimens available for study by complementary techniques.

The present XRM [4] uses continuum radiation from an arc magnet as the excitation source and an energy dispersive detector (30 mm^2 Si(Li)). With this apparatus, minimum detection limits are typically several ppm for 20 micrometer spots. The purpose of this paper is to summarize the initial results of geochemistry and cosmochemistry experiments performed on this instrument through collaborations between BNL staff scientists and geoscientists from universities and industry.

2. GEOCHEMISTRY

Trace element geochemical research has emphasized partitioning and migrational behavior in silica- and sulfur-rich systems. Migration and partitioning of trace elements are of fundamental importance in understanding crustal evolution. For example, the refinement of petrogenetic models leads to more accurate predictions of lifecycles for geothermal centers and improved knowledge of geochemical migration can aid assessment of radioactive waste repository integrity. The value of trace element signatures as indicators of geological provenance is also under investigation. Improved models of the formation of sedimentary basins may lead to refinements in hydrocarbon exploration strategies, for example. A study of the chemical evolution of fluids in coal-forming environments complements this energy-related research.

2.1 Precious Metal Partitioning in Ores

Preliminary trace element determinations of common sulfides from a Ni-Cu deposit (Sudbury, Canada) and two platinum-group element deposits (Merensky Reef, South Africa; Stillwater, Montana) were made [5]. A synthetic pyrrhotite doped with 900 ppm Se and 1100 ppm Pd was used as a standard. The results show that Merensky and Stillwater pentlandites contain 100-300 ppmw Pd whereas those from Sudbury contain <3 ppmw; Merensky chalcopyrite contains ~ 50 ppmw Pd whereas all other chalcopyrites and pyrrhotites lack measurable Pd. Carlin-type ores have proved to be important sources of gold from an economic standpoint but little has been known regarding the sighting of the gold, its chemical/mineralogical speciation and petrogenesis. Such information is essential for developing extraction techniques and identifying new deposits. Studies of the microdistribution of gold in this material showed that pyrite crystals contained Au below the detection limit of about 5 ppm while concentrations up to 40 ppm were observed in the siliceous matrix [6].

2.2 Elemental Diffusion in Silicate Melts

Diffusion of Al, Si, Cl, Mn, Fe, Zn, Y, Zr and Nb in halogen-bearing rhyolites was measured at 10 MPa and 900 to 1000 °C using electron microprobe for major elements and the synchrotron XRF microprobe for trace elements [7]. The experiments used the diffusion couple technique with one half of each couple being a high Al, low Fe, F-bearing rhyolite and the other half being a low Al, high Fe, Cl-bearing pantellerite. Measured diffusion rates for major and trace elements were similar varying from 2×10^{-11} cm^2/sec (Si at 1000°C) to 1×10^{-10} cm^2/sec (Zn at 1000°C) and exhibited no correlation with ionic radii. The results are consistent with trace element diffusion controlled by diffusional gradients associated with the major elements Si and Al rather than transport by large polyatomic complexes.

2.3 Trace Element Signatures of Quartz

Trace element contents were measured for quartz from volcanic, plutonic and low-temperature rocks to test the value of trace element signature as an indicator of geological provenance [8]. Such an approach would be valuable in determining the origin of individual detrital grains in sedimentary and metamorphic rocks. Because of the generally low impurity content of quartz, high spatial resolution to avoid inclusions is essential in this work. Preliminary x-ray spectra show the presence of Ca, Ti, Mn, Fe, Zn and Ge in most specimens and Cr, Ni and Cu in some. A positive correlation exists between the concentrations for Ca, Ti, Mn and Fe and those obtained by ion microprobe, indicating that the trace elements reside in the quartz structure and not in mechanical impurities. A long term goal is to unravel the early structural and geochemical history of the Earth's crust by studying quartz grains from Archaen rocks.

2.4 Evolution of Coal Sulfides

Sulfide minerals, common components of coal, are potential carriers of environmentally-hazardous chemical species. Analyses have been made on pyrite

(FeS_2), marcasite (FeS_2), and chalcopyrite ($CuFeS_2$) originating from a series of localities within the Yorkshire and East Midland Coalfields in Great Britain [9]. Iron, Ni, Cu, Zn, Se, As, Mo, Tl and Pb were detected with minimum detection limits of 5 ppm for 10 minute acquisitions with a 40 micrometer synchrotron beam. The initial results argue against the theory that marine incursions are the only sources of chemical input to coal-forming environments. A major goal of this research is to study the evolution of fluids in the coal-forming environment by analysing sulfides of different temporal association.

3. COSMOCHEMISTRY

Trace elements in extraterrestrial minerals are valuable indicators of geochemical and geophysical history [10]. Chemical signatures are useful in classifying meteorites and unraveling the histories of the mineral components of breccias. At a fundamental level, trace element compositions provide tests of formation models. Improved knowledge of solar system evolution leads to a better understanding of the Earth's history. High spatial resolution and elemental sensitivity are essential primarily because of the minute mass (typically nanograms) of the most interesting specimens which includes refractory condensates in primitive meteorites and cometary/asteroidal dust.

3.1 Iron Meteorites

Iron meteorites represent core analogs and therefore offer the opportunity to study the geochemistry of planetary differentiation and infer the present state of the Earth's interior. Trace element analyses were performed on Fe/Ni metal and troilite (FeS) from 9 iron meteorites representing 5 geochemical groups, IAB, IIICD, IIB, IIIA, and IVA [11]. Cu and Ni distribution coefficients (troilite/metal) correlated positively with kamacite (α-iron) band width and correlated negatively with bulk Ni content. The observation of Cu distribution coefficients less than 1 for Ni-rich meteorites provides evidence for re-equilibration under sub-solidus conditions.

3.2 Cosmic Dust

Micrometeorites (about 10 micrometers in size) collected in the stratosphere derive from comets and asteroids and possess compositions near those of carbonaceous meteorites for most detected elements. The notable exception is Br which was enriched by factors of 8 to 37 [12]. Since Br is detrimental to the Earth's ozone layer, its high content in these particles may have important implications to atmospheric chemistry. The trace element signatures of meteoritic ablation spheres (about 100 micrometers in size) extracted magnetically from deep sea sediments were depleted relative to carbonaceous meteorites for all detected elements suggesting that ablation/melting has significantly modified the chemical composition of the parent material. Compositional clustering observed for the deep sea particles may be useful in identifying the parental objects.

3.3 Refractory Inclusions in Primitive Meteorites

Calcium- and aluminum-rich inclusions in primitive meteorites, so-called CAIs, represent some of the first solid matter to condense from the early solar nebula. Initial experiments concerned the determination of melilite/liquid partition coefficients for refractory lithophiles in synthetic melts of CAI composition [13]. Measured coefficients were between 0.2 and 0.1 for three rare earth elements (Sm, Y, and Yb), unity for Sr and less than .002 for Zr. The rare earth element values are slightly less than rough estimates based on neutron activation analyses of mineral separates.

4. CONCLUSION

Instrumental upgrades (installation of focussing optics, monochromator, crystal spectrometer) now in progress will provide greater elemental sensitivity, higher spatial resolution and access to K-lines for additional elements. The direct impact on geochemistry and cosmochemistry research will be the ability to study (1) elemental distributions in finer-grained samples, such as finely intergrown sulfides and fine-grained refractory inclusions in meteorites, and (2) geochemical behavior of elements present at concentrations below 1 ppm.

5. ACKNOWLEDGMENTS

We thank the staff of the National Synchrotron Light Source, Brookhaven National Laboratory, for beam time to perform these experiments. Research supported in part by the US Department of Energy, Division of Chemical Sciences, Office of Basic Energy Sciences, Contract No. DE-ACO2-76CH00016; NSF Grant No. EAR-8618346; NASA Grant No. NAG 9-106.

6. LITERATURE REFERENCES

1. M. L. Rivers, S. R. Sutton, J. V. Smith and K. W. Jones: abstract, Geol. Soc. of Amer. Annual Meeting, Phoenix (1987).
2. A. L. Hanson, K. W. Jones, B. M. Gordon, J. G. Pounds, W. M. Kwiatek, M. L. Rivers, G. Schidlovsky, and S. R. Sutton: Nucl. Instrum. Methods, B24/25, 400-404 (1987).
3. M. L. Rivers: this proceedings, Intern. Symp. on X-Ray Microscopy, Upton, NY (1987)
4. B. M. Gordon, A. L. Hanson, K. W. Jones, W. M. Kwiatek, G. J. Long, J. G. Pounds, G. Schidlovsky, P. Spanne, M. L. Rivers, S. R. Sutton and J. V. Smith: this proceedings, Intern. Symp. on X-Ray Microscopy, Upton, NY (1987).
5. M. L. Rivers, L. Cabri and J. V. Smith: in preparation (1987).
6. J. R. Chen, E. C. T. Chao, J. A. Minkin, J. M. Back, W. C. Bagby, M. L. Rivers, S. R. Sutton, A. L. Hanson and K. W. Jones: Nucl. Instrum. Methods, B22, 394-400 (1987). Also, Chen et al.: this proceedings, Intern. Symp. on X-Ray Microscopy, Upton, NY (1987).
7. D. Baker, et al.: Jour. Non-crystalline Solids, submitted (1987).
8. J. V. Smith, C. M. Skirius, M. L. Rivers, and K. W. Jones: EOS, Trans. Am. Geophys. Union 66, no. 46, 117 (1985).
9. R. N. White, D. A. Spears, and J. V. Smith: abstract, Annual Geol. Soc. of Amer. Meeting, Phoenix (1987).
10. S. R. Sutton, M. L. Rivers, and J. V. Smith: Nucl. Instrum. Methods, B24/25, 405-409 (1987).
11. S. R. Sutton, J. S. Delaney, J. V. Smith and M. Prinz: Geochim. Cosmochim. Acta, in press (1987).
12. S. R. Sutton and G. J. Flynn: Proc. 18th Lunar Planet. Sci. Conf., in press (1987).
13. D. S. Woolum, D. S. Burnett, M. L. Johnson and S. R. Sutton: abstract, 50th Meteoritical Society Meeting, Newcastle Upon Tyne, England (1987).

Part V

Summary of Session on
Future X-Ray Microscopy Facilities

Report on Special Session on Future X-Ray Microscope Facilities

M.R. Howells

Center for X-Ray Optics, Lawrence Berkeley Laboratory,
University of California, Berkeley, CA 94720, USA

1. INTRODUCTION

A special evening session was held to discuss ways of developing
x-ray microscopy facilities in the future with particular emphasis on
new synchrotron radiation x-ray sources such as the Berkeley Advanced Light
Source. Almost all the conference participants were present and a lively
discussion was held. This report is a distillation of the material
presented both by the 14 speakers and members of the audience. The speakers
in order of presentation were:

S. Rothman, University of California, San Francisco, USA.
T. Ford, Royal Holloway and Bedford New College, UK.
A. Stead, Royal Holloway and Bedford New College, UK.
J. Elliott, London Hospital Medical College, UK.
G. Schmahl, University of Göttingen, FRG.
P. C. Cheng, State University of New York at Buffalo, USA.
D. Shinozaki, University of Western Ontario, Canada.
F. Polack, LURE, Universite Paris-Sud, France.
H. Rarback, Brookhaven National Laboratory, Upton, USA.
E. Lattman, Johns Hopkins University, Baltimore, USA.
S. Fan, State University of New York at Stony Brook, USA.
S. Richards, University of Keele, UK.
C. Buckley, State University of New York at Stony Brook, USA.
G. Nyakatura, University of Göttingen, FRG.

2. GENERAL PRECONDITIONS

Biological research was a different kind of information gathering than
physical science (Rothman) and required different conditions. A comfortable
low-noise environment was important (Schmahl, Michette) and there should be
an awareness of the difficulties which physics-based instrumentation
created for biologically trained scientists. Efforts should be made to
provide clear and complete information about instruments (Richards)
and consideration should be given to providing training sessions for
would-be users (Panessa-Warren).

It would be impractical to design a facility to please everyone
(design by committee). It was important to aim for an achievable
goal (Knotek). For effective research a reliable microscope was a high
priority and this also argued for modest technical goals (Schmahl).

The characteristics of storage rings for high resolution imaging
were of great importance and special attention to system stability
was required (Rothman, Rarback, Shinozaki). Electron beam steering by feedback
would be needed (as at NSLS) and even with such a system the Brookhaven group

444

had experienced picture artifacts that were traceable to a 12 Hz vibration lying outside the feedback system bandwidth.

Although the majority of foreseeable applications seemed to be in biology the opportunities in material science should not be overlooked (Shinozaki, Burge).

3. TYPES OF MICROSCOPE

There would be demand for a wide variety of techniques but the most popular in 1987 appeared to be contact microscopy. The interesting wavelengths would include the "water window" region (23-44 Angstroms) but much importance also attached to the region on the low wavelength side of this (Burge). There were biologically important edges there (K,Cl,S,P,Mg, Na,Zn,Cu,Mn,Co,Fe) (Rothman) plus which the transparency of water at wavelengths 8-15 Angstroms was the same as in the water window. The element mapping capability of x-ray imaging would be very attractive and to utilise it one would need a monochromator (Richards) which had a convenient way to "dial up" a wavelength (Rothman) and proper indication of its current setting (Stead). Efficient suppression of higher diffracted orders was important in certain cases (Shinozaki). Broad band radiation would also find application (Pannessa-Warren) to increase imaging speed.

There were cases where a horizontal specimen stage was important (Stead, Skinner) and this would involve some technical development work. It was also important to be able share the x-ray beam conveniently (Polack). The standard conveniences of the optical microscope such as ease of focussing, (Rothman) and ability to search a large field (Shinozaki) would provide similarly attractive payoffs for the x-ray microscope and should be pursued as far as possible. An ancillary optical microscope should be combined with the x-ray microscope for set-up purposes (Shinozaki, Buckley). The optical microscope should have as wide a range of imaging modalities as the situation allowed (Cheng). An automated facility for characterisation of the resolution and applied specimen dose would be highly valuable (Rothman).

It would be essential for applications to have a microscope that was maintained in a standard operating condition. This would imply that a separate instrument be available for technique development (Thompson, Sayre, Newberry).

It would be interesting to explore a variant of scanning X-ray microscopy in which the detected signal was visible light flourescence. This would allow the use of antibody labelled fluorochrome to map the distribution of specific proteins (Cheng).

4. SAMPLE ENVIRONMENTS

An important advantage of the x-ray microscope was its potential for flexibility in the choice of sample environments. It would be important to image intact cells and other objects without modification of their normal water and air surroundings (Ford) or the gases and ions in solution (Cheng). The possibility of manipulating the content of the sample fluids offered many interesting opportunities (Lattman). For imaging procedures that were slow, some compromise of the ideal natural condition of samples such as fast freeze, high pressure freeze, (Fan, Cheng), critical point drying (Ford) or the use of drugs or chemical fixatives (Ford, Cheng) might be necessary. Such interventions would certainly destroy some significant features (Stead) but, as demonstrated in other forms of biological imaging,there were some types of research that would not be invalidated (Rothman). Some of these activities would require some technique development (Richards).

5. RADIATION DAMAGE

Damage by the illuminating radiation was a factor in all imaging studies including x-ray microscopy. Biological effects of radiation could occur for very small doses effecting a very small proportion of molecules. However such effects might be slow as shown by the dose rate dependence of x-ray crystallography. They might also leave many of the interesting features unchanged from the point of view of an image at the resolution values expected in x-ray microscopy. The role of damage by soft x-ray illumination would need to be understood on a case-by-case basis and was an important area for research and development (Rothman). Valuable insights would be provided by studies of the damage issues in other microscopies (Stead).

The most promising way to circumvent the damage problem would be to make an image in a sufficiently short time. This could be done using the nanosecond pulses from flash sources (Ford) such as laser heated plasmas or by the use of intense synchrotron radiation. In the latter case the question of how fast would be fast enough had to be addressed. For some experiments 1 second (Stead) would be attractive, for others something nearer 1 millisecond (Ford) would be more appropriate. The kinetics of x-ray damage processes were not sufficiently well understood to make general statments.

6. TIMESCALES

For biology, the timing of the preparation-measurement sequence was important. The choice of the type of preparation and image formation depended on feedback from the previous image (Lattman, Ford). It was necessary for the biologist to have control of a shutter which started and stopped the exposure (Stead). If pumpdown was required it was important for it to be fast (Ford). For contact microscopy it was vital to have a capability to process the resist and see the picture immediately after exposure (Ford) in order to decide the next move.

The standard arrangement of long (2-3 week) time slots that was typically used at synchrotron radiation sources was not well suited to biology because the large effort needed for sample preparation often led to wastage of beam time. A more suitable approach would be to have two or more groups sharing the beam with interleaved preparation time and beam time (Stead).

7. ORGANIZATION

A number of speakers had experienced frustrations in trying to do biology research at the kind of large central facilities that were needed for x-ray microscopy. A key requirement was the right kind of support personnel (Cheng, Pannessa-Warren, Richards, Shinozaki). Both the Canadian Synchrotron Radiation Facility at Stoughton and the Daresbury facility had appointed beamline managers which were considered very effective (Shinozaki, Duke). Both in-house physicists (with good communication skills (Richards)) and biologists (Schmahl) were needed. Some examples of how these organisational issues were being addressed could be seen at other types of large facility, such as high voltage electron microscopes (Lattman).

The creation of major biological facilities in the United States was currently a priority with the National Science Foundation (Lattman). The timescale on which applications for beam time were processed should be fast enough that one could get beam time and do an experiment in the time frame of biological research grants (Panessa-Warren).

8. ANCILLARY FACILITIES

The provision of adequate support facilities was the centerpiece of a user-friendly environment for biological research. The totality of those needed would amount to a substantial assignment of resources in terms of people, money and physical space. The chief items were a well-equipped microscope room, a wet biology laboratory and small-scale x-ray sources.

A microscope room was needed for comparative studies paralleling those carried out by x-ray microscopy (Rothman). It was necessary to have both scanning and transmission electron microscopes and an up-to-date range of optical systems including phase-contrast, Nomarski and epi-fluorescence attachments and a confocal microscope. All of these would require a modern capability for digital image display and processing (Cheng, Elliott, Stead, Shinozaki, Richards, Lattman, Rothman, Pannessa-Warren). These devices would also play a necessary role in the processing and readout of resist recordings.

The wet-biology laboratory should be located close to the x-ray microscope and should allow samples to be unpacked, mounted, oriented and developed (Panessa-Warren). Such a laboratory should contain sinks, a fume hood, two refridgerators (one for organic/biological materials and one for reagents), a freezer, incubators, a binocular stereo microscope, an inverted microscope equipped for photomicrography, modern types of microtome, a table top centrifuge, a scale and accessories for making solutions, a resist spinner, storage space for users and general consumable supplies such as glassware, paper towels, distilled and high purity water and a few reagents. Additional desirable features were positive pressure clean space, capability for handling live animals and plants, a darkroom, and an evaporator/sputterer for metallisation of samples and resists (Panessa-Warren, Rothman, Lattman, Cheng).

Additional x-ray sources were valuable for pilot experiments and for technique development. They could also play a role when synchrotron radiation was not delivered on the desired timescale. An x-ray tube source with interchangeable anodes would be useful and a flash source of some kind would be even more valuable (Richards, Pannessa-Warren).

Subject Index

Note: Page references are to the first page of the relevant article

Index of Contributors